T0254617

Representing Scientific Knowledge

Chaomei Chen · Min Song

Representing Scientific Knowledge

The Role of Uncertainty

 Springer

Chaomei Chen
College of Computing and Informatics
Drexel University
Philadelphia, PA
USA

Min Song
Department of Library and Information
 Science
Yonsei University
Seoul
South Korea

ISBN 978-3-319-87336-7   ISBN 978-3-319-62543-0   (eBook)
https://doi.org/10.1007/978-3-319-62543-0

© Springer International Publishing AG 2017
Softcover re-print of the Hardcover 1st edition 2017
This work is subject to copyright. All rights are reserved by the Publisher, whether the whole or part
of the material is concerned, specifically the rights of translation, reprinting, reuse of illustrations,
recitation, broadcasting, reproduction on microfilms or in any other physical way, and transmission
or information storage and retrieval, electronic adaptation, computer software, or by similar or dissimilar
methodology now known or hereafter developed.
The use of general descriptive names, registered names, trademarks, service marks, etc. in this
publication does not imply, even in the absence of a specific statement, that such names are exempt from
the relevant protective laws and regulations and therefore free for general use.
The publisher, the authors and the editors are safe to assume that the advice and information in this
book are believed to be true and accurate at the date of publication. Neither the publisher nor the
authors or the editors give a warranty, express or implied, with respect to the material contained herein or
for any errors or omissions that may have been made. The publisher remains neutral with regard to
jurisdictional claims in published maps and institutional affiliations.

Printed on acid-free paper

This Springer imprint is published by Springer Nature
The registered company is Springer International Publishing AG
The registered company address is: Gewerbestrasse 11, 6330 Cham, Switzerland

Preface

The 2014 Ebola outbreak in West Africa raised many urgent concerns about public health and safety, as well as legal and administrative implications. The high mortality rate of the Ebola virus heightened the tension between the public, healthcare providers, patients, and local authorities. In the United States, the White House expressed concerns about possible unintended consequences of quarantine policies enforced on doctors and nurses returning from Ebola-stricken countries. Governors of some states defended their quarantine policies, while the White House worried that the policies might not be grounded in science. Some contractors were deeply concerned about the safety of handling Ebola patients' medical wastes and whether they should discard expensive instruments just because they were used to analyze Ebola patients' blood. Some people firmly believed that people without symptoms of Ebola would not transmit the disease. However, the Center for Disease Control and Prevention (CDC) revising its own guidelines was enough reason for others to take extra prudent measures to minimize the risk.

Charles Haas, an environmental engineering professor at Drexel University, specializes in water treatment and risk assessment. He started his comprehensive search in the literature for any information on how long the Ebola virus might be able to survive in water. He did not find a clear answer in the literature. Instead, he found reports of nonzero probabilities of infection after 21 days, which was the basis for the recommended 21-day quarantine. Similarly, a group of researchers did a deep search in the literature but did not find a clear picture either. The implications of these findings on public health policies, public understanding of science, and information science are striking.

Semantic MEDLINE is a great resource for developing a good understanding of scientific knowledge in terms of semantic predications as well as their original unstructured texts. The complexity of scientific writing is strikingly high. It is common to see long and complex sentences. Studying semantic predications and the contexts in which they appear has revealed how frequently uncertainties go hand in hand with the very knowledge one aims to achieve. Knowledge that is free from uncertainty probably has no value in a research field. Understanding the epistemic status of scientific knowledge is so important that we want to claim that expertise is the knowledge of uncertainty!

The profound and integral role of uncertainty in science, especially in research fronts of a scientific field, has become the core interest of our research. In April and December 2016, two workshops in association with the National Center for Science and Engineering Statistics (NCSES) of the NSF catalyzed the focus on uncertainty further. At the April workshop with the NCSES, Chen presented some of the initial ideas and preliminary results of uncertainties associated with scientific publications in a white paper on the fidelity of visualizing scientific uncertainty.

The preparation and launch of a new open access journal, *Frontiers in Research Metrics and Analytics* (RMA), in midsummer of 2016 provided another boost to the idea. While many have pointed out the shortcomings of overly relying on simplistic and often single metrics of research productivity and quality, evaluators and policymakers are currently limited to only a few options. As a result, it is difficult to compensate the lack of semantic, diagnostic, and analytic reasoning due to over-simplifications of scientific inquiry as a complex adaptive system. The mission of RMA is therefore to bridge the currently loosely coupled research communities. The theme of improving the clarity of the epistemic status of science emerges again in the five grand challenges for accessing and communicating scientific knowledge more efficiently and effectively.

The idea of creating a Visual Analytic Observatory of Scientific Knowledge (VAO) becomes a unifying framework to stimulate and accommodate tools, resources, and applications toward meeting the five grand challenges and beyond. The research project led by Chen is supported by the NSF Science of Science and Innovation Policy (SciSIP) program (Award Number 1633286). The VAO aims to enable researchers to find the epistemic status of scientific knowledge and its provenance of evolution efficiently and effectively. With the worldwide user community of our CiteSpace tool, we believe that the VAO will substantially advance the state of the art. This book introduces the theoretical foundations of how scientific fields develop, which the reader can then use as a referential framework to guide subsequent explorations of scientific knowledge. We also introduce science mapping tools and demonstrate how these tools can help us develop a better understanding of the history and the state of the art of a scientific domain. More importantly, we want to share our methods and principles, both theoretical and practical, with our reader so that we can empower ourselves with computational techniques and analytic reasoning. In particular, creativity comes from competing, contradictory, and controversial views. Reconciliations of existing discrepancies may lead to creative solutions at a higher level. We hope our reader can benefit from the analytic and methodological value of the materials presented in the book.

Chen spent his sabbatical leave at Yonsei University in the Spring semester of 2017 and taught two courses on Yonsei University's beautiful campus on visual exploration of scientific literature. Students from these two classes eagerly and diligently explored and applied the science mapping tools we introduced in this book, namely, CiteSpace, VOSviewer, and CitNetExplorer.

In our previous work, we emphasized the pitfalls and biases of mental models in our reasoning and decision-making. In this book, we aim to demonstrate that uncertainty plays a fundamental role in representing and communicating scientific knowledge.

We are truly grateful for the encouragement and support from many people at various stages of our research and the production of the book. Chen would like to take this opportunity to thank our coauthor Min Song, researchers in his Text and Social Media Mining Lab (TSMM), and students at Yonsei University for collaborative research and the hospitality during Chen's sabbatical in Seoul. Chen is also grateful to Jiangen He and Qing Ping as graduate research assistants at Drexel University, Sergei V. Kalinin at Oak Range National Laboratory for exploring applications of science mapping in material sciences and for organizing a tutorial in Boston, Maryann Feldman for encouragements and guidance, Gali Halevi, Henk Moed, and Mike Taylor for their valuable contributions toward research on tracking emerging trends, Caroline S. Wagner for organizing one of the workshops with NCSES and serving as a guest editor for a Research Topic with RMA, and Jie Li at Shanghai Maritime University for his extensive efforts in disseminating science mapping tools in China. We would like to say thank you to Beverley Ford at Springer for her initiative, encouragement, persistence, and patience.

As always, to the members of Chen's loving family, Baohuan Zhang, Calvin Chen, and Steven Chen, thank you for everything.

Acknowledgements

Chaomei Chen wishes to acknowledge the support of the NSF SciSIP Program (Award Number 1633286) and industrial sponsorship in the past from Elsevier and IMS Health.

Min Song acknowledges the support of the Ministry of Education of the Republic of Korea and the National Research Foundation of Korea (NRF-2015S1A3A2046711).

Philadelphia, USA   Chaomei Chen
Seoul, South Korea   Min Song
August 2017

Endorsement

Chaomei Chen and Min Song have written an important book that opens up a new area in the study in scientometrics and informetrics as well as information visualization, namely the study and measurement of uncertainty of scientific knowledge and how uncertainty is expressed in scientific texts. At the same time the book is a tutorial and review of relevant methods in natural language processing and gives step by step instructions on how they can be implemented. What I like most about the book, however, is how it integrates this new approach with existing theories in the history and sociology of science. In my view, uncertainty is key to understanding the development of scientific knowledge.

Henry Small
Senior Scientist
SciTech Strategies

Contents

About the Authors

Chaomei Chen is Professor in the College of Computing and Informatics at Drexel University and Professor in the Department of Library and Information Science at Yonsei University. He is the Editor in Chief of Information Visualization and Chief Specialty Editor of Frontiers in Research Metrics and Analytics. His research interests include mapping scientific frontiers, information visualization, visual analytics, and scientometrics. He has designed and developed the widely used CiteSpace visual analytic tool for analyzing patterns and trends in scientific literature. He is the author of several books such as Mapping Scientific Frontiers (Springer), Turning Points (Springer), and The Fitness of Information (Wiley).

Min Song is Underwood Distinguished Professor at Yonsei University. He has extensive experience in research and teaching in text mining and big data analytics at both undergraduate and graduate levels. Min has a particular interest in literature-based knowledge discovery in biomedical domains and its extensions to a broader context such as the social media. He is also interested in developing open source text mining software in Java, notably creating the PKDE4J system to support entity and relation extraction for public knowledge discovery.

List of Figures

List of Tables

Chapter 1
The Uncertainty of Science: Navigating Through the Unknown

Abstract Accessing the state of the art of scientific knowledge timely and precisely remains a profound challenge. The vast majority of scientific articles are transient in nature. They may never receive attention from the scientific community or other relevant stakeholders. Advances of science must deal with controversial, conflicting, incomplete, and discrepant information. Uncertainties are an integral part of scientific inquiry and scholarly communication, but their essential role in understanding scientific knowledge as a whole has been considerably underestimated. We introduce a conceptual framework for the study of uncertainties associated with the creation, validation, and communication of scientific knowledge. We utilize science mapping techniques and approaches to illustrate the evolution of a particular body of scientific literature in terms of intellectual landmarks, critical paths, turning points, and boundary spanning bridges.

Introduction

The world's top players of the ancient board game *Go* were handily defeated by AlphaGo, Google's stunning piece of work. The emotionless AlphaGo has quietly unfolded a new world. What most people found amazing (and amusing) was how a machine could make moves so beyond those of a human's, after all it learned from human players' games in the first place. The advances of science and technology have been transforming our lives quietly and firmly through voice recognition, handwriting recognition, face recognition, and self-driving cars. And now, here comes AlphaGo, which seems to be able to teach itself at an alarmingly fast speed.

Despite the seemingly boundless increases in machine capabilities, however, there are many short but penetrating moments that allow us to glimpse another side of science. One man's medicine is another man's poison. Regulatory agencies such as the Food and Drug Administration (FDA) and Centers for Disease Control and Prevention (CDC) know it too well how hard it is to handle uncertainties from

© Springer International Publishing AG 2017
C. Chen and M. Song, *Representing Scientific Knowledge*,
https://doi.org/10.1007/978-3-319-62543-0_1

perceived risks to surprises. We are not far from powerless when facing the natural course of aging and perhaps cancer. We may feel desperate when we realize that modern medicine seems to have neglected us. Our goal here is to address a series of fundamental issues concerning our understanding of the vast, complex, and changing body of scientific knowledge, and how we can make better decisions and deal with more challenging situations.

One of the most fundamental issues is how we can communicate scientific knowledge reliably, accurately, and effectively. What does it take for scientists in the same domain of research to communicate effectively to each other? What more is needed for communication between scientists and the public, or anyone who simply lacks the desirable domain knowledge? Is that even possible? What can we learn from our attempts to communicate to aliens who might happen to be out there?

Scientific knowledge is not a settled painting. It may change in many ways. What are the plausible driving forces? What is pushing the boundary of what we know? Are there potentially critical processes and mechanisms of which we may still lack a good understanding? Would it be at all possible to consider unknowns as part of the big picture?

Mount Kilimanjaro

Africa's highest mountain is Mount Kilimanjaro. It is 5895 m above the sea level. Imagine for a moment that the weight of the entire mountain comes from a small area, with its size negligible to the whole. It may be easier to imagine the uneven distribution of the mass and the vast, mysterious void out there in the universe. It is not always obvious that such uneven distributions also exist in a mountain of scientific articles published by scientists.

On October 29, 2014, Nature celebrated the 50th anniversary of the Science Citation Index (SCI). SCI was originally invented by Garfield (1955). When he passed away earlier this year, he left the world with his increasingly utilized but significantly controversial legend of citation indexing, especially in the realm of research assessment.

In many ways, citations to scholarly publications are similar to the mechanisms of the popular social networking website Twitter. Citations to publications are analogous to retweets of tweets. The more a tweet has been retweeted, in general, the more popular the original tweet was likely to be. In the realm of scholarly publications, a citation refers to an explicit mention of a previously published article in a scientific publication. The article that makes the citation is also known as the citer or the citing article, which is the source of a citation. In contrast, the cited article is also known as a cited reference, which is the target of a citation.

Apart from citations made by researchers in their publications, there are at least two other types of citations worth mentioning. One is the citations in legal documents and the other is the citations in patent applications as well as granted patents.

Citations in legal documents refer to the reasoning and rulings of previous decisions made by judges. For example, the U.S. Supreme Court opinions record documents not only of the decisions made, but also (and perhaps more importantly) the reasoning that led to those decisions. The reasoning and decision-making processes in such contexts are essentially battles between different lines of arguments put forth by different parties. It is in each party's own interest to formulate the best possible argument to win over its audience, typically including the judge, grand jury, opponent lawyers, and defendant. This boils down to the effectiveness of an argument as a form of communication. Different lines of argumentation are likely to collide. Different parties may present different sources of evidence to support their own arguments and undermine their opponents' arguments. More importantly, different parties may offer drastically different interpretations of the same piece of "evidence." In other words, one lawyer's positive evidence could be another lawyer's negative evidence. What qualifies a particular piece of evidence as viable evidence is not entirely subject to the host of the evidence. Rather, the qualification depends on whether it can turn an otherwise unconvincing story to a compelling and plausible one.

Citations in patents can be made either by the inventors or by patent examiners. Considering that inventors and patent examiners differ in their roles and perspectives, their citations are usually made with different motivations. For example, one may expect that citations made by patent examiners are likely to over-estimate their relevance to the invention that is being examined. After all, the reason that a patent examiner is there in the first place is to vet the novelty and the utility of an invention with reference to the prior art. In contrast, the ultimate goal of an inventor is to convince the patent examiner that their invention is indeed novel by showing that other seemingly relevant inventions are in fact not relevant enough.

As we can see, citations in scientific publications share a lot of situational factors with citations in legal decisions and in determining the novelty of an invention. Scholars need to demonstrate and justify their lines of reasoning and differentiate how exactly their work differs from what has been done so far. On the other hand, citations play different roles in different contexts. For example, establishing the novelty of an idea is more central to research and innovations than in practicing law.

Furthermore, citations are not created equal. Citations are made by judges, patent examiners, and researchers, who all differ as much as one can imagine. Their decisions and judgements are determined by their knowledge of the domain in questions, their past experiences, and their perceptions and interpretations.

The simplest way to estimate the value of a scholarly publication is the total number of times it has ever been cited worldwide across all languages in non-scientific as well as scientific publications. The more a reference has been cited, the more valuable it is likely to be. This is a simple assumption. It has numerous exceptions and it has been criticized and misinterpreted since it was introduced. Admittedly, this assumption is based on many more simplifications. For example, any citations take place within the space that we have been observing. Additional citations take place beyond the boundary of our observation would not be reflected.

This is an issue about the scope. Google Scholar, Scopus, and the Web of Science are widely known sources of scholarly publications and they cover the largest scopes.

Another issue is more complex. As we have mentioned earlier, citations are not created equal for epistemological reasons as well as sociological and philosophical reasons. We cannot determine whether two references with the same number of citations will have the same degree of importance or value. We cannot determine whether a reference with 100 citations is worth as twice as more than a reference with 50 citations, even if they appear to be equal otherwise, for example, with the same date of publication in the same journal. Even then one could argue that it is not possible to ensure they can get exposed equally to their audience. Indeed, there are researchers who suspect that the running order of articles in the same issue may advantage some and disadvantage others. The third issue, which is even more complex, is the question what exactly citation counts represent.

Before we turn to these questions in detail, let us take a look what *Nature* did to convey the disproportionally distributed citations on the backdrop of the Mount Kilimanjaro.

The height of 100 sheets of paper is about one centimeter, or 1/100 of a meter. In 2014, the coverage of the Science Citation Index (SCI), or more commonly known as the Web of Science, was about 58 millions of cited references. If we take one sheet from each of the 58 million articles and the height of 100 sheets is approximately 1 cm, then the core of scientific literature will be a mountain of papers as high as Mount Kilimanjaro (Fig. 1.1).

Unlike the more or less evenly distributed mass on Mount Kilimanjaro, the distribution of citations on the mountain of scientific literature is rather disproportional. The vast majority of the mountain, from the sea level to the 4400 m level, would be composed of articles that have never been cited at all or cited just once. Though taking up about 85% of the mountain's volume, their impact at this scale is indifferent. Above the 4400 m level and all the way up to the level that is 1 m below the peak (5894 m), the mountain is occupied by articles having 10–1000 citations. Articles with over 1000 citations would climb within 1 m to the top, but the last one centimeter of the mountain would be positioned by 100 articles most cited since 1900.

What kinds of articles can make it to the top one centimeter? In other words, using a simplistic measure of impact as perhaps naïve as the citation count, what kinds of articles tend to attract the attention of the scientific community? What kinds of articles are more likely to relatively stable positions on the mountain than others? Perhaps more importantly, what kinds of articles that can move quickly from the bottom of the mountain to the top?

While the mountain metaphor nicely illustrates the distribution of citation frequencies, there are many questions that cannot be answered with the mountain metaphor. For example, the position of an article on the mountain does not correspond to its current position in a latent and potentially gigantic structure of scientific knowledge. Two equally cited articles at the same level on the mountain may be light-years apart in the conceptual universe of the scientific knowledge

Fig. 1.1 Mountain of scientific publications indexed in the Web of Science (as of 2014). *Source* Reproduced with permission from http://www.nature.com/news/the-top-100-papers-1.16224# mountain

conveyed by published scientific articles. Furthermore, if a highly cited article was retracted, where would it reposition itself? Would it remain to be at its position secured purely by its citations? Or should it be pulled out of the mountain altogether and put it on a separate mountain? From the point of view of individual researchers, being cited negatively or critically is probably much better than not being cited at all.

We are interested in how scientific knowledge should be represented so that we can characterize the growth and decay of current research areas. For our purpose, a more suitable metaphor is the universe of concentrations of knowledge separated by voids of unknowns. The notion of a knowledge space is not uncommon. Many have explored various aspects of such frameworks. However, we will take the universe of scientific knowledge for a different and unique spin.

Heilmeier's Catechism

There are many guidelines and advices on how to do research properly. The simplest and most well-articulated one is from George Heilmeier, known as Heilmeier's Catechism. He was the director of DARPA. His catechism is organized as a series of 8–9 straightforward questions, depending on minor variations of the detail. By answering these questions, or simply by considering these questions, one would be able to gain a clear understanding of what a research project should accomplish. The emphasis on the simplicity of the entire communication process is the key to clear communication and an understanding of extraordinarily complex that one may give up the idea to explain it to someone who may not even have the necessary domain knowledge to begin with. Heilmeier's Catechism tells us that no matter how complex a research topic is, it is always possible to explain it in such a degree of clarity that so anyone can understand. Next time, before we are going to tell someone "you wouldn't understand", think about Heilmeiier's Catechism.

Heilmeier's Catechism is a series of straightforward questions. The straightforwardness of a question does not mean it is easy to answer. Rather, it requires that the answer to the question needs to be at the same level of clarity—easy to understand without any prior knowledge of any domain. The ultimate goal is to communicate our ideas, no matter how complex, to someone else in the simplest possible way so that they can understand at a higher level of abstraction. Once we set the stage for the question-and-answer dialog, we can get started with the questions.

The first question is "what are you trying to achieve"? The question should be answered without using any jargons so that anyone can understand it. Possible answers could include "to make a car that can drive itself smoothly in the busiest road in rush hours" or "to make computer software that can fix its own problems."

The second question is to figure out where you are now and how hard it is to arrive where you want to be. If our goal was to make a self-driving car that can navigate through rush-hour traffic, we would need to know the capability of today's best self-driving cars. We would also need to know critical obstacles between us and our destination. Developing the degree of an awareness needed is typically time-consuming and cognitively demanding, especially when we have to construct a big picture of the situation we are in from scratch all by ourselves. Given a domain of research, there are several ways potential sources where one can obtain structural information at a macroscopic level. Notably, if a review or survey of the domain is readily available, it could save a lot of time and effort. The best review is certainly a comprehensive review that is relevant, timely, and highly recognized. In the scientific field, therefore, the best review is one published in relevant journals or conference proceedings, using the same set of keywords, published as recent as possible, with many citations. If we want to obtain a big picture of our domain of research, we can perform a meta-analysis, which is a study of studies. There are, in fact, meta-meta-analysis,[1] i.e. a meta analysis of meta-analyses.

Heilmeier's second question is critical. As far as the researcher is concerned, in order to answer this question adequately, one needs a good understanding of not only what has been done in the past, but more importantly, how they are related to the original research goal. Research articles routinely include sections on "related work." However, the quality of how related work is written varies considerably. A good summary of related work should focus on the relatedness between a previous study and the research question at hand. The nature of the connection should be made explicit. If, as it turns out, a previously published study is not related to the research question after all, then the writer should eliminate the previous study from the discussion of related work. A study becomes part of related work only if it qualifies.

In order to answer the second question thoroughly and comprehensively, the best option is to do so with a systematic review of the domain that covers relevant research topics. Now it becomes clear that Heilmeier's second question could be really challenging to answer.

The third question is about what is new and what is special in your plan. After all, what is the key that may bring you the cutting edge that those relevant efforts made so far do not have? To answer this question, one needs to figure out not only what has been done and what falls short, but also what may unlock the current situation and resolve the problem that didn't seem to have a better solution until now. This is where the creativity comes in. There are numerous theories of discovery. The more theories we know of, the more likely we will be able to recognize opportunities that could be otherwise overlooked. In the study of social networks, for example, research has shown that those who have came across the concept of structural holes are more likely to be able to recognize them than those who have never been exposed to the concept. It is often the case that our imagination is rather limited by our prior knowledge and experience. Once a teacher asked a class of students to draw what aliens would look like in their wildest imagination. The resultant drawings, as it turned out, still show strong influences of the sources of their inspiration—what we see in our everyday life or what we see in museums. On the other hand, there are strategies that can be applied to a broad range of situations, for example including ones such as divide and conquer, recombination, thinking of

---

[1]https://doi.org/10.1016/0005-7967(79)90011-1.

the opposite, and many others. The fundamental value of such generic strategies is their potential role as a good icebreaker—when we are in a situation in which there is no clear clue of the next move we should make. It is commonly seen in science when a new approach is proposed in the opposite direction that would be recommended by a traditional approach. As a matter of fact, in this book you will see several this type of approaches, including the seemingly odd combination of uncertainty and scientific knowledge.

Other questions in Heilmeier's Catechism include questions that are more concerned with project management issues such as milestones and budgets. How will we know your project is on track? How much will it cost?

If one can confidently answer these straightforward questions from Heilmeier's Catechism, he or she is very likely to obtain the expertise required to carry out a research project. The less jargon used in one's answers, the deeper understanding he or she has reached. In this book, what we are concerned is what options are available for us to obtain such levels of expertise effectively. As more scientific articles are published, we will be dealing with a mountain of papers that is not only increasingly higher and larger than Mount Kilimanjaro, but also undertaking fundamental changes. The answers we worked out yesterday may be no long valid with today's new landscape.

How Does Science Advance?

Science produces scientific knowledge. Scientists are engaged in scientific activities. From the viewpoint of an individual researcher, Heilmeier's Catechism suggests a conceptual framework to characterize what research is about. Simply speaking, the researcher will select a topic of interest and formulate a research question to answer. Then the researcher will need to understand where he or she is with reference to where he or she wants to be. The next critical step for our researcher is to figure out how to get there from here. Along the way, new problems may emerge, which need to be solved in order to move on. Sometimes new problems become so significant that they may overshadow the original research question. Individual researchers may sidetrack and pursue new lines of research.

Individual researchers are not alone. How do their research inquiries influence others? How are they influenced by others' work? What can we learn from scientists as a group or as a community?

The research community that is active at the forefront of a research area has been referred to as an invisible college. On the one hand, the existence of an invisible college underlines the fact that the publically accessible scientific literature only represents the body of scientific knowledge partially. Beyond what is published in peer reviewed journals, books, and conference proceedings, there are numerous publications in what is known as the gray literature, including self-archived preprints, technical reports, and rejected manuscripts. Compared to the gray literature, the vast majority of published articles positioned below 4400 m on Mount

Kilimanjaro should be considered the lucky ones. At least in the traditional pub-lication model, published articles should be more accessible than documents that have not obtained the blessing of anonymous reviewers. The rapid growth of the open-access journal industry is presenting significant challenges that traditional journal publishers have to face. Many traditional journals now offer authors to choose whether to publish their articles in the open-access track in an otherwise traditional subscription-based journal.

Figure 1.2 illustrates the evolution of the research of scientometrics in terms of a progressively synthesized network of references cited by articles published in the journal Scientometrics. The colors of the areas indicate the average age of a cor-responding area, starting with the navy blue area from the left, passing through the light green, dark green, and reaching the most recent yellow/orange area on the right. The red circles are cited references that have been viewed frequently over the last 6 months when the dataset was retrieved from the Web of Science. This holistic view of the development of a research field, as reflected by a single journal in this case, underlines a feature that is commonly seen with many other fields of research —the research field is constantly moving ahead. In general, as time goes by, researchers' attention moves along. The state of the art of scientific knowledge changes over time.

What makes the scientific community work so hard to make new discoveries? What makes the universe of scientific knowledge transform itself? There are dif-ferent answers from different perspectives, ranging from epistemological, socio-logical, practical, historical and evolutionary.

Paradigm Shift and Gestalt Switch

The most widely known philosophical theory of the development of scientific knowledge is Thomas Kuhn's structures of scientific revolutions (Kuhn 1962). The key to the development of science is a transformative process characterized by a paradigm shift. According to Kuhn, the development of science is characterized by a process that moves from a normal science stage, to a stage of crisis, a paradigm shift, and a scientific revolution. The process is iterative in nature, with a scientific revolution leading to another cycle of normal science, new crises, and new revolutions.

The central concept in Kuhn's theory is a paradigm. The life and death of a paradigm at a particular time represents the status of scientific knowledge at the time. As the scientific community adopts a new paradigm and abundant a previ-ously predominating paradigm, a scientific revolution takes place. Kuhn's theory has had a profound impact on the scientific community and beyond, but also drawn many criticisms. For example, the incommensurability between competing para-digms means the lack of an accommodating framework to fit all competing para-digms so that one can make comparisons between them. Thus, one may wonder whether the development of science is moving to a position that is somewhat better

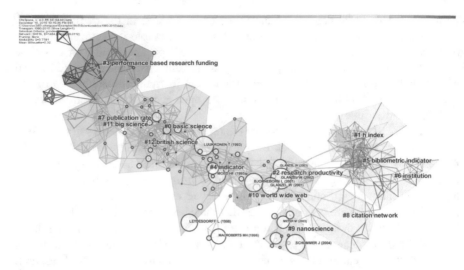

Fig. 1.2 This network of co-cited references reveals rapid changes of research topics in Scientometrics (1980–2015)

than before, and how paradigmatic improvements can be measured. Another common criticism is that ground-shaking revolutions demonstrated by Kuhn are rare events after all; therefore, Kuhn's paradigm shifts may never come for a particular field of research. Other criticisms accuse Kuhn's historical perspective misinterpret what is really happening.

Suppose for a moment Kuhnian paradigms do exist and do behave like how they are described in Kuhn's theory, then how would they show up in the scientific literature? Are they observable? Each paradigm has its own unique set of concepts and its own terminology. What can we expect to see at each stage of the Kuhnian process? At the stage of normal science, we can expect a well-established nomenclature. Variations of terms would have converged. Landmark cases and exemplar studies would have been widely recognized. At the stage of a crisis, we can expect the frequency of descriptions and discussions of abnormality that causes the crisis would stand out. New evidence and new discoveries would be identified. We assume there are ways one can differentiate discussions associated with a crisis from other types of discussions. Later in the book, we will introduce computational techniques for such purposes. Finally, let's imagine what would we see if a scientific revolution does take place. A new paradigm would emerge with reference to other existing paradigms. New concepts and distinct terminologies would appear.

How can we computationally identify the emergence of new concepts and any markers of a new paradigm? If we can detect key signatures of a paradigm, then we can systematically study the rise and fall of a paradigm. Furthermore, if we can identify the constructs predicted by other theories, then we might be able to gain deeper insights into the development of scientific knowledge through a wider spectrum of perspectives than what a single perspective could possibly offer.

We should note that there are different interpretations or even misinterpretations of Kuhnian paradigm shifts. In our view, Kuhnian paradigm shifts can take place at multiple levels of granularity. In other words, a field of research in its normal science stage may contain areas or branches that are in their own paradigmatic stages. A paradigm shift in its broadest sense can define an individual's pursuit of a specific research question as well as a community of researchers. The key value to the paradigm shift model is the role of gestalt switch in our course of discover and creative problem solving. There is no reason for us to limit the scope or the scale for gestalt switch to become valid or invalid. Rather, the extended form of gestalt switch characterizes an important mechanism for us to model the world.

Kuhn's theory is philosophical and historical. Many scholars have tackled the question from different disciplinary viewpoints. Stephen Fuchs, for example, offers his theory from a sociological perspective (Fuchs 1993). Naturally, in his theory, social contexts play a fundamental role.

Competing for Recognition

To Stephen Fuchs, the fundamental force that pushes scientific knowledge forward is not coming from the inside of knowledge per se, at least, that is not the most important reason. Rather, he argues, the advances of science benefit from scientists' competition in the social environment where they belong. Scientists compete for reputations and many more things that tend to follow once you have established a remarkable reputation, for example, more resources such as funding and research students and more opportunities that would further reinforce your reputation such as invited talks, advisory roles and memberships of a board of directors. To win such competitions, scientists would need to stand out despite waves and waves of new publications that flourish the fast-growing and often equally fast forgotten scientific literature.

Productivity certainly draws attention, especially publications on high-profile and prestigious journals. Publications in those journals have a better chance to position your articles near to the top of Mount Kilimanjaro. More fundamentally speaking, however, more publications may not necessarily give us the right recipe. The key to establish and maintain the reputation of a researcher is to make high-impact and novel ideas. To Fuchs, the novelty-centered research is the consequence of competition for recognitions and reputations in a social environment. Awards and prizes are given to scientists who made the original breakthroughs. In scientific writings, we often see claims of the novelty and the originality—we are the first to introduce this idea and we are the first who thinks of this solution.

In this sociological theory of scientific change, Stephen Fuchs emphasizes the influence of social structures on the severity of competition and the perceived value of outcomes from a particular area of research. The interaction between social structures and the severity of competition would, in his view, explain the differences between natural sciences and social sciences. In natural sciences, scientists work in a dense social-cognitive environment in the sense that one's work is more likely influenced by others who work in the same area than in social sciences. One scientist's win may be other scientists' loss. Once a novel discovery is made by one or one group of scientists, the efforts and investments of other scientists that have made in searching for the solutions to the same research problems will be substantially discounted. Recognitions, additional funding, and other resources will be disproportionally allocated to the scientists who beat the others in the intellectual race. And the others who had worked equally hard if not even harder will be lumped into the category of the others. Therefore, the stake is high. As Fuchs mentioned in his writing, there are scientists who decided to pursue other career paths. There are also scientists who migrate to other research areas as they see a better chance to exercise their competitive edges from their original professions.

If the scientific community works as Stephen Fuchs has described, what signs, features, patterns, or trends should have been observable and traceable in scholarly publications? How would different groups of scientists in the same subject domain interact through their publications, or more precisely, through specific claims and declarations made through their publications? If we apply the competition theory to the study of scientific collaboration, what can we expect to uncover?

What does it imply if we consider Kuhnian paradigms and Fuchs' competition simultaneously? What would it mean to scientists to compete for recognition in a field of research that is considered in a period of normal science? How would one win a competition in an environment that is predominated by an established paradigm? Perhaps one can win by a high productivity or by scalability. If one can accomplish the same task a lot faster, cheaper, or simpler, then it should be clearly seen as a competitive potential or an actual competitive gain. Our competitors can be more ambitious. They may induce a new scientific revolution. That would definitely fulfill their competitive motive, although the path would be dangerously stiff. First of all, they would have to be able to poke a hole in the foundation of the currently well-accepted paradigm. Then they would have to convince the scientific community that the problems with the current paradigm are beyond the repair and that there is a better alternative and indeed a life-saving new paradigm to accomplish the mission that bounds the community in the first place. The most cost-effective Kuhnian stage for competing scientists would be the crisis stage. The stage is set to search for a solution that would settle the crisis. Some scientists are likely to be better positioned than others to take the lead because they have adequate resources and abilities to adapt new patterns of research.

As we can see, considering two theories that are supposed to characterize the same phenomenon could be productive because it may lead to many ways to verify expected features or signs empirically. This is a strategy that we will recommend and follow in our own book so that we can understand the complexity of the growth of scientific knowledge from different perspectives. Next, we will introduce yet another model and we will attempt to use these theories to interpret each other so as to identify the underlying concepts that are in fact common across different theories and unique concepts that might explain the differences between distinct perspectives.

The Evolution of a Specialty

The third theory of how a specialty evolves is proposed by Shneider (2009). A specialty refers to a community of researchers who share some common interests. Usually there is no clear way to define the membership of a specialty except that members should have interests in some common research issues. Such interests could be demonstrated in publications, participations in conferences, and other engagements in relevant research activities. It is quite possible that someone who is deeply interested in the research, but has no relevant publications or is not yet visible on the new stage; for example, doctoral students in their early years or a seasoned research who is recently attracted to this part of the world would belong to this group. Since our primary focus is on scientific knowledge that is publically available, researchers without publications may be temporarily invisible from the specialty.

In Shneider's evolutionary theory, the development of a scientific discipline can be characterized by a four-stage evolution of a specialty. The first stage marks the birth of a new specialty. The new specialty is conceived and the central research questions are identified for researchers to answer. The scientific community would probably know very little about these new research questions. Thus the initial formulation of these research questions may still lack of critical details or the level of precision and rigor that one may wish to have, but the attraction would be strong enough. Researchers would be enthusiastic, curious, or otherwise intrigued by a possibly broad-brushed sketch of a research agenda. The next stage is the next logical step—building enabling or augmenting tools so that one can study the original research problems thoroughly and systematically. The second stage is therefore a tool-construction stage. Several good examples come from astronomy. An initial conceptualization could be motivated by the idea to scan the entire sky so that researchers can streamline their research with the large amount of detailed and easy to access data. A notable example is the Sloan Digital Sky Survey (SDSS), which sparked even large scale surveys later on. The tool construction stage would come naturally for such projects. Powerful telescopes and more sensitive cameras would be built to meet the requirements of such tasks.

The third stage of the evolution is to apply the tools that are made available in the second stage to carry out the original plan. The overall productivity of the specialty would be increased sharply as researchers are now equipped with purposefully designed powerful tools. The threshold for entering the research would be substantially lowered due to now routinely available instruments and readily accessible high-quality data. A considerably lowered threshold is likely to introduce new people and trigger new ideas that one might have considered them infeasible. In other words, with easier to use and more powerful tools in hand, researchers would be eager to find new targets to make unexpected use of their tools. Consequently, as more and more applications of these tools go beyond the original scope of research, we bound to come across findings that will surprise us. Surprising findings may trigger a Kuhnian crisis in the specialty, which in turn may lead to a paradigm shift. Some researchers have long been interested in how a new specialty may branch off from an existing specialty. The original specialty may continue to go on with its original research agenda, while the new spinoff specialty may take different turns to pursue something that might not be possible within the original specialty's framework.

The fourth stage is characterized by reflections and summarizations of what have been achieved by the specialty. The lessons learned and the knowledge gained are codified into comprehensive reviews and textbooks. The specialty is matured and its structure becomes relatively stable. As a specialty, it may be sustainable if it continuously draws inputs from its environment; otherwise, it may be dissolved as its members leave for other specialties where they can find more fulfilling roles to play, especially those that are in their earlier stages of the evolution.

Now we have three theories of the development of scientific knowledge and the evolution of the associated research community. Each of these theories casts a unique perspective on the complex and dynamic matter. Each of these theories aims to characterize macroscopic properties of scientific research that involves not only individual scientists and scholars but also loosely or tightly-coupled groups. Each of these theories gives a different explanation of how a research field advances. We are now in a situation that is very similar to what blind men must have gone through when they realize that other people seem to have a different idea about what an elephant is. Is it like a tree trunk, a snake, a spear, or a rope? Are we even talking about the same elephant?

Perhaps a period of Kuhnian normal science overlaps with later stages in Shneider's theory, namely the tool construction stage, the application stage, and perhaps the reflection stage. A specialty in these stages would be governed by a relatively stable paradigm. In contrast, the crisis stage in Kuhn's theory shares many characteristics of Shneider's stages in which the dominant paradigm is questioned and challenged or an alternative path has emerged as a promising route to pursue. There is a greater chance to see such alternatives when researchers in one specialty reach out beyond the original scope of their research or when they encounter perspectives or models of the world that profoundly differ from theirs. Research in creativity and the history of scientific discovery has noticed the connection between the clutch of conflicting or contradictory views and making revolutionary discoveries. Great discoveries are often connected with resolving contradictory conceptualizations at a new level of abstraction. Contradictions that seem to be irreconcilable at one level may be resolved surprisingly from a fresh perspective at another level.

To Kuhn, the fresh perspective may represent a promising new paradigm. To Fuchs, researchers who make novel contributions would increase their competitiveness, whereas researchers who can create a new field of research would be immensely recognized. To Shneider, applying research methodologies beyond their originally intended scope increases the chance to uncover something unexpected, which in turn may lead to the emergence of a new specialty. In short, the development of scientific knowledge is so complex that each of the existing theories may capture only a small part of the underlying phenomenon. It would be valuable if we can learn from the discrepancies, controversies, and contradictions of scientific knowledge as highlighted by different theories from different angles.

Searching for the Unknown

Some of the most fundamental challenges for scientists to make novel leaps are related to the fact that when we know little about the target of our search, it is hard to work out a search strategy that would take us to the target. It may not occur to us to look for something that is beyond the radar of our mind in the first place.

A study in the late 1960s from the National Science Foundation (NSF) named TRACES[2] revealed long-term patterns of several high-impact inventions such as a video tape recorder. Some common patterns emerged from the diverse range of inventions studied by experts in corresponding domains. First, each of the inventions is built on multiple lines of research that may be traced back 50 years ago or even a longer period of time in the past. Second, different lines of research did not communicate with each other and did not converge until much later in the overall course of the development. In other words, individual lines of research did not have any idea of the roles it might play so many years down the road. Thus, for a considerably long period of time, there would be no reasons for different lines of research to be aware of the existence of others. For two tribes separated by a mountain, what would trigger them to think beyond their own tribe? And what would make them wonder what the other side of the mountain may look like?

Let's imagine that there is a river passing along one side of the mountain but none on the other side. The cultures of the two tribes would be different; at least one is evolved from what one can do with the river, for example, fishing, and the other is resulted from how to make a living by the mountain, for example, hunting for rabbits in the woods. How can the two tribes establish the first contact? Perhaps one tribe could suddenly realize there may be someone else out there except themselves. One can think of many scenarios of what could happen, for example, one tribe hearing voices from the other side of the mountain at a very quiet night, members of one tribe chasing a rabbit around the mountain and accidently discovered the other tribe, or some traveler from far away who just wanted to explore what is around the mountain. However, it appears that these scenarios share some common elements—the initial contact between different tribes is likely to be accidental rather than planned.

---

[2]Illinois Institute of Technology (1969). TRACES.

Scientific Controversies

Controversus is a Latin term and means "turned in an opposite direction." Controversies are the clashes of opposing opinions. Controversies are equivalent to conflicts and contradictions. A controversy needs to last long enough to become a controversy. It needs to involve a considerable number of scientists.

Conflicts and Contradictions in Science

Studies of science have long realized the role of conflicts in science, especially in research areas with significant social political implications such as climate change. Mazur (1987) compared the role of scientific controversies in science with the role of wars in history: "Just as historians used to chart the course of empires by tracing the links from one war to another, one could write a passable history of modern science by linking the great theoretical and experimental controversies." Making sense of conflicting findings is an integral part of science (Endrikat et al. 2014).

Brante and Elzinga (1990) proposed a theory of scientific controversies. They identified a few reasons why scientific controversies have not been studied adequately. One reason is that controversies are seen as abnormal episodes in the course of science and they are something to be eliminated. Brante and Elzinga noted that the frequent occurrence of controversies is an indicator of tensions rooted deeply in the very heart of science.

A controversy implies the existence of contradictions at a deeper level. A scientific controversy is primarily concerned with contending knowledge claims. In general, controversies are likely due to contradictions of a theoretical, social, or other type in nature. A controversy can be understood as a structural breaking point or rupture as Brante and Elzinga argued.

Different kinds of conflicts define different types of scientific controversies. McMullin (1987), for example, classifies controversies with respect to facts, theories, and principles. Controversies of fact dispute what is observed. The famous example is when Galileo saw mountains on the Moon, the experts of the Church contested or even refused to look through Galileo's telescope. Theoretical controversies are at a higher level of abstraction because they are resulted from discrepancies in explanations of two or more different theories. Controversies of principle are more general, involving the foundations of scientific disciplines, as in the example of Heisenberg's uncertainty principle.

Brante and Elzinga identified three different approaches to controversies, namely, epistemological, descriptive, and political approaches. Epistemological approaches focus on philosophical issues such as how one should validate knowledge. Epistemological approaches commonly compare the argumentative structure of the contending controversies. The concept of incommensurability is a key to epistemological approaches. Controversies between rival paradigms are incommensurable. Incommensurable controversies may lead to Kuhnian scientific revolutions because there is no overriding rational criterion. Thus, it is important to distinguish incommensurable and commensurable controversies.

A descriptive approach is a historical one in nature. It studies the course of development of a controversy in terms of three overall phases: emergence, development, and termination. Political ones emphasize the importance of context.

Studying scientific controversies brings several benefits to science. Scientific debates disclose and clarify otherwise hidden premises and tacit assumptions. Contending parties need to strengthen their arguments more carefully because their opponents will examine and exploit any weaknesses in their arguments.

Research in related areas include a conflict view of intellectual change. For example, Collins (1989) presented a conflict view of intellectual change. New ideas are generated through a conflict process in which intellectual rivals negate each other. He also introduced the notion of intellectual attention space. He is one of the few influential researchers who are particularly interested in how intellectual conflicts shape the intellectual world.

The Mass Extinction Debates

A good example of a length scientific debate is on the causes of dinosaurs' extinctions. Five mass extinctions have occurred in the past 500 million years on earth, including the greatest ever Permian-Triassic extinction 248 million years ago and the Cretaceous-Tertiary extinction 65 million years ago, which wiped out the dinosaurs among many other species. The Cretaceous-Tertiary extinction, also known as the KT extinction, has been the topic of intensive debates over the last twenty years, involving over 80 theories of what caused the mass extinction of dinosaurs. Paleontologists, geologists, physicists, astronomers, nuclear chemists, and many others are all involved.

Five mass extinctions occurred in the past 570 million years on earth. Geologists divided this vast time span into eras and periods on the geological scale. The Permian-Triassic extinction 248 million years ago was the greatest of all the mass extinctions. However, the Cretaceous-Tertiary extinction, which wiped out the dinosaurs from the earth 65 million years ago within a short period of time along with many other species, has been the most mysterious and hotly debated topic over the last two decades.

Dinosaurs' extinction occurred at the end of the Mesozoic. Many other organisms either became extinct or were reduced greatly in abundance and diversity. Among these were the flying reptiles, sea reptiles, and ichthyosaurs, the last disappearing slightly before the Cretaceous-Tertiary boundary—known as the K-T boundary. Strangely, turtles, crocodilians, lizards, and snakes were not affected or were affected only slightly. Whatever factor or factors caused it, there was a major, worldwide biotic change at about the end of the Cretaceous. But the extinction of dinosaurs is the best-known change by far and has been a puzzle to paleontologists, geologists, and biologists for two centuries. Many theories have been offered over the years to explain dinosaur extinction, but few have received serious consideration. Proposed causes have included everything from disease, heat waves and resulting sterility, freezing cold spells, and the rise of egg-eating mammals, to X rays from a supernova exploding nearby. Since the early 1980s, attention has focused on the impact theory by the American geologist Walter Alvarez, his father, the physicist Nobel Prize winner Luis Alvarez, and their colleagues.

There have been over 80 theories of what caused the extinction of dinosaurs, also known as the KT debate. Paleontologists, geologists, physicists, astronomers, nuclear chemists, and many others all have been involved in this debate (Alvarez 1997). Throughout the 1980s the KT debate was largely between the impact camp and the volcanism camp. The impact camp argued that the KT extinction was due to the impact of a gigantic asteroid or comet, suggesting a catastrophic nature of the KT extinction. The volcanism camp, on the other hand, insisted that the mass extinction was due to massive volcanism over a much longer period of time, implying a gradual nature of the KT event. The impact camp had evidence for the impact of an asteroid or a comet, such as the anomalous iridium, spherules, and shocked quartz in the KT boundary layer, whereas the volcanism camp had the Deccan Traps, which was connected to a huge volcanic outpouring in India 65 million years ago.

Catastrophism

In their 1980 Science article, Alvarez et al. (1980), a team of a physicist, a geologist, and two nuclear chemists, proposed an impact theory to explain what happened in the Cretaceous and Tertiary extinction. In contrast to the widely held view at the time, especially by paleontologists, the impact theory suggests that the extinction happened within a much shorter period of time and that it was caused by an asteroid or a comet.

In the 1970s, Walter Alvarez found a layer of iridium sediment in rocks at the Cretaceous-Tertiary (K-T) boundary at Gubbio, Italy. Similar discoveries were made subsequently in Denmark and elsewhere, both in rocks on land and in core samples drilled from ocean floors. Iridium normally is a rare substance in rocks of the Earth's crust (about 0.3 parts per billion). At Gubbio, the iridium concentration was found more than 20 times greater than the normal level (6.3 parts per billion), and it was even greater at other sites.

There are only two places one can find such high concentration of iridium: one is in the earth's mantle. The other is in extra-terrestrial record. Iridium can be found in the earth's mantle and in extra-terrestrial objects such as meteors and comets. Scientists could not find other layers of iridium like this above or below the KT boundary. This layer of iridium provided the crucial evidence for the impact theory. However, the impact theory has triggered some of the most intense debates between gradualism and catastrophism. The high iridium concentration did not necessarily rule out the source could not be from the Earth.

Gradualism

Gradualists believed that mass extinctions occurred gradually instead of catastrophically. The volcanism camp is the leading representative of gradualism. The volcanism camp had a different explanation of where the iridium layer in the KT boundary came from. They argued that this iridium layer may be the result of a massive volcanic eruption. The Deccan Traps in India was dated 65 million years ago, which coincided with the KT extinction; the Siberia Traps was dated 248 million years ago, which coincided with another mass extinction—the Permian-Triassic mass extinction, in which as many as 95% of species on Earth were wiped out. The huge amount of lava produced by such volcanic eruptions would cause intense climatic and oceanic change worldwide.

Another line of research has been focusing on the periodicity of mass extinctions based on an observation that in the past there was a major extinction about every 26 million years. The periodicity hypothesis challenged both the impact theory and the volcanism to extend the explanation power of their theories to cover not only the KT extinction alone but also other mass extinctions such as the Permian-Triassic mass extinction and other major extinctions. Some researchers in the impact camp were indeed searching for theories and evidence that could explain why the Earth could be hit by asteroids or comets every 26 million years.

A watershed for the KT impact debate was 1991 when the Chicxulub crater was identified as the impact site on the Yucatan Peninsula in Mexico (Hildebrand et al. 1991). The Signor-Lipps effect was another milestone for the impact theory. Signor and Lipps (1982) demonstrated that even for a truly abrupt extinction, the poor fossil record would make it look like a gradual extinction. This work in effect weakened the gradualism's argument.

In 1994, proponents of the impact theory were particularly excited to witness the spectacular scene of the comet Shoemaker-Levy 9 colliding into Jupiter because events of this type could happen to the Earth and it might have happened to dinosaurs 65 million years ago. The comet impacts on Jupiter's atmosphere were spectacular and breathtaking.

In the controversy between the gradualist and catastrophist explanations of the dinosaurs' extinction, one phenomenon might not exclude the other. It was the explanations of the highly concentrated layer of iridium that distinguished two competing paradigms. The catastrophism was one of the major beneficiaries of the periodicity paradigm because only astronomical forces are known to be capable of producing such a precise periodic cycle. There were also hypotheses that attempted to incorporate various terrestrial extinction-making events such as volcanism, global climatic change, and glaciations. There was even a theory that each time an impact triggered the volcanic plume, but supporting evidence was rather limited. A few landmark articles in the periodicity frame addressed the causes of the periodicity of mass extinctions using the impact paradigm with a hypothesis that asteroids or comets strike the earth catastrophically every 26 million years.

The initial reaction from the impact camp was that the periodicity hypothesis completely conflicted with the impact theory. What can possibly make asteroids hit the earth at such pace? The impact paradigm subsequently came up with a hypothesis that an invisible death star would make it possible, but the hypothesis was still essentially theoretical.

Public Understanding of Science

Public understanding of science involves a broad range of fundamental and practical aspects, from public health policies that may influence everyone's daily life immediately to science policies that may lead to profound impacts in a long run but for many of them the differences are simply too soon to tell. One can organize various issues in some loosely defined categories:

1. Communicating scientifically settled knowledge to the public
2. Communicating something that even scientists do not have a sufficiently clear understanding, including controversial, conflicting, or contradictory findings, to the public

Communication in the first category itself can be a very challenging task, especially if one needs to communicate complex scientific knowledge to the public or to anyone who does not have any domain knowledge. As what George Heilmeier advocates, the most effective communication is to convey the idea clearly without using technical jargons. It is a tremendous challenge for both scientists and the public to handle communications in the second category effectively and efficiently when we navigate between the interplays of established scientific knowledge, new discoveries in the making, and the uncertainty of the unknown (Corbett and Durfee 2004; Fischhoff 2013). For instance, government regulatory agencies such as the US Food and Drug Administration (FDA) need to assess, regulate, and communicate risks of new medical interventions that may involve a variety of uncertain, incomplete, and potentially contradictory information from scientific inquiries in the making (Institute of Medicine 2014).

The Tower of Babel

Ancient Mesopotamians believed that mountains were holy places and gods dwell on top of mountains and such mountains were contact points between heaven and earth, for example, Zeus on Mount Olympus, Baal on Mount Saphon, and Yahweh on Mount Sinai. But there were no natural mountains on the Mesopotamian plain, so people built *ziggurats* instead. The word *ziggurat* means a "tower with its top in the heavens." A ziggurat is a pyramid-shaped structure that typically had a temple at the top. Remains of ziggurats have been found at the sites of ancient Mesopotamian cities, including Ur and Babylon.

The story of the *Tower of Babel* is in the Bible, Genesis 11: 1–9. The name Babylon literally means "gate of the gods." It describes how the people used brick and lime to construct a tower that would reach up to heaven. According to the story, the whole earth used to have only one language and a few words. People migrated from the east and settled on a plain. They said to each other, "Come, let us build ourselves a city, and a tower with its top in the heavens, and let us make a name for ourselves, lest we be scattered abroad upon the face of the whole earth." They baked bricks and used bitumen as mortar. When the Lord came down to see the city and the tower, the Lord said, "Behold, they are one people, and they have all one language; and this is only the beginning of what they will do; and nothing that they propose to do will now be impossible for them. Come, let us go down, and there confuse their language, that they may not understand one another's speech." So the Lord scattered them abroad from there all over the earth, and they left off building the city. Therefore its name was called Babel, because there the Lord confused the language of all on the earth; and from there the Lord scattered them abroad over the face of the earth. Archaeologists examined the remains of the city of Babylon and found a square of earthen embankments some 300 ft on each side, which appears to be the foundation of the tower. Although the Tower of Babel is gone, a few ziggurats survived. The largest surviving temple, built in 1250 BC, is found in western Iran.

The Tower of Babel has been a popular topic for artists. Pieter Bruegel (1525–1569) painted *the Tower of Babel* in 1563, which is now in Vienna's Kunsthistorisches Museum Wien. He painted the tower as an immense structure occupying almost the entire picture, with microscopic figures, rendered in perfect detail. The top floors of the tower are in bright red, whereas the rest of the brickwork has already started to weather. Maurits Cornelis Escher (1898–1972) was also intrigued by the story. In his painting in 1928, people were building the tower when they started to experience the confusion and frustration of the communication breakdown caused by the language barrier.

While the Tower of Babel is unlikely to bring us a good understanding of the state of the art of a scientific discipline, linguistic and semantic resources such as BabelNet (Navigli and Ponzetto 2012) play an increasingly active role in resolving discrepancies in processing unstructured text from heterogeneous sources. We introduce several widely known text mining and knowledge representation resources shortly.

Communicating with Aliens?

Space probes Pioneer and Voyager are travelling into deep space with messages designed to reach some intelligent forms in a few million years. If aliens do exist and eventually find the messages on the spacecraft, will they be able to understand? What are the assumptions we make when we communicate our ideas to others?

Pioneers 10 and 11 both carried small metal plaques identifying their time and place of origin for whatever intelligent forms might find them in the distant future. NASA placed a more ambitious message aboard Voyager 1 and 2—a kind of time capsule—to communicate a story of our world to extraterrestrial.

Pioneer 10 was launched in 1972. It is now one of the few most remote man-made objects. Communication was lost on January 23, 2003 when it was 80 AU[3] from the Sun. It was 12 billion kilometers or 745.6 million miles away. Pioneer 10 was headed towards the constellation of Taurus (The Bull). It will take Pioneer over 2 million years to pass by one of the stars in the constellation. Pioneer 11 was launched in 1973. It is headed toward the constellation of Aquila (The Eagle), Northwest of the constellation of Sagittarius. Pioneer 11 may pass near one of the stars in the constellation in about 4 million years.

According to "*First to Jupiter, Saturn, and Beyond*" by Fimmel et al. (1980), a group of science correspondents from the national press were invited to see the spacecraft before it was to be shipped to Kennedy Space Center. One of the correspondents, Eric Burgess, visualized Pioneer 10 as mankind's first emissary beyond our Solar System. This spacecraft should carry a special message from mankind, a message that would tell any finder of the spacecraft a million or even a billion years that planet Earth had evolved an intelligent species that could think beyond its own time and beyond its own Solar System. Burgess and another correspondent Richard Hoagland approached Director of the Laboratory of Planetary Studies at Cornell University, Dr. Carl Sagan. A short while earlier, Sagan had been involved in a conference in the Crimea devoted to the problems of communicating with extraterrestrial intelligence. Together with Dr. Frank Drake, Director of the National Astronomy and Ionosphere Center at Cornell University, Sagan designed a type of message that might be used to communicate with an alien intelligence.

Sagan was enthusiastic about the idea of a message on the Pioneer spacecraft. He and Drake designed a plaque, and Linda Salzman Sagan prepared the artwork. They presented the design to NASA; it was accepted to put on the spacecraft. The plaque design was etched into a gold- anodized aluminum plate 15.25 by 22.8 cm (6 by 9 in.) and 0.127 cm (0.05 in.) thick.

This plate was attached to the antenna support struts of the spacecraft in a position where it would be shielded from erosion by interstellar dust. The bracketing bars on the far right are the representation of the number 8 in binary form (1000), where one is indicated above by the spin-flip radiation transition of a hydrogen atom from electron state spin up to state spin down that gives a characteristic radio wave length of 21 cm (8.3 in). Therefore, the woman is 8×21 cm = 168 cm, or about 5' 6" tall. The bottom of the plaque shows schematically the path that Pioneers 10 and 11 took to escape the solar system—starting at the third planet from the Sun accelerating with a gravity assist from Jupiter out of the solar system. Also shown to help identify the origin of the spacecraft is a radial pattern

---

[3]Astronomical Unit: one AU is the distance between the Earth and the Sun, which is about 150 million kilometers (93,000 million miles).

etched on the plaque that represents the position of our Sun relative to 14 nearby pulsars (i.e., spinning neutron stars) and a line directed to the center of our Galaxy. The plaque may be considered as the cosmic equivalent to a message in a bottle cast into the sea. Sometime in the far distant future, perhaps billions of years from now, Pioneer may pass through a planetary system of a remote stellar neighbor, one of whose planets may have evolved intelligent life. If that life possesses the technical ability and curiosity, it may detect and pick up the spacecraft and inspect it. Then the plaque with its message from Earth may be found and deciphered.

Pioneer 10 will be out there in interstellar space for billions of years. One day it may pass through the planetary system of a remote stellar neighbor, one of whose planets may have evolved intelligent life. If that life possesses sufficient capability to detect the Pioneer spacecraft—needing a higher technology than mankind possesses today—it may also have the curiosity and the technical ability to pick up the spacecraft and take it into a laboratory to inspect it. Then the plaque with its message from Earth should be found and possibly deciphered. Due to the loss of communication, we may never hear from it again unless one day it could be picked up by intelligent aliens in the deep space.

Voyager 1 and 2 were launched in the summer of 1977. They have become the third and fourth human built artifacts to escape our solar system. The two spacecraft will not make a close approach to another planetary system for at least 40,000 years.

The Voyager carried sounds and images to portray the diversity of life and culture on Earth. These materials are recorded on a 12-in gold-plated copper disk. Carl Sagan was responsible for selecting the contents of the record for NASA. They assembled 115 images and a variety of natural sounds, such as those made by surf, wind and thunder, birds, whales, and other animals. They also included musical selections from different cultures and eras, and spoken greetings from Earth-people in 55 languages, and printed messages from President Carter of the United States of America and United Nation's Secretary General Waldheim. Each record is encased in a protective aluminum jacket, together with a cartridge and a needle. Instructions, in symbolic language, explain the origin of the spacecraft and indicate how the record is to be played. The 115 images are encoded in analog form. The remainder of the record is in audio, designed to be played at 16-2/3 revolutions per second. It contains the spoken greetings, beginning with Akkadian, which was spoken in Sumer about six thousand years ago, and ending with Wu, a modern Chinese dialect. Following the section on the sounds of Earth, there is an eclectic 90-min selection of music, including both Eastern and Western classics and a variety of ethnic music. It will be forty thousand years before they make a close approach to any other planetary system. In Carl Sagan's words, "The spacecraft will be encountered and the record played only if there are advanced space-faring civilizations in interstellar space. But the launching of this bottle into the cosmic ocean says something very hopeful about life on this planet."

Controversies in the Ebola Crisis

Most people in the U.S. had probably not heard of Ebola until the extensive news coverage of the deadly virus and a series of high-profile incidences surrounded the virus and its consequences. The public's impression of Ebola was that Ebola is highly infectious, that there is no reliable treatment, or no treatment at all, and that an incubation period could be as long as three weeks without patients showing any symptoms.

With such an understanding, there was no surprise to see the level of panic when news flashes confirmed and reinforced people's beliefs. A nurse was infected while handling an Ebola patient's body fluids. Another nurse who had spent long enough of her time with Ebola patients was returning to the U.S. She didn't have any symptoms of Ebola when she entered the U.S. Should she be allowed to travel freely? The World Health Organization (WHO) and the CDC had their recommended quarantine protocols. The CDC did revise its protocols. The protocol then required a 21-day quarantine. The public was concerned if the nurse was traveling freely in the U.S., she might get others infected. Of course, that was the worst scenario. The nurse insisted that she was not showing any symptoms, then there was no need to enter a quarantine. Two governors disagreed. The public needed a reliable scientific source to determine which side's argument is scientifically sound. Furthermore, it was also a strong argument that given the potentially deadly consequences, it would seem to be reasonable to ask for an individual to inconvenient or even sacrifice for the interest of the public. Miles (2015) specifically analyzed public health policy implications and politics of fear surrounded the Ebola nurse Kaci Hickox.

What was the best scientific knowledge available at that time? A group of scientists and their students combed the scientific literature for answers, but they didn't find answers (Bibby et al. 2015). More precisely, the answer did not exist. The state of the art at the time was not capable of giving a reliable answer. Charles Haas, a Drexel University professor, was searching for answers to a seemingly simple question: how long can Ebola virus survive in water? He didn't find answers either. He went on to search for sources in the scientific literature concerning the minimum duration of an Ebola quarantine. There was no clear-cut answers. There were non-zero probabilities associated with the longest incubation period. Although in theory the change of getting infected could be well below 1% or 100th of 1%, the fact the infection was almost an irreversible process. Once one is infected, it may be just too late for the patient to backtrack and start it all over. On the one hand, when the perceived risk is high and the stake is high, we tend to think of ways to avoid taking the risk at any price as long as it is justifiable. To many, perhaps except the nurse, taking a potentially over-predictive quarantine does not seem too much to ask for given a potentially life-and-death risk for not taking an adequate quarantine.

Figure 1.3 depicts a geographic distribution of researchers who have published articles on Ebola in journals indexed in the Science Citation Index (SCI). An area

Fig. 1.3 The geographic distribution of authors who published on Ebola

with a strong concentration of authors is shown as an area of red or orange. In contrast, areas in green have a much lower degree of concentration.

Argary (2015) discussed the accountability and public health policies in the Ebola crisis. Johnson and Slovic (2015) explained why keeping the public in the dark about the risk beyond the 21 day quarantine could backfire. Researchers found that people with higher levels of education tend to have a lower level of perceived risk than people with relatively lower levels of education. The Ebola crisis underlines a few issues of concern. First, this is a communication issue. The state of the art of science concerning Ebola was not communicated to the public accurately and precisely. The communication between domain experts, administrators, policy makers, and the public is inadequate. Second, this is an issue of the understanding of science, especially when involving areas where the uncertainty is relatively high. The uncertainty may be due to incomplete, missing, or conflicting information.

A major point that we are going to make in this book is concerning the role of uncertainty in science. To our knowledge, this is a much under-represented area. A major exception is a public workshop organized by the FDA on uncertainties in regulating the approval of new drugs (Institute of Medicine 2014). The second public workshop is scheduled on September 18, 2017.[4] For example, the profound uncertainty exists in anticipating how a new drug will affect a population that is much larger and significantly more diverse than the samples of individuals tested in clinical trials.

---

[4]https://www.fda.gov/forindustry/userfees/prescriptiondruguserfee/ucm378861.htm.

Laypersons Explanations of Conflicting Scientific Claims

Contradictions between different scientists are quite common. From an epistemological point of view, criticizing each other's knowledge claims is considered a typical part of research.

However, if a layperson searches for information in order to help them make a realistic decision, it is unlikely for them to be able to resolve contradictory information by themselves. In a situation that involves conflicting information from different scientists or scientific literature they publish, laypersons or non-experts usually do not have the specific knowledge to reach their conclusions. Besides, there are many topics even scientists cannot resolve. Thus one may not be able to reach a decision in time.

In an interesting study published in 2015, how laypersons would react to contradictory information put forward by scientists were investigated by a group of German psychologists (Thomm et al. 2015). Building on prior qualitative research, they developed an instrument to measure laypersons' assumptions on concrete reasons for scientific conflicts. For example, they first investigated how laypersons would react in a scenario in which the role of cholesterol in the buildup of plaques in the arteries—arteriosclerosis—was associated with contradictory views. In particular, participants of their study were given a scenario in which university professors from two different teaching hospitals made contradictory statements on their websites. One claims that the most important of the factors is cholesterol, whereas the other suggests that cholesterol plays an insignificant role.

They asked the participants of their study express the degree to which they agree or disagree with a variety of potentially valid explanations of conflicting scientific claims. To make sure they are indeed qualified as laypersons, the study excluded anyone who might have any relevant domain knowledge, which led to 285 valid participates (158 female participants). Participants rated the extent to which they would agree or disagree with a particular statement that offers a plausible explanation. Participants' answers were analyzed by using Principal Component Analysis (PCA), which is one of the most widely used techniques to characterize a complex set of results. PCA is typically used to identify a small number of the most significant underlying factors. In this study and a follow-up study conducted by the same group of researchers, four-component solutions were taken as the most representative solutions.

The four representative categories of explanations are:

- researchers' motivations
- differences in research process
- differences in competences, and
- thematic complexity.

Here are some examples: External factors such as competitive pressure, rivalry, marketing, or advertising influence the scientists' work. The scientists' work is influenced by their personal goals (e.g., recognition, money, promotion). The desire for reputation and recognition influences the work of the scientists. The reasons grouped under the differences in research process include that different answers to this question are based on different methodological approaches (e.g., different measurement instruments) and that the research methods differ from one another. The third component contains explanations focusing on competences. For example, scientists differ from one another because each of them invests a different amount of work into research on this specific topic. If the scientists would only engage in practical work and not theory, these contradictions would not appear. Finally, the fourth component cites the thematic complexity. There is still not enough research on the subject to classify the results. The topic is too complex to deliver clear results. There is probably another factor that could explain the differing opinions; however, this factor has yet to be discovered.

In order to test the stability of their quantitative measurement, the researchers replicated their method on two more areas of research: one is still in the medical and healthcare and the other is on climate change. The stability of their four-component solution was established in terms of the PCA results obtained. They named their questionnaire as the Explaining Conflicting Scientific Claims (ECSC).

As we shall see in our book, one can easily relate the four common dimensions shared by three areas of research to some of the well-known theories of scientific change. For example, the differences in research process and thematic complexity are epistemic beliefs in that they focus on how knowledge is discovered and justified. The other two dimensions, namely, researchers' motivations and their competences essentially focus on their roles in a social environment and how the behaviors of fellow researchers in the same area of study may act or react in a shared environment. If we label the first two as epistemic beliefs, then we could label the latter two as social beliefs. We can see similar distinctions in Thomas Kuhn's structures of scientific revolutions and Stephen Fuchs' sociology of scientific change. The former primarily focuses on the maintenance of holistic views, i.e. paradigms, whereas the latter is socially oriented. Although laypersons may not be aware of any of the theories of scientific change, the stability of such dimensions across different subject matters may represent some common perceptions of why scientists would want to compete for their often apparently controversial claims.

For example, Stephen Fuchs in a sociological theory of scientific change emphasizes the role of competing for reputation and resources in making novel discoveries. When facing scientific controversies, contending parties' reputations are likely to be at a higher stake than in a situation where explicit contradictions may be lacking.

Grand Challenges

In 2016, when we launched a new journal Frontiers in Research Metrics and Analytics,[5] we identified five grand challenges to underline our long-term goal to bring together several currently loosely coupled research communities so that the development of research metrics and in-depth analytic studies of scientific disciplines, fields, or sub-fields can reinforce each other (Chen 2016). Research metrics alone is never meant to provide the only source of information for complex sense making and decision making needs, which typically involve multidimensional heterogeneous factors (Hicks et al. 2015). Quantitative indicators of scholarship should be what we begin with in an assessment process as opposed to what we end up with. The role of an indicator is to draw our attention so that we can decide what examinations are in order. Physicians do not diagnose their patients with their vital signs; rather, they begin with the vital signs and proceed with a series of tests so as to diagnose the underlying problems.

When physicians diagnose underlying problems with their patients, analytic reasoning plays a critical role in developing a theory that may explain much of the symptoms. The nature of a diagnostic procedure is to establish plausible theories of an underlying problem. The role of indicators is critical but not diagnostic. In contrast, theories, or equivalently hypotheses and mental models, play an essential role in making the right decisions and taking the right course of action. Theories of scientific change, such as the ones we outlined earlier from Kuhn, Fuchs, and Shneider, provide the basis of what we can expect to see when we study the evolution of a scientific field. The five grand challenges below are further revised and extended based on the original version, especially emphasizing the fundamental importance of understanding uncertainties in advances of science as well as in scientific knowledge.

Accessibility

Scientific literature is increasingly volatile. *PLoS One* alone published 30,000 articles in 2014, an average of 85 articles per day. The Web of Science has accumulated over 1 billion cited references. The scale of retraction has stepped up —in one incidence publishers retracted 120 gibberish papers simultaneously.[6] While it is easy to locate a paper that we are looking for, keeping abreast of the advances of scholarly work is a constant challenge.

In addition to the common focus on documents, more efficient and incrementally maintainable approaches should enable researchers recognize and match information of interest beyond the constraints of the form or the language. The appropriate

---

[5]http://journal.frontiersin.org/journal/research-metrics-and-analytics.

[6]https://www.nature.com/news/publishers-withdraw-more-than-120-gibberish-papers-1.14763.

scope of a subject should be naturally and automatically expanded to attract documents through a wide variety of intellectual linkage, from semantic, linguistic, to social, and economic linkages, just as how an experienced expert would expand his/her own oeuvre of domain expertise. In addition, the self-organized and updated oeuvre of knowledge should help us understand the significance of research at the same level of clarity as *Heilmeier's Catechism*.

Clarity on Uncertainty

Scientific knowledge is never free of uncertainty. A good understanding of the underlying landscape of uncertainty is essential, especially in areas where information is incomplete, contradictory, or completely missing. For instance, there is no information on how long Ebola virus can survive in the water environment (Bibby et al. 2015). If surrogates with similar physiological characteristics can be found, then any knowledge of such surrogates would be valuable. Currently, finding such surrogates in the literature presents a real challenge.

It is difficult to communicate uncertainty clearly, especially on issues with widespread concerns, such as climate change (Heffernan 2007) and Ebola (Johnson and Slovic 2015). The way in which the uncertainty of scientific knowledge is communicated can impact many aspects of our life.

Another form of uncertainty rises when new inputs alter the existing structure of scholarly knowledge. New discoveries may strengthen a previously weak link as well as undermine or eliminate a previously considered strong connection. Distortions may be introduced by citations and reinterpretations (Greenberg 2009; Horn 2001) or false claims made by retracted studies (Chen et al. 2013). In many areas, damages may remain unnoticed for a long time due to the lack of efficient and systematic mechanisms.

Active researchers are aware of such uncertainties in their areas of expertise. They choose words carefully and use hedging and other rhetorical mechanisms to convey their findings in the context of uncertainty. These common practices in scholarly communication have further increased the complexity of understanding science, especially for those without relevant expertise and for computational approaches. Future developments should enable stakeholders to access scholarly knowledge with a great degree of clarity on uncertainty as well as the domain knowledge itself. The knowledge of the history and the current status of a scientific frontier is in nature the knowledge of knowledge. Researchers have identified situations where the notion of metaknowledge can be valuable (e.g., Evans and Foster 2011).

The goal of a scientific inquiry is to improve our understanding of what is previously unknown or not fully understood. It is intuitive that the overall level of uncertainty associated with a research topic is likely to decrease over time. The level of uncertainty is likely to be the highest when a new topic is emerging or when it is experiencing a scientific crisis. As scientists improve the understanding of a research topic, the uncertainties relevant to the topic are likely to decrease. Thus, if one can quantify the level of uncertainty associated with the domain knowledge, especially if one can measure the reduction of uncertainties as a result of a particular course of research, the notion of uncertainty may play a more revealing role in detecting emerging trends and assessing the scholarly impact.

Connecting Diverse Perspectives

The vast body of scholarly knowledge is a gold mine for making new discoveries. Pioneering efforts have demonstrated the value of connecting disparate bodies of knowledge discovery (Swanson 1986) and a recombinant search in technology landscapes (Fleming and Sorenson 2001). More recent attempts to enhance the process of scientific discovery with publicly available knowledge include detecting transformative potentials based on structural variations (Chen 2012), atypical combinations (Uzzi et al. 2013), diversity in interdisciplinarity (Rafols and Meyer 2010), and connecting different scientific domains through analogy (Small 2010).

What influential ideas have in common is their rich connections with other ideas (Goldschmidt and Tatsa 2005). The value of divergent thinking in scientific discovery, decision making, and creative problem solving is widely recognized. Reconciling multiple perspectives is critical by exposing competing views on the same issue and resolving seemingly contradictories at a new level (Linstone 1981; Chen 2014).

To meet this challenge, new computational and analytic tools should enable researchers and evaluators work with multiple perspectives directly. The unit of operation and analysis should focus on perspectives and paradigms as well as their premises, evidence, and chains of reasoning.

Benchmarks and Gold Standards

Repositories of well-documented exemplar cases analyzed from multiple perspectives should be created, maintained, and shared with the research community so as to enable researchers test and calibrate their metrics and analytic tools. Such repositories should include the full coverage of high-impact scientific breakthroughs, the most complex cases of retracted studies, and the longest scientific debates in the history of science. As shared resources, they will provide valuable benchmarks and testbeds for the development and validation of new metrics and analytic capabilities.

The role of readily available gold standards is crucial for a wide variety of scholarly activities. Biomedicine research is a good example. The development of biomedical and genomic analysis tools has profoundly benefited from information accessible through the array of portals maintained by the National Center for

Biotechnology Information (NCBI), including genes, proteins, and many more other biomedical and genomic information.[7]

Integrating Scholarly Metrics and Analytics

Scholarly metrics and qualitative studies of scientific discoveries and long-range foresights need to work together. The value of experts' opinions has been widely recognized. The challenge is in soliciting and synthesizing a wide variety of views from a diverse range of experts (Linstone and Turoff 1975). As strongly advocated in the Leiden manifesto, scholarly metrics should serve the supporting role to qualitative and in-depth analytics of scholarly content and activities (Hicks et al. 2015).

Numerous scholarly metrics have been proposed, ranging from the widely known h-index, citation counts with or without field normalization, to altmetrics. Scholarly metrics are meant to be universal, quantifiable, field independent, and easy to communicate (Bollen et al. 2009; Radicchi et al. 2008; King 2004). They convey extrinsic characteristics of research.

A profound challenge to integrate the indicative power of research metrics and the insight-seeking analytic approaches is the difficulty in linking two perspectives that differ in so many ways at so many levels. A single perspective is not capable of characterizing and conveying the breadth and the depth of scholarly activities. Aggregation is often necessary but important details may be lost.

A problem of great challenge in one perspective may become resolvable in another. Field normalization, for example, has been intensively studied in order to achieve the universality of research metrics across different disciplines. Drawing the boundary of a discipline is notoriously hard. More effective methods may approach the issue from multiple perspectives. Until we are able to move back and forth between distinct perspectives efficiently and effectively, our ability to measure and utilize the value of scholarly knowledge would be rather limited.

The Organization of the Book

This book is written for anyone who is interested in how a field of research evolves and the fundamental role of understanding uncertainties involved in different levels of analysis. We introduce a series of computational and visual analytic techniques from research areas such as text mining, deep learning, information visualization, and science mapping such that readers can apply these tools to the study of a subject matter of their choice. In addition, we set the diverse set of methods in an

---

[7]http://www.ncbi.nlm.nih.gov/.

integrative context that draws upon insights from philosophical, sociological, and evolutionary theories of what drives the advances of science such that the readers of the book can guide their own research with their enriched theoretical foundations.

Scientific knowledge is complex. A subject matter is typically built on its own set of concepts, theories, methodologies, and findings discovered by generations of researchers and practitioners. Scientific knowledge, as known to the scientific community as a whole, experiences constant changes. Some changes are long-lasting, whereas others may be short lived. How can we keep abreast of the state of the art as science advances? How can we effectively and precisely convey the status of the current science to the general public as well as scientists across different disciplines?

The study of scientific knowledge in general has been overwhelmingly focusing on scientific knowledge per se. In contrast, the status of scientific knowledge at various levels of granularity has been largely overlooked. This book aims to highlight the role of uncertainties in developing a better understanding of the status of scientific knowledge at a particular time and how its status evolves over the course of the development of research. Furthermore, we demonstrate how the knowledge of the types of uncertainties associated with scientific claims serves as an integral and critical part of our domain expertise.

The organization of the book aims to introduce the overarching theme of representing scientific knowledge logically and intuitively. The contents of individual chapters are outlined as follows.

Chapter 2 presents macroscopic theories of scientific change, from philosophical, sociological, and problem solving perspectives. The chapter begins with the notion and major characteristics of generalized types of mental models. These macroscopic theories share many of the characteristics of mental models. It is important to bear in mind serious pitfalls and biases associated with these theories, or any theories for that matter. These theories collectively provide a conceptual framework for the analysis, reasoning, and interpretation in subsequent chapters.

Chapter 3 introduces science mapping tools and a few exemplars of science mapping applications. The design principles and major functionalities of popular science mapping tools such as CiteSpace, VOSviewer, and CitNetExplorer are explained in detail such that the reader can utilize these freely available tools to study a scientific domain of their own choice. CiteSpace provides a variety of metrics and indicators concerning trends and patterns in scientific literature. Many of these metrics are explained further with illustrative examples from applications of CiteSpace. CiteSpace also includes extensions that are particularly used to generate examples in the book. Three examples of systematic scientometric reviews using CiteSpace are included to illustrate relevant concepts and analytic functions.

Chapter 4 summarizes an array of fundamental and widely used concepts and computational methods for measuring scholarly impact as well as identifying more generic properties such as semantic relatedness, burstness, clumping, and centrality. Normalizations of metrics across scientific fields and the year of publication are discussed with concrete examples.

Chapter 5 explains the structures and features of several widely known and inspirational resources for representing concepts and semantic relations in biomedical knowledge, namely MeSH, ULMS, SemRep, and Semantic MEDLINE. Many examples in subsequent chapters make use of these resources.

Chapter 6 reviews a series of techniques and resources of natural language processing and text mining and their roles in handling unstructured text. In particular, topics such as information extraction, topic modeling, and deep learning are introduced along with illustrative examples.

Chapter 7 focuses on literature-based discovery, which is a major area of research in its own right with applications of the principles and computational techniques introduced in previous chapters. Pioneering works, landmark systems, and more recent developments and challenges are outlined. In particular, the design of PKD4 J, a scalable and flexible engine for literature-based discovery is demonstrated in detail.

Chapter 8 demonstrates a series of studies of semantic predications from Semantic MEDLINE, including the detection of semantic predications with burstness and in association with conflict, contradictory, or other sources of uncertainties of scientific knowledge. Semantic networks of predications are analyzed within the framework of structural variations. Examples in this chapter represent scientific knowledge at a level of granularity that differs from those in Chap. 3.

Chapter 9 presents a conceptualization of further research on uncertainties in scientific knowledge. Several common sources of uncertainties in scientific literature are characterized, notably including retracted scientific publications, hedging, conflicting or contradictory findings. Semantically equivalent uncertainty cue words and their connections with semantic predications are identified and visualized as the first step towards a systematic study of uncertainties in accessing and communicating the status of scientific assertions.

We do not assume our readers to have any particular prior knowledge of the topics discussed in this book. We expect that the book is accessible to a diverse range of readers from a wide variety of backgrounds and pragmatic needs. Representing scientific knowledge itself is a concept that requires a high order of abstraction as well as analytic thinking across multiple levels of granularity. Thus, abstraction, critical thinking, and analytic reasoning are all valuable for you to get the most out of the book. Readers with a first-hand research experience are likely to find easier to make connections than readers who may not yet have such experiences. We include numerous examples based on citation data from bibliographic records retrieved from the Web of Science and semantic predications from Semantic Medline. We include examples along with MySQL queries. Replicating some of the examples may require you to operate basic Java programs. Nevertheless, it is not essential to have any prior knowledge or experience in association with these resources or tools. The recommended sequence follows the order of the chapters. Of course, you may also reader the book in any order you like. You can always look up the relevant concepts and examples when you need to.

References

Alvarez LW, Alvarez W, Asaro F, Michel HV (1980) Extraterrestrial cause for the Cretaceous-Tertiary extinction. Science 208(4448):1095–1098

Alvarez W (1997) T. rex and the Crater of Doom. Vintage Books, New York

Asgary R (2015) Accountability and public health policies impacting proper Ebola response: time for a bioethics oversight board. Am J Bioeth 15(4):72–74. doi:10.1080/15265161.2015. 1010695

Bibby K, Casson LW, Stachler E, Haas CN (2015) Ebola virus persistence in the environment: state of the knowledge and research needs. Environ Sci Technol Lett 2(1):2–6. doi:10.1021/ ez5003715

Bollen J, Sompel HVd, Hagberg A, Chute R (2009) A principal component analysis of 39 scientific impact measures. PLoS ONE 4(6):e6022. doi:10.1371/journal.pone.0006022

Brante T, Elzinga A (1990) Towards a theory of scientific controversies. Sci Stud 3(2):33–46

Chen C (2012) Predictive effects of structural variation on citation counts. J Am Soc Inform Sci Technol s63(3):431–449. doi:10.1002/asi.21694

Chen C (2014) The fitness of information: quantitative assessments of critical evidence. Wiley, New York

Chen C (2016) Grand challenges in measuring and characterizing scholarly impact. Front Res Metr Analytics. doi:10.3389/frma.2016.00004

Chen C, Hu Z, Milbank J, Schultz T (2013) A visual analytic study of retracted articles in scientific literature. J Am Soc Inform Sci Technol 64(2):234–253. doi:10.1002/asi.22755

Collins R (1989) Towards a theory of intellectual change: the social causes of philosophies. Sci Technol Human Values 14(2):107–140

Corbett JB, Durfee JL (2004) Testing public (un)certainty of science: media representations of global warming. Sci Commun 26(2):129–151

Endrikat J, Guenther E, Hoppe H (2014) Making sense of conflicting empirical findings: a meta-analytic review of the relationship between corporate environmental and financial performance. Eur Manag J 32:735–751

Evans JA, Foster JG (2011) Metaknowledge. Science 331(6018):721–725

Fimmel RO, Allen JV, Burgess E (1980) Pioneer: first to Jupiter, Saturn, and beyond. Scientific and Technical Information Office, NASA, Washington, DC

Fischhoff B (2013) The science of science communication. PNAS 110(suppl. 3):14033–14039

Fleming L, Sorenson O (2001) Technology as a complex adaptive system: evidence from patent data. Res Policy 30(7):1019–1039. doi:10.1016/s0048-7333(00)00135-9

Fuchs S (1993) Three sociological epistemologies. Sociol Perspect 36(1):23–44

Garfield E (1955) Citation indexes for science: a new dimension in documentation through association of ideas. Science, 122:108–111

Goldschmidt G, Tatsa D (2005) How good are good ideas? Correlates of design creativity. Des Stud 26(6):593–611. doi:10.1016/j.destud.2005.02.004

Greenberg SA (2009) How citation distortions create unfounded authority: analysis of a citation network. BMJ 339:b2680. doi:10.1136/bmj.b2680

Heffernan O (2007) Clarity on uncertainty. Nature Reports Climate Change 5. https://www.nature. com/climate/2007/0710/pdf/climate.2007.57.pdf. doi:10.1038/climate.2007.57

Hicks D, Wouters P, Waltman L, Rijcke Sd, Rafols I (2015) Bibliometrics: The Leiden Manifesto for research metrics. Nature 520(7548):429–431. doi:10.1038/520429a

Hildebrand AR, Penfield GT, Kring DA, Pilkington M, Carmargo ZA, Jacobsen SB, Boynton WV (1991) Chicxulub crater: a possible Cretaceous-Tertiary boundary impact crater on the Yucatan Peninsula, Mexico. Geology 19(9):867–871

Horn K (2001) The Consequences of Citing Hedged Statements in Scientific Research Articles: When scientists cite and paraphrase the conclusions of past research, they often change the hedges that describe the uncertainty of the conclusions, which in turn can change the uncertainty of past results. BioScience 51(12):1086–1093. doi:10.1641/0006-3568(2001)051 [1086:tcochs]2.0.co;2

Institute of Medicine (2014) Characterizing and communicating uncertainty in the assessment of benefits and risks of pharmaceutical products: workshop summary. The National Academies Press, Washington, DC. doi: 10.17226/18870

Johnson BB, Slovic P (2015) Fearing or fearsome Ebola communication? Keeping the public in the dark about possible post-21-day symptoms and infectiousness could backfire. Health Risk Soc. doi:10.1080/13698575.2015.1113237

King DA (2004) The scientific impact of nations. Nature 430(6997):311–316

Kuhn TS (1962) The structure of scientific revolutions. University of Chicago Press, Chicago

Linstone HA (1981) The multiple perspective concept: With applications to technology assessment and other decision areas. Technol Forecast Soc Change 20(4):275–325. doi:http://dx.doi.org/10.1016/0040-1625(81)90062-7

Linstone HA, Turoff M (eds) (1975) The Delphi method. Addison-Wesley Publishing Co., Reading, MA

Mazur A (1987) Scientific disputes over policy. In: Engelhardt, Caplan (eds) Scientific controversies. Cambridge University Press, Cambridge

McMullin E (1987) Scientific controversy and its termination in In: Engelhardt, Caplan (eds) Scientific controversies. Cambridge University Press, Cambridge

Miles SH (2015) Kaci Hickox: public health and the politics of fear. Am J Bioeth 15(4):17–19. doi:10.1080/15265161.2015.1010994

Navigli R, Ponzetto SP (2012) BabelNet: The automatic construction, evaluation and application of a wide-coverage multilingual semantic network. Artif Intell 193:217–250

Radicchi F, Fortunato S, Castellano C (2008) Universality of citation distributions: toward an objective measure of scientific impact. Proc Natl Acad Sci 105(45):17268–17272

Rafols I, Meyer M (2010) Diversity and network coherence as indicators of interdisciplinarity: case studies in bionanoscience. Scientometrics 82(2):263–287

Shneider AM (2009) Four stages of a scientific discipline: four types of scientists. Trends Biochem Sci 34(5):217–223

Signor PW, Lipps JH (1982) Sampling bias, gradual extinction patterns, and catastrophes in the fossil record. Geol Soc Am Spec Pap 190:291–296

Small H (2010) Maps of science as interdisciplinary discourse: co-citation contexts and the role of analogy. Scientometrics 83(3):835–849

Swanson DR (1986) Fish oil, Raynaud's syndrome, and undiscovered public knowledge. Perspect Biol Med 30:7–18

Thomm E, Hentschke J, Bromme R (2015) The Explaining Conflicting Scientific Claims (ECSC) Questionnaire: Measuring Laypersons' explanations for conflicts in science. Learn Individ Differ 37:139–152. doi:10.1016/j.lindif.2014.12.001

Uzzi B, Mukherjee S, Stringer M, Jones B (2013) Atypical combinations and scientific impact. Science 342(6157):468–472

Chapter 2
The Dynamics of Scientific Knowledge: Macroscopic Views

Abstract Macroscopic theories of scientific change are holistic views of what drives the creation and acceptance of scientific knowledge. At the grand scale of scientific communities, such theories offer a conceptual framework for analyzing the development of a scientific discipline through philosophical, sociological, and problem solving perspectives. As mental models, however, these models of scientific processes are subject to pitfalls and biases that may hinder our analytic reasoning. Integrating theoretical and empirical studies has the potential to help us reach a new level of understanding the dynamics of scientific knowledge. Three major theories of scientific change are presented from philosophical, sociological, and problem-solving perspectives to highlight distinct concepts and expectations as well as shared characteristics.

Introduction

A complex adaptive system consists of numerous interconnected components. The state of the system as a whole may change due to a variety of reasons. Components can be added or removed, and their internal state and links to other components are also subject to modification changed internally. Perhaps the most widely known concept is the butterfly effect—small changes may cause large effects. A butterfly flapping its wings is usually considered a minor perturbation to the scale of any weather system. However, if such minor perturbations may cause magnificent effects later on, then we are dealing with a complex adaptive system.

There are many complex adaptive systems around us—forests, societies, and the growing body of scientific knowledge. A macroscopic view of a forest focuses on the ecosystem of a wide variety of individual trees, plants, and flowers that come with different shape, size, color, and growth patterns. A sociological perspective of the population focuses on relationships between individuals of different personalities and backgrounds. A philosophical view of science focuses on how different

© Springer International Publishing AG 2017
C. Chen and M. Song, *Representing Scientific Knowledge*,
https://doi.org/10.1007/978-3-319-62543-0_2

disciplines of science create and organize their knowledge of the world, and what forces advance of scientific knowledge. These various perspectives share a deep interest in a complex adaptive system.

Is a tree falling down a minor event to a forest that would survive a threatening fire many years later? Is the publication of a scientific article a groundbreaking revolutionary event, or simply a minor perturbation to the growth of scientific knowledge as a whole? In this chapter, we set the platform of our analysis and discussions at a macroscopic level. Our primary focus is not only on the trees, but also the forest in which all the trees play a role. We will focus not only on individual researchers but also the groups and the communities in which individual researchers trailblaze their pathways and leave their footprints.

Mental Models

We borrow the idea of a mental model from psychology (e.g., Johnson-Laird 1983). A mental model is what we believe how a system operates or what is going on. For example, we may believe that the earth is the center of the universe. Given this mental model, we would be able to interpret what we see—galaxies further away from the earth seem to receding from us faster than galaxies nearby. In addition, we would be able to make predications on things that we have never seen. Many accidents occurred because of wrong mental models (Chen 2014). Individuals' mental models may differ due to social, organizational, cultural, and individual differences (Markus and Kitayama 1991). Mental models play critical roles in a wide variety of situations and activities such as situation awareness (Endsley 1995), intelligence analysis (Heuer 1999), creativity (Csikszentmihalyi 1996), discourse comprehension (Kintsch 1988), public understanding of science (Bostrom et al. 1994), and practice of law (Sutton 1994).

Easy to Form

Mental models are associated with a few very intriguing properties that may have profound implications on what we do or not to do with our own mental models. First of all, it doesn't take much of input for us to form our mental model. A glance may be all we need to generate a vivid and convincing mental model, or a story. We can fill up the details effortlessly with our imagination and with our prior experience and knowledge. Many cultures share a similar story: Someone had some of his properties missing and he started to suspect his neighbor. He decided that he would give a 'neighbor's watch' to his neighbor's behavior. The more closely he watched, the more seriously he was convinced: it must be him—it all fits! Before he got any chance to do anything about his neighbor, his properties were recovered but it had nothing to do with his neighbor. When he neighbor-watched his neighbor next day, he thought the neighbor didn't behave like a thief after all. He didn't bother to think what made him so convinced earlier on. This story explains why it is pointless to attempt to prove a hypothesis in science and why it is far more revealing to disprove a hypothesis.

The creation of our mental model is a very subjective process. It is influenced by our past experience, our education, cultural values, and expectations of others. We may unconsciously fill in the gaps by adding details that are not found anywhere in initial observations. Hallucinations and Pareidolia can be considered as extreme cases of mental models.

Pareidolia is an interesting psychological phenomenon in which one may see something in an image that does not exist. Widely known examples include seeing a human face on Mars in a 1976 image, seeing a female figure on Mars in a 2007 image, seeing Jesus and a number of celebrities in the Pillars of Creation, and seeing various objects in the Shroud of Turin. Sometimes amusing and creative sightings make entertaining news headings, whereas sometimes contradictory interpretations sustain lengthy debates between scientists with a substantial degree of domain expertise. In our 2014 book on the fitness of information, we have discussed the subjectivity of evidence in detail. It is important to bear in mind that our perspective is determined by our mental model. We can only see what our mind sees.

Designers may use various design metaphors to help the user to develop a mental model that would fit to a given design metaphor. For example, with a desktop metaphor, all our understanding of and experience with a desktop in the real world instantly become transferable and applicable. We can open a file. We can save a file. We can drag a file to the trashcan. We would be able to figure out how we are supposed to react to some of the actions even we haven't learned how to use them specifically. The greatest value of using a design metaphor is that we can save a lot of efforts in communicating various details. As long as we get the right mental model, we should be able to transfer a lot of our knowledge gained in one cir-cumstance to another. The downside is obviously when a mental model departs from its original source, or its prototype, it is very hard for us to detect when two belief systems fail to match one another. When we just start to learn a foreign language, it may be difficult or inconvenient, but it is rarely dangerous. The most dangerous stage, however, arrives much later when we have convinced ourselves that we fully understand what is expressed in the foreign language. In other words, it is much easier for us to realize whether we understand something than whether we misunderstand anything. When we have a convincing and self-explained story, it would be much harder for us to even pay any attention and ourselves whether we get the story right in the first place. Stereotypes and prejudices are among some of the most common examples of a mental model that is formed too fast, especially when what we see is no more than the tip of an iceberg.

Hard to Change

The second property of a mental model is that it is hard to change. In the neighbor-watch story, additional observations reinforced the initial perception of a suspiciously behaving neighbor. Until he found his misplaced properties, the incorrect perception was reinforced incorrectly. Conclusive and undisputable evidence is essential to break the reinforcement loop. Intelligence analysis of the lack of evidence on Iraqi's mass destructive weapons resembles some of the issues discussed here. If we held a mental model that Iraq concealed mass destructive weapons, then Iraq's denial can only reinforce the belief. After all, if they were hiding the mass destructive weapons, of course they would deny the existence. The questions in such situations are not the answers to which would convince us our initial guess was right. Rather, a more valuable line of inquiry would question how we would know that our initial guess was not wrong.

The notion of diagnostic evidence is a key in making a decision in a complex and dynamic setting, especially when available information is incomplete, conflicting, or contradictory. Diagnostic evidence is the information that is capable of differentiating alternative interpretations and thus advancing the diagnostic process. A tuberculin skin test is a common test to see if someone has ever been exposed to tuberculosis (TB). If someone has ever been exposed to the TB bacteria, the skin test will see a firm red bump on the arm where a small amount of TB protein is injected under the skin. However, a positive skin test still cannot tell whether the infection is inactive (latent) or active (contagious). Thus, a skin test is not diagnostic if we need to know whether there is a risk for the TB to be passed to others.

Intelligence analysis has identified procedures that may help us to avoid some of the pitfalls and biases because of these properties of our mental models. For example, we are advised to brainstorm as many hypotheses as possible at the initial stage of investigation and refrain ourselves from diving into the evaluation of individual hypotheses. Evaluating a hypothesis would inevitably enrich the mental model that would justify the hypothesis with concrete and vivid details. We are particularly vulnerable to arguments that come with concrete and vivid details. "I know someone who was exactly in the same situation, ..." and "Look, here is a photo to prove it." In general, we tend to believe what we see and it is easier for us to be convinced by specific details. A thorough brainstorm step earlier on in the process will help us to expand the horizon of our consideration because it will become increasingly hard to do so later in the process.

Once we have brainstormed as many hypotheses as possible, we need to eliminate hypotheses or the mental models that can be possibly disproved by the available evidence. Science is full of examples in which one question may have numerous possible answers. For example, 65 million years ago, at the K-T boundary, hundreds of species or even more became extinct within a relative short period of time, including dinosaurs that once seemed have dominated the earth. While the consequences are evident, what caused the massive extinction was far from clear. Researchers proposed as many as over 80 theories to explain what happened. Some theories direct our attention to forces from the inside of the earth such as massive and continuous flows of lava. Some theories draw our attention to forces from the deep space, including one-shot asteroid to periodical visits of astronomical objects orbiting around invisible stars. Many pieces of evidence are subject to alternative interpretations. The debates between scholars from different schools of thought or schools of beliefs lasted over a decade until diagnostic evidence was found in the Mexico Bay.

The best way and the most valuable way to validate a mental model is to challenge its very foundation with an alternative or competing mental model (Chen 2011, 2014). We use the term mental model, theory, and hypothesis exchangeably in this context. As researchers such as Randall Collins (1998) have studied in detail, the creativity typically arises from the intellectual confrontations of competing schools of thought. Collins noticed that great philosophers in history are likely to be the ones that fight for and defend their own schools of thought against the attacks of other schools of thought. The highest form of victory is not to defeat the opponent; rather, it is to accommodate the opponent in a higher level of order. In science, a powerful theory would adopt two contradictory theories as its special cases.

We know that the quality of our idea increases as we continue to come up with more ideas. In other words, we tend to generate increasingly better ideas. This is another reason that we should maintain a brainstorming process long enough to see good ideas. A scientific theory may not be formed as quickly as a mental model in our everyday life, but it shares some of the most fundamental properties of a mental model—any theory is a simplified and thus incomplete abstraction of the reality or the underlying phenomenon. Any abstraction may be locally accurate but globally wrong. Contradicting theories may not be good news to each individual theory, but holistically, it is likely to be the most valuable sign to these contradicting theories as a whole.

Theories of Scientific Change

We will introduce three accounts of scientific change from rather different perspectives. The first one is a philosophical theory of scientific revolutions from Kuhn (1962). The second one is a sociological theory of scientific change from Fuchs (1993a, b). The third one is an evolutionary theory of a generic problem solving process by Shneider (2009).

Scientific Revolutions

The central idea of Kuhn's theory is that science advances through a series of scientific revolutions. Each scientific revolution takes place when the predominant position of one paradigm is taken by a new paradigm. A paradigm is a view of the world, or a belief of what the world is and how it works. In other words, a paradigm is a mental model of the world shared by a community of scientists. As a mental model, a paradigm provides a basic framework for researchers to investigate a set of research questions with recognized methodologies.

Gestalt psychology believes that our mind is holistic. We see the entirety of an object before we attend to its parts. And the whole is greater than the sum of its parts. In terms of information theory, the way that individual parts form the whole gives us additional information about the system as a whole. In his *Patterns of Discovery*, Norwood Russell Hanson argues that what we see is influenced by our existing preconceptions (Hanson 1958).

Kuhn further developed the view how a *gestalt switch* is involved in scientific discovery and explained the nature of a paradigm shift in terms of a gestalt switch. Kuhn cited an experiment in which psychologists showed participants ordinary playing cards at brief exposures and demonstrated that our perceptions are influenced by our expectations. For example, it took much longer for participants to recognize unanticipated cards such as black hearts or red spades than recognize expected ones. Kuhn quoted one comment: "I can't make the suit out, whatever it is. It didn't even look like a card that time. I don't know what color it is now or whether it's a spade or heart. I'm not sure I even know what a spade looks like. My God!"

Paradigm Shift

A paradigm may go through a process that has the following stages: normal science, crises, and a paradigm shift, which defines a scientific revolution. At the normal science stage, the research in the field of study is well defined by the predominating paradigm. There is a consensus in terms of the kinds of research questions that should be investigated. Since the research agenda is set by the paradigm, research in this period is largely incremental as opposed to disruptive or revolutionary.

At the crisis stage, anomalies become inevitable and they challenge the very foundation of the currently predominating paradigm. Patchwork on the current paradigm is no longer adequate to resolve the crises. Researchers may propose drastically different paradigms to resolve the immediate crises. In addition, as the candidate of an alternative view of the world, the newly proposed paradigm should appear to have the potential at least as good as the current paradigm, although at this point, researchers would tolerate the lack of a thorough examination of the new paradigm because everyone knows it will take time to accomplish.

At the revolutionary stage, a critical mass has reached for the new paradigm to claim the predominant position from the once leading paradigm. Researchers in the scientific community start to re-examine the world through the perspective of the new paradigm. Once the new paradigm has established its predominant position, the science will repeat the process that the previous paradigm has gone through. One day there will be another crisis to emerge and challenge the foundation of the currently young and healthy paradigm. There will be another paradigm to emerge and take the leading position from the current paradigm. There will be an endless series of scientific revolutions.

The notion of paradigm shift has become a household name. Researchers from almost every discipline of science have embraced the idea. In the Web of Science, we have found as many as 567 variants of The Structure of Scientific Revolutions. One of the 567 variants alone has been cited 12,101 times. Among its numerous citing articles, 23 are in the category of highly cited articles in their own field.

Kuhn himself vividly described what we would expect to see in terms of citations. During the normal science, there should be a few highly cited groundbreaking articles that serve as exemplars of the predominating paradigm. Researchers routinely draw their inspirations from these groundbreaking articles. During a period of crisis, researchers are likely to cite articles that originally revealed the crisis. During the paradigm shift period, researchers are expected to cite the new paradigm.

Criticisms

Kuhn's paradigm shift theory has also drawn extensive criticisms to itself too. One is the suggestion that researchers on each side of competing paradigms may never fully understand the ideas of those on the other side. This is the famous incommensurability issue. Since each side has a mental model that is so different from its competing paradigms, a paradigm may make little sense to those who are occupied by different paradigms. It was believed that some of the hardcore beholders of a paradigm may never make that Gestalt switch and they may never adopt an incommensurable paradigm in their lifetime. Incommensurability refers to the communicative barrier between different paradigms; it can be taken as a challenge to the possibility of a rational evaluation of competing paradigms using external standards. If that was the case, the argument may lead to the irrationality of science.

Masterman (1970) examined Kuhn's discussion of the concept of paradigms and found that Kuhn's definitions of a paradigm can be separated into three categories:

1. Metaphysical paradigms, in which the crucial cognitive event is a new way of seeing, a myth, a metaphysical speculation
2. Sociological paradigms, in which the event is a universally recognized scientific achievement
3. Artefact or construct paradigms, in which the paradigm supplies a set of tools or instrumentation, a means for conducting research on a particular problem, a problem-solving device.

She emphasized that the third category is most suitable to Kuhn's view of scientific development. Scientific knowledge grows as a result of the invention of a puzzle-solving device that can be applied to a set of problems producing what Kuhn has described as "normal science."

While we believe it is hard for one to see the world from different perspectives, we can all learn to see the world through a fresh perspective. The key is the information that is subject to multiple alternative interpretations. If multiple mental models hinge on such information, the hinges seem to be a good point to start. In fact, as we will see, Kuhn's philosophical account and the sociological account that we will introduce next may appear to be incommensurable. Can we find a point that the two theories differ in their interpretations of the same thing we can all observe?

A common criticism of the notion of a Kuhnian paradigm shift is that it doesn't seem to be fully consistent with the history of science. Some argued that Kuhnian revolutions are rare events. The Copernican revolution is a classic example of a paradigm shift. It marked the change from the geo-centric to the solar-centric view of our solar system. Another classic example is Einstein's general relativity, which took over the authoritative place of Newtonian mechanics and became the new predominant paradigm in physics. How often are we experiencing scientific revolutions at the Kuhnian scale? When was the last time a scientific field was turned upside down?

A Kuhnian paradigm may correspond to a cluster of co-cited references, or a group of references that are frequently cited together. We can verify that distinct paradigms are behind different clusters of co-citations because of the conceptual frameworks they work with, which are determined by their paradigms. van Raan (1990) reported that co-citation clusters appear to be scale free. It means that there may be no such thing as a typical size of such clusters. In other words, a cluster of any size seems to be possible.

A domain of any size may be represented by a corresponding network. The dynamics of the domain can be largely characterized by the network, which can be further decomposed into clusters or specialties at a finer level of granularity. The notion of Kuhnian Gestalt Switch, i.e. paradigm shift, can be applied to each of these specialties as well as to the domain as a whole. Researchers in a particular specialty may work with their own paradigmatic research agenda. Some specialties may last longer than others, but they are all driven by a paradigm of their own, or a world view of their own. Researchers may remain in a specialty, if they continue to follow the same paradigm. In contrast, researchers may leave for a different specialty, if they want to branch off to a different paradigm. Thus, we believe that the process of scientific revolutions is not limited to rare and once-in-a-life-time revolutions. Rather, scientific revolutions take place all the time at different scales. For example, at a disciplinary level, computer science is relatively stable overall. However, at the level of one of its components, for example, artificial intelligence, it is inevitable to notice that a scientific revolution is taking place in at least in one area—neuro networks, notably, deep learning.

Explanation Coherence

Thagard (1992) noted although historians and philosophers of science have recognized the importance of scientific revolutions, there has been little detailed
explanation of such changes. He proposed a computational approach to explain
what he called conceptual revolutions with a special focus on how the conceptual
structure changes in a scientific revolution.

He introduced the concept of explanation coherence of a theory and argued that
the acceptance of a scientific theory is essentially due to its explanation coherence.
Suppose we have two theories A and B. Both of them can explain the same set of
phenomena, but theory A has fewer assumptions than theory B. In such situations,
theory A is considered superior. Similarly, if two theories have the same number of
assumptions, but one can explain more phenomena than another, the one with more
explanation power is considered superior.

Thagard examined examples of scientific revolutions such as the conceptual
development of plate tectonics in the latest geological revolution and Darwin's
natural selection theory. A conceptual revolution may involve structural and
non-structural changes. For example, the continental drift was transformed to
modern theories through a structural change, whereas the change of the meaning of
the evolution concept in Darwin's origins of species is non-structural.

Thagard suggests that we should focus on rules, or mechanisms, that govern how
concepts are connected. For example, we should consider the variation of strengths
of links over time. Adding a link between two concepts can be seen as strengthening an existing but possibly weak link between the two concepts. Removing an
existing link can be seen as a result of a decay of its strength; they no longer have a
strong enough presence in the system to be taken into account. Thagard identified
nine steps to make conceptual changes:

1. Adding a new instance, for example that the blob in the distance is a whale.
2. Adding a new weak rule, for example that whales can be found in the Arctic
   Ocean.
3. Adding a strong rule that plays a frequent role in problem solving and explanation, for example that whales eat sardines.
4. Adding a new part-relation, also called decomposition.
5. Adding a new kind-relation, for example that a dolphin is a kind of whale.
6. Adding a new concept, for example narwhale.
7. Collapsing part of a kind-hierarchy, abandoning a previous distinction.
8. Recognizing hierarchies by branch jumping, that is, shifting a concept from one
   branch of a hierarchical tree to another.
9. Tree switching, that is, changing the organizing principle of a hierarchical tree.

Branch jumping and tree switching are much rare events associated with conceptual revolutions. Thagard examined seven scientific revolutions:

1. Copernicus' solar-centric system of the planets replacing the earth-centric theory
   of Ptolemy

2. Newtonian mechanics, synthesizing celestial and earth-bound physics, replacing the cosmological views of Descartes
3. Lavoisier's oxygen theory replacing the phlogiston theory of Stahl
4. Darwin's theory of evolution by natural selection replacing the prevailing view of divine creation of species
5. Einstein's theory of relativity replacing and absorbing Newtonian physics
6. Quantum theory replacing and absorbing Newtonian physics
7. The geological theory of plate tectonics that established the existence of continental drift

Thagard's central claim is that it is best to explain the growth of scientific knowledge in terms of *explanation coherence*. The power of a new paradigm must be assessed in terms of its strength in explaining phenomena coherently in comparison with existing paradigms. He demonstrated how the theory of *continental drift* gained its strength in terms of its explanation coherence.

Competition Leads to Scientific Change

Fuchs (1993a, b) proposed a sociological theory of scientific change after he criticized the Kuhnian paradigm shift as an oversimplification of the complex reality. Fuchs argues that advances of science are driven by sociological reasons. Scientists compete for recognition and reputation. Fuchs explains why a few types of scientific change may result from competitions when two factors interplay, namely mutual dependence and task uncertainty.

Mutual dependence refers to the social and organizational dependencies between scientists and their competing peers. Task uncertainty refers to the level of uncertainty involved in the course of scientific inquiry. The task uncertainty is high in scientific frontiers where research is essentially exploratory in nature and there is a high amount of tacit knowledge involved, for example, scientific discoveries of high creativity. In contrast, the task uncertainty is low in areas where tasks are routinized. A combination of high task uncertainty and high mutual dependence will lead to original scientific discoveries, which will bring a substantial degree of recognitions and reputations such as Nobel Prizes. A research area with intensified competitions is also likely to have a high retraction rate (Chen et al. 2013). A combination of low task uncertainty and high mutual dependence will result in specialization to maintain the tension between scientists with high mutual dependence while they work on routinized research.

According to Fuchs, Kuhn's theory does not account for the many possible reactions to perceived anomalies. Apart from switching to a different world view altogether, one can choose to ignore them, explain them away, try to accommodate them into established knowledge or make minor modifications of the theory. In the terminology of our mental models, one could choose to keep the existing mental model or ignore the anomalies. The question is what if anomalies are too prominent to ignore or too fundamental to patchwork the existing mental model.

Fuchs further attacked Kuhn's theory by arguing that Kuhn's theory expects only two basic types of scientific activity: normal science and revolutionary science. "Revolutions, however, are as rare in science as in other areas of society. Most scientific change appears to be nonrevolutionary." Fuchs quoted sociological studies that questioned the idea of sudden and holistic gestalt shifts. Instead, the sociological studies argued that even the few revolutions are dramatic culminations of a long series of smaller incremental changes and that the normal-revolutionary dichotomy in Kuhn's theory is too simple to provide an adequate account for the complexity of how science may change.

Fuchs proposed a sociological theory of scientific change named the Theory of Scientific Organization (TSO). His theory views scientific specialties as reputational work organizations in which material resources and social structures shape how scientists do research. A specialty is a group of researchers who have similar training, attending the same conferences, reading and citing the same set of literature. Specialties usually have a small core of highly productive and visible researchers, a semi-periphery of researchers with much less visibility, and a large periphery of inactive or transient researchers.

Fuchs' theory is built on three components: (1) liberal or conforming cognitive styles: how we think and what we perceive are shaped by social structure. Knowledge is social imagery because it reflects an underlying social organization. In a cohesive and homogeneous group, one is under pressure to conform to its cognitive standards. In contrast, loosely coupled and heterogeneous groups tend to have more liberal cognitive styles. (2) centralized or decentralized social structures: the nature of the work influences the social structures and cognitions of a group. Routine and predictable work is likely to emphasize formal rules, codified procedures, and administrative hierarchy. In contrast, uncertain, exploratory, and creative work is likely to have more informal, flexible, and decentralized social structure. (3) the materialist theory of consciousness: those who control instrumental resources and organizational facilities also control how ideas are generated.

Randall Collin's theory of the intellectual world (Collins 1998) and Richard Whitley's comparative typology of scientific fields (Whitley 1984) are two theories that have direct impact on TSO. Collins suggests that a specialty's structure is determined by two factors: how much scientists need to coordinate and how certain the research agenda is. Similarly, Whitley suggests two parameters: task uncertainty and mutual dependence between scientists. Physics, for example, has very high reputational autonomy and highly centralized resources. Researchers in such structures heavily depend on those who control reputations and resources. Task uncertainty is low because of the tight controls. In comparison, the mutual dependence between sociologists is low, while task uncertainty is high because of decentralized resources and a variety of options to gain reputations. Building on these concepts, Fuchs proposed to explain how and why various types of scientific change take place in various areas of science in terms of mutual dependence and task uncertainty as two organizational variables.

Table 2.1 Four categories of scientific change

		Task uncertainty	
		Low Routine, repetitive, predictable	*High* Information is incomplete, ambiguous, controversial, unpredictable
Mutual dependence	*Low* Loosely coupled networks, decentralized means of production	Stagnation	C: Fragmentation
	High Tightly coupled networks, concentrated means of production	B: Specialization Teaching and textbook writing, follow-up research; Competition is low	Permanent Discovery Research fronts, invisible colleges, highly competitive

The key idea is that competition drives change because scientists compete for attention, reputation, and resources. Scientists who are seen to advance the state of knowledge will receive the highest rewards. Thus competition drives scientists to produce novel findings. Nevertheless, the pressure to produce something new is not equally distributed in various areas of science. The core of Fuchs' theory is summarized in Table 2.1. The combinations of mutual dependence and task uncertainty define three major types of scientific change:

- Permanent Discovery
- Specialization
- Fragmentation
- Stagnation, or the lack of activity

Permanent Discovery

The most productive, most visible, and most impactful groups are research fronts or invisible colleges. The mutual dependence and task uncertainty are both high. These are the small and tightly coupled core groups of highly productive researchers. The task uncertainty is high as their work belongs to the frontiers of science. The competition here is the highest with a very short half-life of research papers. Since the research topics they are working on are so advanced, the members of the group cannot rely on published literature. Instead, they rely on the invisible colleges to maintain their leading positions. Small and Crane (1979) showed that this type dense and highly interactive research fronts are reflected in the scientific literature as highly interactive co-citation clusters. Due to the density of the network, changes happen in one part of the network will be quickly spread to the rest of the group. If there is one important lesson that we can learn from this type of scientific change, it is about how a discovery finds its way in the network. The significance of a discovery cannot be materialized until people start to pay attention to it. More importantly, the spread of a new discovery needs the attention from the leaders and members of the core; otherwise, the new discovery is unlikely to get very far.

Scientific knowledge at the research fronts has not reached the status of knowledge that is certain enough to be accepted by the research community. Research frontiers are areas where the uncertainty is the highest. There is no textbook that can teach us what is going on at research frontiers. This is the primary reason that one should stay in touch with the core group of the specialty to pursue the highly exploratory research.

Specialization

The sociology of science and scientific knowledge differentiates two types of science: (1) controversial, conflictual, and uncertain science and (2) objective, consensual, and authoritative science. The former is what is happening in science in the making, whereas the latter is after the research has settled and rationalized in hindsight. We will continue to discuss the topic of uncertainty of scientific knowledge later in the book. Uncertainty is an integral part of scientific inquiry.

Competition under the conditions of dense networks and low uncertainty of task will lead to specialization. Research that follows the specialization is very similar to the normal science stage in Kuhn's theory.

Specialization can be seen as an extension, refinement, application, or expansion of the pioneering work that has been done by researchers at the frontiers of the specialty. Fuchs described the tasks as "handed down" from the research fronts. The novelty is relatively low for doing routinized tasks. The prestige is relatively low and the competition is not as fierce as in research front groups.

Specialization is also seen as an option to create a shelter from fierce competition by branching off the core specialty and establishing an area where researchers may reduce the competition. A specialty's general pattern of growth and decline starts with a tentative and exploratory search at the beginning, followed by a period of fast growth and then a gradual decline. As the chance of making significant new discoveries is decreasing, the core scientists would leave the increasingly routinized area and search for new areas.

Fragmentation

The third type of scientific change is fragmentation. From a sociological point of view, natural sciences and social sciences differ because scientists and social scientists work in different organizational structures. The differences in the strengths and objectiveness of scientific statements are resulted from the type of networks they are in. The subject matter does not matter.

Social sciences and the humanities are soft because they have fewer, weaker, and more dispersed resources. "Strong and closely coupled organizations produce science; loosely coupled and textual organizations produce hermeneutics." Because researchers are in a loosely coupled network, their work is unlikely to travel far and wide as quickly as researchers in a tightly coupled network. The high uncertainty associated their tasks means that it is generally difficult to determine the significance of change and obtain a sense of direction in which such changes may direct. Thus scientific change under such conditions would be mostly unstructured with an unclear direction or sense of progress.

Fuchs' theory attempts to explain scientific change from a sociological perspective. Its central premise is that competition leads to scientific change. A specialty starts with a tightly coupled core group of scientists. The most innovative and advanced research is likely to appear in this part of the specialty. The research fronts laid down the groundwork, which would typically lower the uncertainty and probably the cost of accomplishing similar tasks in new areas. As a result, specialization becomes an attractive option. Scientists carve out an area where they can routinize highly special procedures with increased sophistications to shield their professions from external competitions. Specialized research becomes highly productive. In both of the research fronts and specialized areas, scientists belong to tightly coupled networks. The competition would be considerably weaker in loosely coupled networks, where researchers have more controls of what they do. The task uncertainty divides researchers who are in loosely coupled networks further: those with high task uncertainty and those with low task uncertainty. Task uncertainty in this context reflects whether the research in question is creative and original or codified and trivial. The former leads to fragmentation. The latter leads to stagnation.

What do the theories from Kuhn and Fuchs have in common? How do they differ? Are they describing the same phenomenon from different perspectives or the two sides of the same coin? Are these two mental models compatible?

Kuhn's normal science and Fuchs' specialization share many similarities. Researchers in both cases have an established framework to pursue their research. For Kuhn, the stability is provided by the predominant paradigm. For Fuchs, the routinization is resulted from the pioneering work of the research fronts.

Research frontiers in Fuchs' theory are the most creative, volatile, and uncertain stage of a specialty's growth. In Kuhn's theory, the most creative and unpredictable stage is when the currently predominating paradigm is in crisis. A philosophical point of view may not pay much attention to sociological parameters of researchers behind such crises. Who would be the one to draw our attention to anomalies, researchers in an area of specialization or researchers from the research fronts of a specialty? By the time a paradigm may encounter a threatening crisis, pioneers in the original core of the paradigm have probably already moved on. In Kuhn's theory, researchers may switch to a new paradigm as a scientific revolution. From a sociological point of view, researchers may have many choices in response to crises. There may be pragmatic reasons. For example, if established researchers were to switch to a new paradigm, it is likely that they will have to throw away a lot of domain expertise that they have earned in a hard way.

An Evolutionary Model

A relatively new theory of the evolution of a scientific discipline is proposed by Shneider (2009). He suggests that his theory complements existing theories of scientific process, including Kuhn's structure of scientific revolutions. He devoted much of his attention to the characteristics of scientists that would be most influential and productive at each stage.

Stage I—Conceptualization

The evolution of a scientific discipline has four stages. At the first stage, scientists introduce a new language to describe a new subject matter. To Isaac Newton, the new language was differential equations and the new subject matter was mechanical movements. To Antoine Lavoisier, the new language was chemical equations and the new subject matter is chemistry as we know it today.

Scientists working at the first stage may not be the ones who discover new facts. Lavoisier, for example, did not discover any substances. Nor did he invent any chemical apparatus. However, the new language created by Lavoisier connected many previously isolated pieces together. Watson and Crick are another example of first-stage scientists. They discovered the double helix structure of DNA.

First-stage scientists would focus on essentials and tolerate the uncertainties in many other aspects. According to Shneider, "What might be considered to be incompleteness and inaccuracy is, in reality, the formation of a first-stage hypothesis." Shneider further elaborated the point using Dmitry Mendeleyev as an example. Mendeleyev created the periodic table. He reserved positions on his periodic table for elements that ought to exist but yet to be discovered.

Shneider characterizes first-stage scientists as those with a broad range of interest who tend to make contributions across different fields of science. First-stage scientists have strong confidence and they are able to sustain criticisms from the most reputable colleagues. They are good at making use of philosophical, esthetic and culture perspectives. They are able to connect seemingly unrelated topics to make their arguments. Finally, the most critical trait of first-stage scientists is their ability to generate interests in their ideas and sustain them into the second stage.

Taken together, the most unique defining characteristic of first-stage scientists is their outstanding vision. They are able to see profound connections or properties that others cannot see. Shneider emphasized the role of introducing a new language in the emergence of a new scientific field. Perhaps a more intuitive way to clarify this is to ask what the new language could convey that other existing languages could not. What do the double helix and Mendeleyev's periodic table have in common?

Stage II—Tool Building

The second stage of the evolutionary model is critical for the development of research methods and tools. The value of the instruments developed at the second stage will be ultimately determined by how much they will be used at the next stage. The importance of method is evident in recognitions made by prestigious awards such as Nobel Prize awards. Method papers are among the most cited types of publications. Many powerful research tools typically have their user populations ranging across a wide variety of scientific disciplines. A stage-two field is probably the easiest one to identify. One can check how many tools have been developed and used within a relatively short period of time, for example, within a five-year interval. The basic local alignment search tool (BLAST), for example, is one of the most highly cited papers.

Stage III—Applications of Tools

The third stage of the evolution is characterized by the application of established research methods to new targets. This is the most productive stage in terms of the data and new knowledge. It is not really whether the new target area is ready to accept new methods from their generous provider. Rather, it is whether the research field as the provider has reached its third stage. This realization may have practical implications. Instead of searching for potentially useful techniques in all scientific fields, one only needs to search for them in stage-three disciplines or fields.

Anomalies are most likely to rise at the third stage. The research at the third stage is probably stretched as far away as possible from the original ideas proposed in the first stage. The discrepancies between the source field and the target field may become apparent. Similarly to Kuhn's theory, one way to handle the anomalies is to switch to a new world view. Third-stage scientists are probably application oriented. They believe that a new theory would be useless unless it can solve concrete problems. First-stage scientists, in contrast, are generally less concerned about finding immediate applications for a potentially valuable theory. They are less likely to be bothered by the lack of clear indications of future applications. This type of research is sometimes referred to as basic research as opposed to applied research. Research conducted by the TRACES project found the duration between the initial basic research and the first clear application can last for 50 years and most likely much longer. If future first-stage scientists try to create the first stage of a new field, they may face the resistance from the current third-stage scientists.

Stage IV—Knowledge Codification

The fourth evolutionary stage is marked by a relatively low productivity of new knowledge. Research activities in the fourth stage become increasingly routinized. In addition, the skills and knowledge learned through the first three stages need to be passed on to next generations. Like first-stage scientists, fourth-stage scientists need to provide a holistic view of their discipline. Unlike first stagers, fourth stagers are good at understanding facts and they follow the latest developments. With their complementary strengths, they are the best helper for would-be first stagers to branch off from a stage-three field.

From a more pragmatic point of view, a grant proposal may benefit from a variety of reviewers from scientists who have strengths associated with different stages of scientific disciplines. Perhaps it would make an interesting exercise to look at the evolution of your own field of research. Can you find who the first-stagers are? Who might be second-, third-, or fourth-stagers?

Multiple Perspectives

There are other theories of scientific change (Mulkay 1975). For example, a transition model of Exploration → Unification → Decline/Displacement was proposed by Mulkay et al. (1975). Nevertheless, the theories outlined above are representative. They cover the major characteristics of the development of a scientific field.

These theories evidently overlap. Kuhn's competing paradigms, Fuchs' research fronts, and Shneider's first stagers all represent the initial conceptualization of a new research framework. Shneider's tool building second stagers and Fuchs' specialization in high dependency and low task uncertainty share common features of routinization and codified knowledge. Shneider's third-stage scientists who apply proven techniques to new targets may expand the scope of a field uneventfully or trigger anomalies that may take much more creativity to handle. This division may roughly correspond to Fuchs' classification of task uncertainty. Fuchs played down the chance of a profound scientific revolution. Instead, he drew our attention to the small and gradual changes that eventually build up.

Shneider's first stagers are most likely the members of the invisible colleges or the core groups of research fronts in Fuchs' theory. Shneider's second stagers, tool builders or methodologists, are likely to be associated with Fuchs' specialization. At least some of Shneider's third stagers may become the research fronts of a new field. From Kuhn's point of view, researchers who are most likely to introduce a new paradigm would be those who are visionary pioneers with a broad range of interest across different fields or specialties.

Kuhn's theory is philosophical in nature and focuses on the forest of scientific knowledge. Kuhn's mechanism of change is very clear—a Gestalt switch in response to irreconcilable anomalies. Fuchs' theory offers sociological explanations to scientific change. Fuchs does not accept a sudden revolutionary change such as Kuhnian paradigm shifts. Instead, Fuchs believes that competition for recognition and reputation leads to scientific change. Fuchs provided more details to explain the types of change may be resulted from an interaction between mutual dependence and task uncertainty. To Fuchs, the most likely source of scientific change is the research fronts or a small core of highly creative people. Scientific change boils down to whether one can attract people's attention and for how long. Having something new to say is one way and probably the most effective way to do it. As we can see from another sociological theory—That's interesting!—the best way to get our attention is to challenge what we believe. If we know our audience believes that the earth is the center of the universe, then we would probably get their full attention if we tell them that the earth is out the center of the universe. Similarly, if we believe that there is only one universe, then we would probably be very eager to find out more if our physicist tells that there are in fact multiple universes— multiverse! How come we haven't seen any signs of any other universes for all these years?

Shneider's theory is probably the most complicated among the three theories. Scientists at each stage are mostly profiled by concrete examples rather than declarative definitions. Few of those characterizations are unique. The problem solving plot is probably the most valuable contribution of Shneider's theory. The problem is identified. Tools are developed to solve the problem. Then we apply the tools to new targets, which are likely to trigger new problems. For those who have a holistic vision, they would document the process and what we have learned along the way. The problem solving framework provides a template for us to interpret the growth of a research area that is so generic that it seems to be applicable to many scientific domains. On the other hand, does BLAST alone sustain a field of research? Do other scientists put their work on hold while tool builders construct new instruments? Similarly, when third-stagers enjoy finding new targets, what would other stagers be doing? We are unlikely to find answers to these questions unless we examine the evolution of several scientific domains in detail. We believe thorough cross-examinations of scientific fields should be done to answer these questions. To our best knowledge such studies of multiple fields across multiple theoretical frameworks are currently missing.

Summary

The macroscopic theories of scientific change introduced in this chapter provide a conceptual framework for the study of the dynamics of scientific knowledge. By presenting these theories with distinct perspectives side by side, we hope that their similarities and differences are made clear. We also need to bear in mind that, as mental models of scientific disciplines, these theories provide a valuable reference for the development and validation of computational approaches to the study of scientific knowledge. It may well be true that science advances through all the possible routes described by these theories collectively in that there are indeed paradigm shifts taking place, scientists do actually compete for their recognition, and applications of special-purpose tools really lead us to new discoveries. The collective value of these theories is that they are insightful and inspirational for us to characterize something as complex, dynamic, and abstract as the knowledge of sciences.

References

Bostrom A, Morgan MG, Fischhoff B, Read D (1994) What do people know about global climate-change. 1. Mental models. Risk Anal 14(6):959–970. doi:10.1111/j.1539-6924.1994. tb00065.x

Chen C (2011) Turning points: the nature of creativity. Springer, New York

Chen C (2014) The fitness of information: quantitative assessments of critical evidence. Wiley, New York

Chen C, Hu Z, Milbank J, Schultz T (2013) A visual analytic study of retracted articles in scientific literature. J Am Soc Inform Sci Technol 64(2):234–253. doi:10.1002/asi.22755

Collins R (1998) The sociology of philosophies: a global theory of intellectual change. Harvard University Press, Cambridge

Csikszentmihalyi M (1996) Creativity: flow and the psychology of discovery and invention. HarperCollins Publishers, New York

Endsley MR (1995) Toward a theory of situation awareness in dynamic-systems. Hum Factors 37 (1):32–64. doi:10.1518/001872095779049543

Fuchs S (1993a) A sociological theory of scientific change. Soc Forces 71(4):933–953

Fuchs S (1993b) Three sociological epistemologies. Sociol Perspect 36(1):23–44

Hanson NR (1958) Patterns of discovery: an inquiry into the conceptual foundations of science. Cambridge University Press, Cambridge

Heuer RJ (1999) Psychology of intelligence analysis. Central Intelligence Agency

Johnson-Laird PN (1983) Mental models. Harvard University Press, Cambridge

Kintsch W (1988) The role of knowledge in discourse comprehension—a construction integration model. Psychol Rev 95(2):163–182. doi:10.1037/0033-295x.95.2.163

Kuhn TS (1962) The structure of scientific revolutions. University of Chicago Press, Chicago

Markus HR, Kitayama S (1991) Culture and the self - implications for cognition, emotion and motivation. Psychol Rev 98(2):224–253. doi:10.1037/0033-295x.98.2.224

Masterman M (1970) The nature of the paradigm. In: Lakatos I, Musgrave A (eds) Criticism and the growth of knowledge. Cambridge University Press, Cambridge, pp 59–89

Mulkay M, Gilbert N, Woolgar S (1975) Problem areas and research networks in science. Sociology 9:187–203

Mulkay MJ (1975) Three models of scientific development. Sociol Rev 23:509–538

Shneider AM (2009) Four stages of a scientific discipline: four types of scientists. Trends Biochem Sci 34(5):217–223

Small H, Crane D (1979) Specialties and disciplines in science and social science. Scientometrics 1:445–461

Sutton SA (1994) The role of attorney mental models of law in case relevance determinations—an exploratory analysis. J Am Soc Inform Sci 45(3):186–200

Thagard P (1992) Conceptual revolutions. Princeton University Press, Princeton

van Raan AFJ (1990) Fractal dimension of co-citations. Nature 347:626

Whitley R (1984) The intellectual and social organization of the sciences. Clarendon, Wotton-under-Edge

Chapter 3
Science Mapping Tools and Applications

Abstract We introduce the design and applications of a few influential science mapping tools, namely CiteSpace, VOSviewer, and CitNetExplorer, such that one can utilize these freely available tools to study a scientific domain of interest. CiteSpace provides a variety of metrics and indicators concerning trends and patterns in scientific literature. Many of these metrics are explained further with illustrative examples from applications of CiteSpace. CiteSpace also includes extensions that are particularly made to generate examples in the book. Three examples of systematic scientometric reviews using CiteSpace are included to illustrate relevant concepts and analytic functions.

Keeping Abreast of Scientific Frontiers

Keeping abreast of the development of a scientific domain is challenging for many reasons. It is time-consuming to search and gather relevant information adequately. There are numerous ways to describe the same topic, so it is challenging to come up with a comprehensive list of keywords for a scientific domain. Furthermore, we are most likely unfamiliar with the domain we plan to search for in the first place. How do we maximize the coverage of our search with our limited knowledge of the target?

One of the scenarios that we need to deal with has become less common because we are more likely to find at least some relevant publications now than we were just a few years ago. Imagine that we carefully formulated a search query, but our query didn't lead to any usual articles. In other words, it seems we have to revise our initial query so that we can at least find something. Once we found relevant articles, there are ways to expand the search and find more relevant publications.

An effective way to minimize the risk of missing anything important in a scientific domain is to see the basis that other researchers have built on. Most likely we can learn valuable information from others that we would probably never think of.

© Springer International Publishing AG 2017   57
C. Chen and M. Song, *Representing Scientific Knowledge*,
https://doi.org/10.1007/978-3-319-62543-0_3

Scholarly Publication

Researchers today find themselves with increasingly more options to publish their work, ranging from the traditional peer-reviewed archival publications to self-directed newsflash-like tweets. Although researchers now have many more options than ever before, the essential process remains the same. Briefly speaking, researchers come across a research question that they can do something about. Of course, finding the right research question is currently also more of an art than science. As we have seen in Hilmeier's series of questions, a critical step in research is to understand the status of the research question in a broad context chronologically and domain ontologically.

Scientific discoveries are made rarely in the order that makes the most logical sense; otherwise, science would be reduced to a simple and straightforward logical reasoning process. Scientists and researchers need to publish their work in order to establish or maintain their intellectual impact in the scientific community. Novelty, originality, interesting, and creativity are among the few criteria that are held strongly in scholarly publication, especially through those venues guarded by various forms of peer reviews. Along the line of novelty, reviewers commonly criticize the lack of originally, the inadequacy of a claimed novelty, or an inadequately established connection to prior work by others in the field. The strongest argument for a novelty is almost certain not the one that simply claims no one has ever done it before or the equivalent cliché that we are the first who did so and so.

Sociologists have noticed that the easiest way to attract people's attention is to challenge the beliefs of your audience to the extent that they would be curious enough to listen to what you have to say. On the other hand, going too far in this direction may put off your audience altogether if it starts to sound ridiculous to their current mindset. In fact, sociologists suggest a few strategic moves that may boost the novelty of your next research question. For example, numerous theories were proposed to explain what happened at the KT boundary that led to the extinction of dinosaurs. The widely known theory was the one that focused on an asteroid impact on the earth and its atmosphere. A competing theory suggested that it was the lava from the insider of the earth rather than what from the sky. Similarly, after the September 11 terrorist attacks, researchers realized that people may still develop post-traumatic stress disorder (PTSD) symptoms even they were never near to a site of trauma, which was previously believed to be impossible. Prior to the September 11 terrorist attacks, PTDS research suggests that a first-hand physical experience of a trauma is essential for developing PTDS. However, researchers found that many people who did not have direct experiences through the trauma because they are thousands of miles away from New York.

The subjectivity of evidence means that the role of a piece of information as evidence is subject to the mindset or the mental model of individuals. The same piece of information can be used by different individuals to support different arguments. As the rest of the universe seems to be redshifted from us, does it mean we are at the center of the universe? As everyone can see the sun rises and sets, they may still come up with different interpretations concerning whether we are at the center of the universe or the earth is orbiting around the sun.

In addition to the subjectivity of information, the uncertainty of the collective knowledge of the scientific community is another fundamental concept that we should bear in mind. We know from sociological perspectives of scientific change that scientists are driven by their desire to establish and consolidate their recognition and reputation in the scientific community or beyond. They seek to attract attention from their peers with novel ideas and astounding findings. The most active areas would be where we know little about the subject. Once we know more and more about a subject area, the level of uncertainty is likely to reduce. Thus, the level of uncertainty is an integral part of our knowledge of an area of research. Indeed, the knowledge of the uncertainty of knowledge is a type of meta-knowledge, which tells us the epistemological status of our knowledge. As we have seen in Shneider's evolutionary model of a scientific discipline, the meta-knowledge of a discipline may tell us which stage of the evolution the discipline is going through (Shneider 2009). Is it still at the first stage when researchers in the specialty are trying to conceptualize a new line of research? Is it at the stage when researchers are concentrated on building the right tools to augment their studies as Galileo was building his telescope?

Citation-Based Analysis

There are many types of scientific publications. We will primarily focus on two of them that are most likely to reveal relevant scientific knowledge of a domain: articles that report original research and reviews of a research topic.

Each formally published scientific article typically consists of the following components:

- A title and sometimes a subtitle.
- A list of authors and their affiliations.
- An abstract, structured or unstructured.
- A list of keywords assigned to the article by the authors.
- A list of keywords assigned to the article by indexing services such as the Web of Science.
- The main body of the article, including text, figures, tables, equations, and other materials.
- An acknowledgement to reviewers, researchers, or research funding or sponsorship.
- A list of references cited in the article.

Terms such as noun phrases appeared in the title of an article can be used to compute how often two terms appear within an article or even at the sentence level. Such terms are called co-occurring terms. For example, the four terms highlighted in yellow in the title are co-occurring terms. Similarly, connections between terms in the abstract can be established in terms of their co-occurrences too.

A citation is an instance in which an article explicitly refers to a previously published article. Eugene Garfield conceived the idea of citation indexing, which taps into association of ideas found in scientific publications (Garfield 1955). The referred article is called a reference or a cited reference, for example, the citations to (Price 1965; van Raan 2000; Abt 1998) in the example shown in Fig. 3.1. In a scientific publication, especially in an original research article, references are cited for specific reasons in connection to an argument of the article. The fact that these references are cited by the same article means that they are co-cited references. In other words, they are cited together. A co-citation relationship between two references implies that, from the point of view of the author of the citing article, the two references are related to one another through the content of the citing article. For instance, one can infer from the text that the co-citation relation between (Price 1965) and (van Raan 2000) is probably because both of them are relevant to properties of transient articles. There may be more instances in which these two references are cited together in the same article later on. Multiple co-citation instances of the same pair of references may strengthen their co-citation relation in terms of the quantity. On the other hand, co-citations in different contexts may increase the diversity of the nature of the co-citation relation.

Traditionally, co-citation relations are established based on the references listed at the end of an article rather than based on an inspection of co-citation instances in the body of the article. In other words, we know that (Price 1965; van Raan 2000) are co-cited by the article because they are both included in the reference list of the

⊕WILEY
InterScience®

title **CiteSpace II: Detecting and Visualizing Emerging Trends and Transient Patterns in Scientific Literature**   terms

author Chaomei Chen
affiliation *College of Information Science and Technology, Drexel University, 3141 Chestnut Street, Philadelphia,*   geolocation
             PA 19104-2875. E-mail: chaomei.chen@cis.drexel.edu

abstract **This article describes the latest development of a generic approach to detecting and visualizing emerging trends and transient patterns in scientific literature. The work makes substantial theoretical and methodological contributions to progressive knowledge domain visualization. A specialty is conceptualized and visualized as a time-variant duality between two fundamental concepts in information science: research fronts and intellectual bases. A research front is defined as an emergent and transient grouping of concepts and underlying research issues. The intellectual base of a research front is its citation and co-citation footprint in scientific literature—** high citations and transient ones with their citations peaked co-citation within a short period of time (Price, 1965). Transient ones ●━┓ are much more common than classics (van Raan, 2000). The ●━┛ average length of time that a research article continues to be cited in the scientific literature is closely connected to the growth speed of the underlying research area (Abt, 1998). ● Understanding the dynamics of how transient articles trans- citation form the intellectual landscape of a scientific field has significant practical implications for scientists in a wide variety of disciplines.

Fig. 3.1 The meta-data of a research article—a 2006 JASIST article on CiteSpace II (Chen 2006). The article is the 2nd of the 10 Google Scholar classic papers in Library and Information Science published in 2006

article rather than we found the sentences that mentioned them on the same page. The simplistic traditional approach is largely due to the accessibility of full text of scientific publications in the mid 20th century when the now famous Science Citation Index (SCI) was conceived by Eugene Garfield. Initial volumes of SCI were themselves created on punch cards and printed on papers. It is not until recent the access to full text articles is gradually taken for granted. As it becomes easier to access the full text of scientific articles, one would wonder what we would miss by deriving co-citation relations from the end-of-article reference list as opposed to pinpoint co-citation instances directly in full text.

The 2006 JASIST paper on CiteSpace is 19-page long, including 18.5 pages of text. The first citation is on the first page to (Price 1965) and the last citation is on page 18 to (Smalheiser and Swanson 1998). Intuitively, the co-citation connection between (Price 1965) and (van Raan 2000), both on the first page, is much more meaningful than the co-citation connection between (Price 1965) and (Smalheiser and Swanson 1998), which spans over 18 pages of text. It seems one should take into account this type of distance between the locations of two references whenever possible. In contrast, due to the limitation of data, co-citation relations in a traditional co-citation analysis cannot further differentiate their strengths within an article. We investigated the effect of the proximity of co-citation locations in the full text articles published in six bioinformatics journals and found that co-citations at the sentence level provide a good approximation to the overall co-citation patterns identified at the article level. It is therefore our recommendation that whenever possible the proximity of co-citation locations should be taken into account, preferably at the citation context level, which is commonly defined as a sentence that contains a citation instance and one or two neighboring sentences before and after the citation sentence.

The Metaphor of a Knowledge Space

An intuitive metaphor of the scientific knowledge is a knowledge space or the universe of entities and relations and various aggregations at higher levels of knowledge representation such as facts, rules, claims, hypotheses, speculations, and other types of elements represented. Stars and quasars in the universe of knowledge would represent concepts and their connections. Each published article would introduce some changes into the existing universe of knowledge. For example, an article may introduce new connections between existing concepts. A more innovative article may introduce a set of new concepts and their relationships all at once. The brightness of a star can indicate how active a concept is. A concept is more active if the concept is being involved in more and more recently published articles. In contrast, if a concept has not been mentioned for a long time, its brightness would become dimmer. Interconnections between concepts can be introduced by an article and subsequently reinforced by additional articles later on. Connections that have not been actively discussed over a long period time may weaken their strengths. Newly published articles may alter the structure of the underlying knowledge space by adding new concepts and new interrelations.

The universe metaphor is not the only one that is intuitive. For example, an alternative metaphor is a neural network that models how human brains function when we learn from a typically large amount of data. The astounding performance of AlphaGo is one of the many impressive applications of artificial intelligence techniques, especially the so-called deep learning techniques. Through deep learning, multiple layers of interconnected neurons are adaptable to recognize patterns at various levels of granularity. For example, tasks that are used to be very challenging for computers but effortless for human beings such as recognizing a human face or a handwriting now can be reliably done through deep learning techniques. We will return to this exciting topic later in our book.

The universe metaphor has the advantage of visualization-congruence, which means it comes natural to derive a visualization design that would fit nicely with what users may expect from their understanding of the universe of astronomical objects such as stars, galaxies, the Milky Way, and the Great Wall. We would expect that many types of changes in the universe of knowledge are as visible and observable as we have seen in the universe of astronomical objects.

We use an interactive visual analytic tool CiteSpace to demonstrate how to generate a systematic review of a scientific field. A traditional systematic review of a field is typically written by someone who has developed a substantial understanding of the field. A typical review article contains over 100 cited references. A systematic review is valuable in the course of the development of a field. Derek Price, a pioneer of the scientometric field, once estimated that a fast-growing field probably needs to have a systematic review paper after every 50 original research articles. Systematic reviews thus serve the role of summarizing what has been achieved by the original research articles since the last review article. However, a field may be too young to have a readily available systematic review. Existing reviews may not give enough attention to the topics that we are particularly interested in. In other words, it is quite possible that our best bet would be simply to review the literature all by ourselves.

Doing the review by ourselves has several distinct advantages:

- We choose the depth and breadth of the topics to cover.
- We choose when it is the time to do it as we need it.
- We develop a deeper understanding of the topics and their connections along the way.

The major challenge is the lack of the knowledge of the domain as a whole or that of a few specific areas of the domain. On the other hand, this would be true to any potential researchers who are planning to review the literature of a scientific domain. Given the scale and the volume of today's scientific literature, it is unlikely for an individual to master the depth and breadth of a subject domain. Many new research students face the challenge when they search for potential dissertation topics. By any standard, it is a time-consuming task to sift through hundreds of hand-picked articles to identify some potential research questions. More importantly, identifying a research question cannot be done meaningfully without a good understanding of what characterizes it in a broader context. In other words, the amount of the effort required to articulate a research question properly is probably about the same as, if not more than, the amount of the effort required to develop a good understanding of a field. As we have seen in Heilmeier's Catechism, researchers need to figure out not only the status of their research problems in a potentially boundless context but also how to articulate the status most effectively to anyone who might concern.

Let's assume that the analyst does not have any special training in the target subject domain, which is most likely the case for many who need to find out more about the domain in the first place. The first thing our analyst needs to find out about the universe of knowledge is its structure or how the various galaxies are organized in the space, how they are related to one another, how long they have been there, what changes are taking place, and what one may expect to see in the future. Within a specific galaxy, our analyst would be interested in stars that stand out in one way or another. What is the brightest star? Which one has the greatest mass? Which one is the most unstable one? Which one is on its way to collapse? Which one is about to collide and merge with another one?

A visualization tool to our analyst is like a telescope to an astronomer or a GPS to a driver. The resolution of a visualization tool is determined by the resolution of the underlying data. A high-resolution GPS would be more useful for us to navigate through a dense and complex road layout than a low-resolution GPS. A high-resolution telescope would be more powerful for us to see finer structures of astronomical objects than a less powerful telescope. The resolution of a visualization of co-cited references in scientific publications can be measured in terms of the number of pages or the distance between the locations of co-cited references. Take the references cited in the 2006 JASIST article on CiteSpace (Chen 2006). In a traditional co-citation analysis, the resolution is the same 18-pages to all the co-cited references. The 18-page resolution is the maximum possible distance between two references cited in the article. If the full text of the article is accessible, then the resolution can be further improved by replacing the 18-page distance with the actual distance between the locations of two citations. If two references are cited multiple times in the same article, then the co-citation proximity can be defined through several options. For example, we can use the minimum distance between the locations of two citations. Alternatively, we can use the median distance to represent the strength of the co-citation link.

CiteSpace: Visualizing and Analyzing a Knowledge Domain

CiteSpace is an interactive visual analytic tool written in Java (Chen 2004, 2006; Chen et al. 2010). It is freely available. The motivation behind the development of CiteSpace is to enable researchers to conduct a systematic review of a scientific field with little relevant domain knowledge or no prior knowledge of the domain at all. It is suitable for a new research student to search for potential dissertation topics, for an experienced researcher to keep abreast of the development of an established field of study (Chen 2017), or for a scientist to explore emergent trends in one or more research areas (Chen et al. 2012, 2014a, b).

CiteSpace is not intended to replace the role of conventional systematic reviews. Rather, CiteSpace aims to provide a computational approach that can be easily applied by the vast majority of researchers to meet their own needs. The procedure is repeatable at practically no cost to the analyst so that the analyst can generate a new review whenever necessary.

It is not our intention to use tools such as CiteSpace to eliminate the role of domain expertise in interpreting the analytic results of a CiteSpace application. On the contrary, we precious the value of domain expertise and we believe any domain expertise is hard to come by. We want to provide a tool for areas where domain expertise is not readily accessible or not available in a timely manner.

CiteSpace is probably the first computer application that is specifically designed to support the visual analysis of scientific literature. The development of CiteSpace has been particularly inspired by a number of pioneering software systems that have been made freely available, notably Pajek for analyzing large networks (Batagelj and Mrvar 1998), information visualization toolkits such as prefuse (Heer 2007), software programs generously shared by Loet Leydesdorff. Many wonderful and relatively new systems are made freely available, including VOSViewer (Van Eck and Waltman 2010), CitNetExplorer (Van Eck and Waltman 2014), and Gephi.

CiteSpace is unique in several ways in comparison with other systems that also take science citation data as the input. First of all, CiteSpace is designed to support the analyst to obtain a good understanding of the development of a scientific domain, or a knowledge domain. The unit of analysis is a subject domain, which means all the landmark publications and articles that have played a critical role in the holistic view of the knowledge domain as a complex adaptive system. With the support of CiteSpace, our analyst should be able to develop a good sense of the fundamental issues and major methods associated with the research domain. Second, the focus on a domain of knowledge is reinforced by various visual encoding that characterizes patterns and features with reference to underlying theories of the development of a scientific domain. For example, a cluster of co-cited references provides a representation of the intellectual base of a research specialty. The nature of inter-cluster relationships is underlined by cited references with strong betweenness centrality scores. The boundary-spanning or brokerage implications of such references are supported by theories such as the Structural Hole Theory (Burt 1992) and as a focal point in a paradigm shift from a Kuhnian point of view. CiteSpace is designed in such a way that the search for critical information of the development of a scientific field is turned to the visual search of patterns and features that standout in an overview of the domain.

Figure 3.2 shows an overview of terrorism research (1996–2003). We will explain the details shortly, but for now let us check what features would draw our attention, assuming we know nothing about this research domain. What we can see

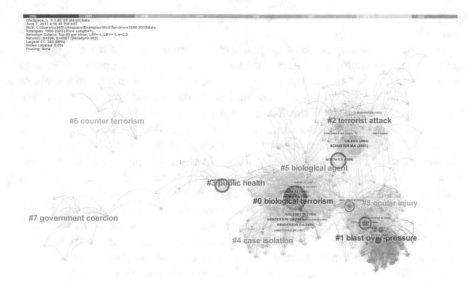

Fig. 3.2 An overview of terrorism research (1996–2003)

effortlessly is a property called preattentativeness, which means they will get our attention within the first 200 ms. It is the time required to redirect our attention. This property is also called standout or popout. In the visualized network of cited references, our attention is likely to be directed towards the few big circles in purple. Then we could also notice some discs in red. Next we would probably explore the surrounding areas of these focal points and perhaps read the text labels in different colors and font sizes. At the highest level of granularity, we could see 3–5 concentrations (clusters) in different colors. The legend above the visualization indicates the navy blue color is associated with 1996 on the left and the orange color is associated with 2003 on the right. The area in the orange color, with a label "#2 terrorist attack," must essentially correspond to the year 2003. In contrast, the area in navy blue, labeled as "#1 blast over-pressure," must be connected to the year 1996. Further inspections would reveal that purple circles seem to be positioned between different areas such as NORTH CS (1999) between #2 terrorist attack and #0 biological terrorism on the upper right region of the display, HOFFMAN B (1998) between the mainland and the peninsula stretching into the west (#7 government coercion), and MALLONEE S (1995) linking the #0 biological terrorism and #8 ocular injury in the lower right region.

As you can see, we are able to identify a small number of elements in the research domain without referencing to any specific domain knowledge. Evidently these elements must play some special roles in further understanding the research domain. What makes these elements standout in the overview of the research landscape is due to the co-citation patterns found in scientific articles written by researchers in the scientific community. In other words, these patterns reflect something profound shared by individual researchers because the emergence of a pattern requires the consensus, endorsements, and reinforcements of many researchers.

Now we can identify some of the major characteristics of a knowledge domain. It may consist of multiple inter-connected topic areas. The development of each of these topic areas is likely to last for a period of time, which may have a variable duration. The key to the inter-relationship between two topic areas is largely hold by the brokerage node or nodes that connect the two topic areas. Our analyst can reach this level of understanding of a knowledge domain through a visual inspection that won't take much longer than a few minutes, although our experience shows that most of the users would eagerly dive into the juggles of specific references before pondering the overall structure of the forest and the implications of the structure on subsequent exploration of the domain's landscape.

Ben Shneiderman, a pioneer in visual information seeking and human-computer interaction in general, proposed a simple mantra that summarizes the strategy of visual information retrieval for designers as well as for end users (Shneiderman 1996). Shneiderman's mantra states "overview first, zoom and filter, then details on demand." The first step—overview first—is to form a hierarchical organization of a domain and its topic areas. The entire research community behind the domain as a whole can be considered as a single specialty. Each topic area corresponds to a subset of the overarching specialty. Such subsets can be considered as distinct specialties in their own rights. The best way to understand the nature of a specialty is not only to see what topic it focuses on but also how it distinguishes itself from specialties associated with its neighboring topic areas. In the terrorism research (1996–2003) example, understanding that bioterrorism is a major concern in the research domain is one thing, but a deeper understanding of practical implications of bioterrorism on healthcare and the preparedness of emergency responders is a significant step towards understanding what a research domain is really about. Achieving an understanding at this strategic level brings numerous advantages to our analyst in subsequent exploration of the knowledge domain. Once we have established an organizing framework based on the writings of many active researchers in the domain, one can easily categorize newly published research and recognize in what sense the new research is novel. Answering Heilmeier's questions is no longer as challenging as they seemed to be before our inspection of the overview of the domain.

Visual Exploration of Scientific Literature

The general procedure of visually exploring the scientific literature of a knowledge domain consists of several basic steps:

- collecting data
- configuring representation models
- generating interactive visualizations of the domain.

Data Collection

The goal of the data collection step is simple: to collect data that can adequately and accurately represent the domain in question. In practice, this is easier said than done. First, we are probably not familiar with the domain of our choice. We may be very interested in the domain, but we are probably not aware of various terminologies or jargons that have been used in scientific writings to describe topics relevant to the domain. Furthermore, concepts, theories, and practices may have evolved over time. The best case scenario is when we are familiar with the vocabulary of the domain, have an easy access to a domain ontology or thesaurus, and have a domain expert on the team. In the toughest scenario we would have none of them. In general, it is likely that we are somewhere in between. A common strategy is to snowball the query-and-refinement process so that as we learn more and more about the domain, we are better able to characterize what the representative data would look like.

More sophisticated search strategies are possible to improve the quality and efficiency of data collection further. For example, if we are familiar with a theory of the development of scientific knowledge, then we may derive a complex set of queries such that we can cover various aspects of the target domain systematically. In a recent example, we found it effective to organize our queries with reference to a theory of the evolution of a scientific discipline. The theory was proposed by Alexander Shneider. It is simple and intuitive. According to the theory, the evolution of a scientific discipline goes through four distinct stages in sequence: conceptualization, tool construction, tool application, and knowledge codification. The conceptualization is the first stage of the evolution. New ideas are conceived, although a lot of details remain unknown. The tool construction stage focuses on developing instruments that would be necessary to investigate the research questions conceived at the conceptualization stage. The tool application stage is when the application of enabling and augmentative techniques to the research questions result in new discoveries and new knowledge. When we formulate a complex query for relevant articles, we can include sub-queries that would cover specific aspects of an evolving scientific domain. For example, we can use one query to specify the basic concepts of the domain, use another query to specify the types of tools that are particularly relevant to the domain, and yet another query to specify applications of the research method.

Configuration of Representation Models

A key concept in CiteSpace is the time slicing technique. The idea is similar to the concept of a sliding window. A long period of time can be time sliced into a series of adjacent time slices. A snapshot of the domain knowledge can be represented by scientific articles published within the corresponding time slice. A time slice can be a one-year window or a multi-year window. The primary effect of time slicing is to enhance the impact of research in a particular year. Adjacent time slices can overlap with each other. One of the effects of overlapping time slices is to smooth the variations over time. Besides, it makes more sense to consider that articles published in December and in January next year should belong to the same group as well as articles published in June and July. The effect of allowing an overlapping sliding window is a smoother transition of various patterns. The traditional network analysis without time slicing is a special case when the width of the window becomes the entire time interval. If we allow the duration of overlapping years to vary from 0 to the entire time interval, then the traditional method is a special case when the overlapping years become the entire duration of the time interval. In the following examples, we use non-overlapping time slices for its simplicity.

The clarity of a network is typically affected by several factors. An excessive number of links in a network would make it harder to differentiate salient patterns from common linkages. There are many strategies for reducing the number of links. Some of them make clear-cut decisions, whereas others follow sophisticated criteria that take into account local structural properties or even global structural properties.

Link Selection

Removing weak ties from a network is a commonly adopted strategy. Weak ties are often associated with a higher level of uncertainty, including underrepresented connections. There are many ways to select the weak ties to remove and they tend to impact the remaining network differently. The simplest way is to rank all the links in a network by their strengths and remove links from the bottom of the list, for example, by removing links with the strength below a cut-off threshold or by removing the 20% of the links with the lowest strengths. The downside of this approach is the risk of removing nodes that do not have strong links to survive. Although one may argue that we do not lose much anyway considering those nodes do not have strong ties with the rest of the network, weak ties may bring us valuable and unanticipated information. According to a famous study entitled the strengths of weak ties in social networks, the value of weak ties lies in their potential role in informing us something that may be unexpected. From an information scientist's point of view, any information that surprises us is a learning opportunity because it shows that our current belief, or our mental model, is inadequate, inconsistent, or even totally invalid. Weak ties in a social network imply a connection between people from different social circles. Information from different social circles is more likely to bring us something new as opposed to information from the same circle of friends.

According to sociologist Burt (1992), the potential value of the information flow is not because the ties are weak; rather, it is because weak ties are more likely in the position to connect different groups of individuals. The more broadly we are exposed to different ideas, diverse perspectives, and alternative interpretations, the more likely we are able to come up with creative solutions and better handle a complex situation. According to Burt, our positions in a social network are not equal because the chance of seeing a diverse range of information flowing by is different. The difference, according to Burt can have profound consequences because one can translate such potential to a competitive edge.

We have learned at least two things from the above discussion: (1) we should avoid removing weak ties simply because they are weak, and (2) some nodes are worth our attention more than others because they may indicate where the competitive edges are or where the creativity is. Let us see if we can meet the two criteria simultaneously. Instead of dealing with all the links in a single list, take all the nodes in the network and consider links that connect each node to the rest of the network. This arrangement makes it possible to retain all the nodes while removing relative weak ties from each node. Along this line of reasoning, we can come up with additional methods to reduce the number of links but preserve global properties of the network. In CiteSpace, the user can tie the number of links proportional to the number of nodes in the network. We know that the least number of links to connect N nodes is N − 1 and the maximum number of links in a fully connected undirected network is N*(N − 1)/2. Turning a network to a minimum spanning tree will give us a network of N − 1 links. However, we have shown in our previous research that using a minimum spanning tree to approximate the original network has several drawbacks despite its advantages such as computationally simple and efficient. Given a network, there may be multiple minimum spanning trees. Arbitrarily picking one of them does not justify the validity of the resultant representation. A more convincing solution is to retain all the minimum spanning trees if we cannot justify selecting one of them only. This is indeed what Pathfinder network scaling can offer.

Pathfinder network scaling is a link reduction technique that can impose a triangle inequality condition across the entire network (Schvaneveldt 1990). Pathfinder network scaling is able to retain the most salient paths in an associative network. A network that satisfies the triangle inequality condition throughout the network is called a Pathfinder network. Comparing with link reduction techniques such as threshold-based methods, Pathfinder network scaling is theoretically sound. Although initial implementations of Pathfinder network scaling are computationally expensive, fast-algorithms have been developed, especially by a group of scientometrics at the University of Granada. The Pathfinder network is the set union of all the minimum spanning trees of the original network. Alternative paths connecting the same pair of source and target nodes are allowed simply because we do not have reasons to discriminate them, just like our travelers can choose a cheaper multi-city flight as well as a faster but more expensive non-stop flight between two cities.

The length of a citation link from the source article published in year Ys to a target article published in Yt provides information that could be useful for understanding the long-term impact of the target article. If we can afford to ignore citations to target articles published long time ago, then we can remove such citations from the network modeling steps. This parameter in CiteSpace is called

Look Back Years (LBY). It is common to cut off by LBY at 5–8 years. Consistent with the universe metaphor, we can choose to focus on our connections within a radius of our choice.

Node Selection

The construction of a network may also impose restrictions on what kinds of scientific publications would qualify to participate in the modeling process. In other words, we need to decide what kinds of publications should contribute a representation of the knowledge of a scientific domain. Why do we assume that we should select a subset of publications to portray the knowledge structure of a scientific domain rather than including all of them in the process?

No matter what data sources we use it is very difficult, if not impossible, to obtain a collection that we can truly claim to have a coverage of 100%. Both Google Scholar and Elsevier have access to tens of millions of publications. The Web of Science and Scopus are representative but not comprehensive. Pragmatically, increasing the current coverage by 10% may cost extra 90% of efforts and resources. More importantly, what can we learn from the extra 10% coverage that we cannot possibly learn from what is currently covered? Besides, if we are going to apply the same methodology to an extended coverage, what would make the extra information standout and avoid becoming sidelined by the existing high-profile features? Thus, it is more important to have the quality data than aiming to collect the data that may cover everything. We need to be selective in data collection as well as in analytic methodologies. As long as we bear in mind the scope of our data sources, we do not have to perfect the dataset before starting analyzing it. In fact, an iterative strategy is likely to work more effectively than perfectionism that focuses on one step of the process alone because each step is a learning process and an opportunity to refine our process.

The node selection process determines not only what is relevant in the sense of information matched by information retrieval models but also evaluative indicators such as citations and altmetrics. Evaluative indicators provide information regarding the perceived value of an entity in the universe of knowledge. The value and the relevant of a piece of information may not necessarily correlate. In other words, a highly relevant piece of information may have little value to our analyst who is constructing a systematic review of a domain. In terms of the value of a citation to a publication, it is probably not so much how many times it has been cited so far; rather, it probably matters a whole lot more if thought leaders in the research domain or potentially relevant domains cited it. The widely known PageRank algorithm follows the same principle—the significance of a webpage should be recursively determined by the significance of pages that refer to the webpage. Along this line of reasoning, we should pay more attention to what an article has to say if it is written by a Nobel Prize Laureate, if it has been cited by Turing Award recipients or by others who have an established reputation in science and technology, or it has been widely cited for reasons that remain unknown.

Citations and altmetric scores are valuable information on how fellow researchers' react upon a scientific publication. Broadly speaking, the more citations an article has received, the more likely that the article has generated an impact on the research community. The more an article has been viewed and downloaded, the more likely the article is interesting. Using citations as an indicator of research impact is controversial. Some argued that since each citation instance may be motivated differently, it may not make sense to add them up as if each of them is equally accountable. Some argued even further that since some citations are supportive, some are neutral, and some are even challenging the original work, lumping these instances of different nature does not make any sense. Others have questioned the assumption that each citation reflects something about the knowledge of a domain because many mistakes or errors in citing a reference evidently show that one cannot assume everyone reads what they cite. Furthermore, researchers have found some citations even distorted the intended meaning of the original source.

There are at least two ways out of these controversies. One is to further classify the types of citations. The other is to clarify the significance of being cited. A good example of the former is the Shepard's Citation Signal in LexisNexis. For each legal case, for example, Miranda v. Arizona, 384 U.S. 436, the Shepard's Citation Signal identifies the types of citations, i.e. signals, that the case has been cited, including warning, questioned, caution, positive treatment, negative treatment, and criticized by. As shown in Fig. 3.3, instances of citations, or citing decisions, are classified into several types. For example, the Miranda v. Arizona (384 U.S. 436) case has been cited in dissenting opinions of the U.S. Supreme Court in cases listed as the items 21–23, i.e. Florida v. Powell, Montejo v. Louisiana, and Dickerson v. United States. Classifications such as the Shepard's Citation Signal are currently rare in scientific literature. The Web of Science and Scopus do not currently provide any citation information below the article level. In other words, we have no other options except assuming the an article cites all its references uniformly even if we know that this is not a good assumption to make. CiteSeer and Google Scholar are probably the most widely known resources of scientific literature where one can find contextual information of a citation. However, we are not aware of any large-scale resources of scientific publications that enable users to search citations by specific classifications of citations.

The latter way to reconcile much of the controversies or the uneasiness surrounding the use of citation counts as an indicator of scholarly impact is to clarify what we mean by impact. Many assume that the impact implies a positive outcome and that one should rule out any negative impact. It is our view that the term scholarly impact should include both positive and negative influences produced by a scholarly contribution. A failure at one level of consideration can be valuable at another level of thinking. Einstein once said the value of his research in his later years is to stop another fool to make the same mistake. If we can learn from a scientific publication either something to follow or something to avoid, it has a

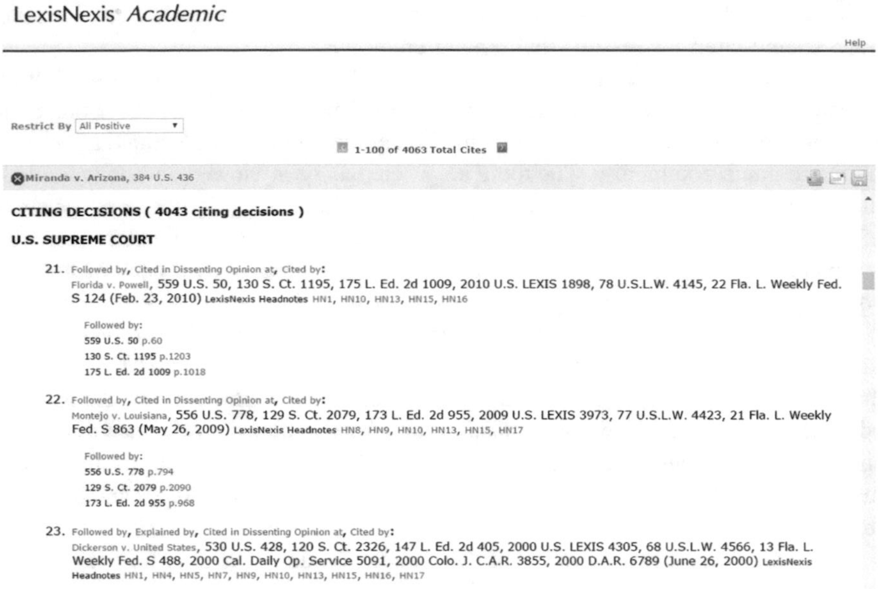

LexisNexis® *Academic*

Restrict By | All Positive ▼ |

☐ 1-100 of 4063 Total Cites ☑

⊗ Miranda v. Arizona, 384 U.S. 436

CITING DECISIONS (4043 citing decisions)

U.S. SUPREME COURT

21. Followed by, Cited in Dissenting Opinion at, Cited by:
     Florida v. Powell, 559 U.S. 50, 130 S. Ct. 1195, 175 L. Ed. 2d 1009, 2010 U.S. LEXIS 1898, 78 U.S.L.W. 4145, 22 Fla. L. Weekly Fed.
     S 124 (Feb. 23, 2010) LexisNexis Headnotes HN1, HN10, HN13, HN15, HN16

     Followed by:
     559 U.S. 50 p.60
     130 S. Ct. 1195 p.1203
     175 L. Ed. 2d 1009 p.1018

22. Followed by, Cited in Dissenting Opinion at, Cited by:
     Montejo v. Louisiana, 556 U.S. 778, 129 S. Ct. 2079, 173 L. Ed. 2d 955, 2009 U.S. LEXIS 3973, 77 U.S.L.W. 4423, 21 Fla. L. Weekly
     Fed. S 863 (May 26, 2009) LexisNexis Headnotes HN8, HN9, HN10, HN13, HN15, HN17

     Followed by:
     556 U.S. 778 p.794
     129 S. Ct. 2079 p.2090
     173 L. Ed. 2d 955 p.968

23. Followed by, Explained by, Cited in Dissenting Opinion at, Cited by:
     Dickerson v. United States, 530 U.S. 428, 120 S. Ct. 2326, 147 L. Ed. 2d 405, 2000 U.S. LEXIS 4305, 68 U.S.L.W. 4566, 13 Fla. L.
     Weekly Fed. S 488, 2000 Cal. Daily Op. Service 5091, 2000 Colo. J. C.A.R. 3855, 2000 D.A.R. 6789 (June 26, 2000) LexisNexis
     Headnotes HN1, HN4, HN5, HN7, HN9, HN10, HN13, HN15, HN16, HN17

Fig. 3.3 Shepard's analysis definitions

direct impact on our thinking. It has an impact! We should not narrowly limit the meaning of impact to positive ones only.

As we can see, the process of selecting qualified sources is iterative in nature. Initially, we may select publications with many citations already or publications that have no citations yet but have been tweeted and retweeted a lot. At the next level, after we analyze the selected publications, we may be able to apply increasingly sophisticated selection criteria. For example, we may select publications that are known to have strong betweenness or eigenvector centrality scores in the network we have analyzed. We may select articles that are known to have sharp increases in their citation counts. We may also want to focus on articles that have been cited by researchers from at least five specialties. Of course, we may want to pay attention to articles that specifically criticized particular publications.

Interactive Visualizations

Vannevar Bush was the head of the U.S. Office of Scientific Research and Development (OSRD) during the World War II. He envisaged how the mankind's knowledge can be collectively organized by association, the same way as how the human memory works (Bush 1945). Highly connected information resources such as the Internet and the Wikipedia are commonly considered as being inspired by Vannevar Bush's visionary MEMEX. Navigating in such a universe of knowledge is called trailblazing. The navigator forges trails that represent new connections. As the metaphor of a universe of knowledge may imply, we need to make inter-galactic travels and study information at different levels of granularity. We may be interested in specific causal relations between a virus and a disease. We may be interested in how similar methods are used in different disciplines of science. Interactive visualization is an integral part of visual analytics. It enables us to explore or forage information at various levels of granularity and trace connections across areas where different perspectives may apply.

Ben Shneiderman's mantra for visual information seeking should be very helpful here. It is intuitive and simple to follow. In addition to the useful mantra, it is a good idea for an analyst to get familiar with a few other theories concerning the process of search and what we may expect to find. As people often say, you will only find what you look for. The better we are theoretically prepared, the better position we will be able to place ourselves to recognize potentially relevant patterns. Otherwise, we may miss important clues even if they are right in front of us. We have discussed a few theories of scientific change at the beginning of the book. We will frame our interpretations with these theories and characterize what we would expect to see if the theory is true.

Structural Variation Analysis

The structural variation theory considers the body of a scientific domain's knowledge as a complex adaptive system (Chen 2012, 2014). Its global structure may be altered significantly by newly published articles, or by semantic predications conveyed by these articles. According to the theory, articles that have the potential to trigger global changes are transformative in nature and they are the ones that are most likely to influence the course of the further development of a scientific field. How do we measure such potentials?

If we represent the domain knowledge as a network, then the modularity measure of the network can be very useful for us to assess the global structure of the network. The modularity of a network is defined with reference to a partition of the network. If we can divide the network into smaller components and minimize inter-component connections, then the modularity quantifies the degree to which the resultant components can be separated from one another. The modularity's value ranges from 0 to 1. The highest value of 1 means that the network is completely modularized by the chosen partition. In contrast, the lowest modularity value of 0 means that these components are tightly coupled and one cannot separate them in any meaningful way.

Figures 3.4 and 3.5 illustrate how the system adapts to the publication of the groundbreaking paper by Watts and Strogatz (1998). The network was derived from 5135 articles published on small-world networks between 1990 and 2010. The network of 205 references and 1164 co-citation links is divided into 12 clusters with a modularity of 0.6537 and the mean silhouette of 0.811. The red lines are made by

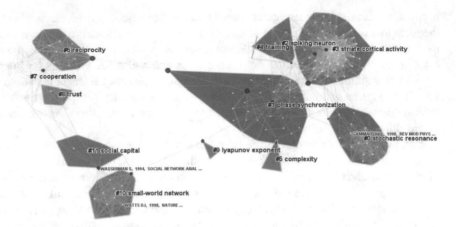

Fig. 3.4 The structure of the system before the publication of the ground breaking paper by Watts and Strogatz (1998)

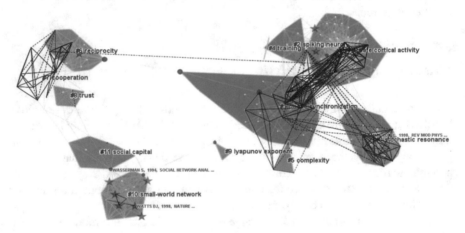

Fig. 3.5 The structure of the system after the publication of Watts and Strogatz (1998)

the top-15 articles measured by the centrality variation rate. Only major clusters' labels are shown in the figure. Dashed lines in red are novel connections made by Watts and Strogatz (1998) at the time of its publication. The article has the highest scores in Cluster Linkage and C_{KL} scores, 5.43 and 1.14, respectively. The figure offers a visual confirmation that the article was indeed making boundary-spanning connections. Recall that the data set was constructed by expanding the seed article based on forward citation links. These boundary-spanning links provide empirical evidence that the groundbreaking paper was connecting two groups of clusters. The emergence of Cluster #8 complex network was the consequence of the impact.

In this view, a network is a system of interconnected blocks. The most fundamental changes for such systems would be changes that alter how existing blocks are connected as well as adding or eliminating participating blocks. Relatively speaking, changes that are essentially limited to the internal state of a block would not be considered as significant as changes that transform inter-block connections. At the article level, each pair of co-occurring semantic predications introduced in the article is potentially an agent of change or a perturbation signal. If the co-occurring connection falls within a single block, then it may generate a local impact without causing any global changes. In contrast, if the co-occurring connection links two blocks in an innovative way or in a surprising or unanticipated way, then it becomes likely that the new link may change not only the local structure but also how the existing continents are organized. In other words, we should particularly pay attention to the predications of the latter kind.

CiteSpace supports structural variable analysis. Given a set of scientific articles, these articles are separated by the year of their publication. For each article published in year Y, CiteSpace will compute all the changes introduced by the article with reference to a network that represents the state of the knowledge prior to year Y. The differences of the networks before and after the publication of the article are used to quantify the likelihood that the article is altering the global structure of the underlying network in a significant way.

Using MySQL Databases in CiteSpace

CiteSpace has a built-in interface with a MySQL database on your localhost. You can upload your data to the database and interact with your data directly as you would with any MySQL database. You can also interact with your data through special-purpose functions provided in CiteSpace (Fig. 3.6).

For each dataset uploaded to MySQL, you can perform some text analysis functions from the Data Processing Utilities interface. The text analysis functions here are slightly different from the network of co-occurring terms in the main interface of CiteSpace. The major difference is that functions here include a selection step based on log-likelihood ratio tests. In theory, the resultant graph visualization should represent the most important patterns of phrases.

VOSViewer and CitNetExplorer

VOSviewer is a popular science mapping software tool developed by Van Eck and Waltman (2010) at the Centre for Science and Technology Studies (CWTS) in Leiden, the Netherlands, for constructing and visualizing bibliometric networks. These networks may include journals, researchers, or individual publications, and they can be constructed based on co-citation, bibliographic coupling, or co-authorship relations. VOSviewer also offers text mining functionality that can be

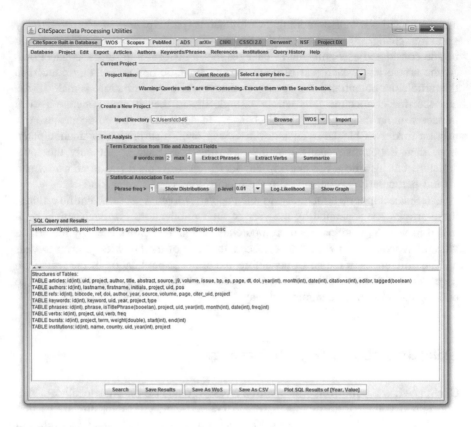

Fig. 3.6 An interface with MySQL in CiteSpace

used to construct and visualize co-occurrence networks of important terms extracted from a body of scientific literature.

VOSviewer maintains a simple workflow from the data to visualization. It is relatively straightforward to generate a visualization from bibliographic records from the Web of Science, Scopus, and a few other sources. Figure 3.7 shows a density map of references cited in the Science Mapping dataset. Comparing with CiteSpace, a noticeable strength of VOSviewer is its nice and simplistic approach to visualizing scientific publications. On the other hand, the strength in occasions may become a weakness of VOSviewer if the analyst needs to conduct in-depth investigations beyond the initial visualization. Perhaps more importantly, to our knowledge, unlike CiteSpace, the visual design in VOSviewer is not driven by theories of scientific change. For example, VOSviewer does not support concepts such as intellectual turning points nor transformative potentials. Although VOSviewer supports the notion of clusters, it does not provide cluster labels. As a result, one has to rely heavily on the assistant of domain experts or on one's own domain knowledge when interpreting VOSviewer visualizations. In fact, the development of CitNetExplorer (Van Eck and Waltman 2014), by the same team of

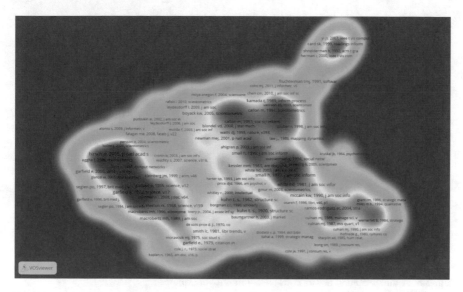

Fig. 3.7 A density map visualization in VOSViewer of references cited in the science mapping dataset

VOSviewer, is primarily motivated by the aim to strengthen the relatively weak support of analytic functionality.

CitNetExplorer supports the visualization and exploration of direct citation networks (Van Eck and Waltman 2014). In a direct citation network, a link pointing from a node n_i to a node n_j represents that the article represented by n_i cites the article of n_j. Figure 3.8 shows an example of a direct citation network of articles in the Science Mapping dataset. Articles are arranged vertically based on the year of their publication with the earliest year on the top of the visualization and the latest year at the bottom. The directed citation link is shown vertically. The colors of nodes indicate their clusters.

CitNetExplorer provides more functions for exploring a visualized network, including the drill down function and the display of the shortest path between two nodes. Figure 3.9 shows the resultant network of performing the drill down function on the Science Mapping network. The user can explore the shortest path between two nodes.

There are an increasing number of computer software programs for analyzing scientific publications. Many of them are freely available. Apart from CiteSpace, VOSviewer, and CitNetExplorer, other widely known systems include HistCite,[1] sci2[2] developed at Indiana University, the growing set of programs developed by

---

[1]https://clarivate.com/products/web-of-science/.

[2]https://sci2.cns.iu.edu/user/index.php.

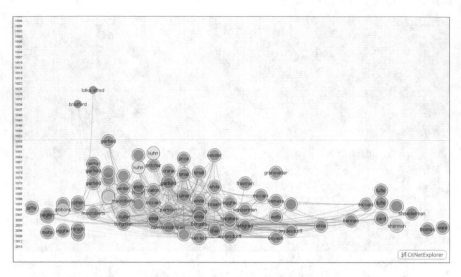

Fig. 3.8 A direct citation network visualized in CitNetExplorer

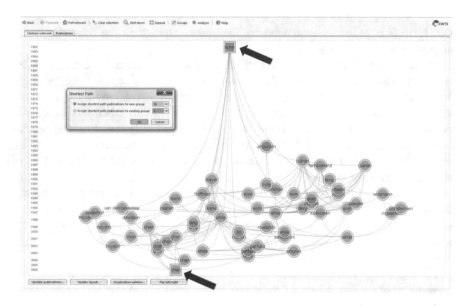

Fig. 3.9 Drill down and the shortest path between two nodes in CitNetExplorer

Loet Leydesdorff in Amsterdam,[3] Alluvial Generator,[4] KnowledgeMatrix Plus,[5] to name the few. A list of tools and resources is accessible from CiteSpace as well as on the web.[6]

Terrorism Research (1996–2003)

Our first example of exploring the knowledge structure of a research domain is the terrorism research (1996–2003). The source of the science citation data is the Web of Science. The dataset comes with the release of CiteSpace as the Demo 1 project for instructional purposes. Although everyone has probably heard of terrorists, terrorist attacks, and terrorism, many may still have no clear idea what terrorism as the subject of research may include. The lack of prior knowledge of the target domain is probably an accurate description of most of the users of the visual analytic procedure.

The data collection was based on a simple query in the Web of Science. If the term terrorist or the term terrorism appears in the title, the abstract, or the keyword list of an article, then the article is considered relevant and it will be included in the dataset to be analyzed further. This type of search is called topic search in the Web of Science. We used CiteSpace to visualize important patterns in the dataset so that we can explore the visualization and learn about the subject domain. At the end of the process, we should be able to obtain a good understanding of the domain in terms of its overall structure, key groups of publications, and critical works in the field.

Citation Bursts

A relatively simple but effective method is to identify publications in the domain that have drawn attention of the research community at various stages of the development. Burst detection is a reliable technique that enables us to accomplish this task (Kleinberg 2002). Given a sequence of frequency values, a burst is an abrupt elevation of the frequencies over a specific time interval. For example, the number of cars crossing a bridge connecting New Jersey and Pennsylvania everyday may experience bursts during rush hours and the number of cars crossing the bridge every month may experience bursts during holidays. As we have discussed, citations received by scientific publications may provide the first-order

---

[3]http://www.leydesdorff.net/software.htm.

[4]http://www.mapequation.org/apps/AlluvialGenerator.html.

[5]http://mirian.kisti.re.kr/km/km_pop_en.jsp.

[6]http://cluster.cis.drexel.edu/~cchen/citespace/resources/.

indicator of scholarly impact. In contrast, bursts of citations provide a higher-order indicator of a scholarly impact in terms of the attention from the research community that is evidently above-and beyond the normally expected level. As a result, publications with strong enough citation bursts during the course of the development of the domain are valuable landmarks for us to navigate the domain further.

Figure 3.10 lists 24 references with the strongest citation bursts between 1996 and 2003. The simplistic diagrams on the right depict the duration of a burst event in red. Overall, the periods of citation bursts drifted over time as new research topics move to the center of the stage. For example, COOPER1983, the second one on the list, has a strong citation burst weight of 5.916. Its citation burst lasted for four years from 1996 till 1999. At this point, although we may not know the specific role played by COOPER1983, we know that this article is evidently

Top 24 References with the Strongest Citation Bursts

References	Year	Strength	Begin	End	1996 - 2003
BRISMAR B, 1982, J TRAUMA, V22, P216	1982	3.6708	**1996**	1999	
COOPER GJ, 1983, J TRAUMA, V23, P955	1983	5.916	**1996**	1999	
ENDERS W, 1993, AM POLIT SCI REV, V87, P829	1993	2.2584	**1996**	1997	
KATZ E, 1989, ANN SURG, V209, P484	1989	4.0719	**1996**	2000	
PHILLIPS YY, 1986, ANN EMERG MED, V15, P1446	1986	2.3415	**1997**	1999	
HADDEN WA, 1978, BRIT J SURG, V65, P525	1978	2.8319	**1997**	1999	
HULLER T, 1970, ARCH SURG-CHICAGO, V100, P24	1970	2.8122	**1997**	1999	
SCHMID AP, 1988, ..., V, POLITICAL TERRORISM	1988	2.3624	**1997**	1998	
FRYKBERG ER, 1988, ANN SURG, V208, P569	1988	4.5731	**1997**	2000	
WATERWORTH TA, 1975, BRIT MED J, V2, P25	1975	2.3415	**1997**	1999	
MALLONEE S, 1996, JAMA-J AM MED ASSOC, V276, P382	1996	3.044	**1998**	2000	
SHALEV AY, 1992, J NERV MENT DIS, V180, P505	1992	2.4654	**1998**	2000	
MESELSON M, 1994, SCIENCE, V266, P1202	1994	2.3691	**1999**	2001	
*CDCP, 1999, MMWR-MORBID MORTAL W, V48, P69	1999	2.4063	**1999**	2001	
SIMON JD, 1997, JAMA-J AM MED ASSOC, V278, P428	1997	3.084	**1999**	2001	
CARTER A, 1998, FOREIGN AFF, V77, P80	1998	2.7523	**1999**	2001	
OKUMURA T, 1996, ANN EMERG MED, V28, P129	1996	2.4063	**1999**	2001	
FRANZ DR, 1997, JAMA-J AM MED ASSOC, V278, P399	1997	3.0233	**1999**	2003	
TOROK TJ, 1997, JAMA-J AM MED ASSOC, V278, P389	1997	2.5347	**2000**	2001	
ALIBEK K, 1999, BIOHAZARD, V, P	1999	2.3249	**2000**	2003	
RICHARDS CF, 1999, ANN EMERG MED, V34, P183	1999	3.1263	**2000**	2001	
INGLESBY TV, 1999, JAMA-J AM MED ASSOC, V281, P1735	1999	2.6297	**2001**	2003	
KOLAVIC SA, 1997, JAMA-J AM MED ASSOC, V278, P396	1997	2.4539	**2001**	2003	
INGLESBY TV, 2000, JAMA-J AM MED ASSOC, V283, P2281	2000	3.2797	**2001**	2003	

Fig. 3.10 Articles with citation bursts in terrorism research (1996–2003)

valuable, especially between 1996 and 1999. We can also tell from the simple depiction that during the same period of time, no other article reached the same level of citation bursts. If we want to invest our precious time on learning more about the research domain, this article should be on our list of a few landmark articles.

In addition to references with the strongest citation bursts, we should also pay attention to references that have the longest duration of citation burst or have the most recent periods of burst. Two of the references have the longest 5-year duration of citation burst, namely KATE1989 and FRANZ1997. We also notice that two of the three most recent bursts from 2001 were authored by INGLESBY in 1999 and 2000, respectively. This is an example of how to identify landmark articles without any prior knowledge of the target domain. This method has a few distinct advantages over using citation counts or altmetrics such as downloads. Usually citation counts are only available as a sum that is accumulated over all the years since the publication of an article. For example, the most cited publication in the Web of Science, the one at the very top of Mount Kilimanjaro, has passed its citation peak many years ago. Citation counts alone cannot tell us whether a highly cited article is still at the center of everyone's attention or its glory is really due to the credit it earned in its golden age that has long gone. Knowing when an article is particularly high performing in drawing the research community's attention is more useful than merely knowing that an article has a lot of citations.

More recent bibliographic records obtained from the Web of Science are likely to include DOIs of cited references. For those references with DOIs, the user can access the full text of a reference through its DOI link, which would be useful for exploring the literature.

Timeline Visualization

CiteSpace supports a few types of visualization, including a cluster view, a timeline view, and a timezone view. A cluster view depicts an overview of a network in a node-and-link diagram. A timeline view still displays the nodes and links but organizes them along multiple parallel timelines. A timeline visualization is intuitive. The analyst can obtain a good overview of the domain with a few simple steps.

Figure 3.11 shows a timeline visualization of the terrorism research (1996–2003). Each line from the left to the right represents a cluster of co-cited references, which in turn reveals the work of a distinct specialty. CiteSpace supports several other types of networks that can be derived from a set of bibliographic records. Here we focus on networks of co-cited references. Studies of this type of networks are called Document Co-Citation Analysis (DCA). Other types of studies include Author Co-Citation Analysis (ACA) and Collaborative Network Analysis.

The timelines are arranged by their size from the largest downwards. The label next to each cluster line summarizes the most likely context in which members of

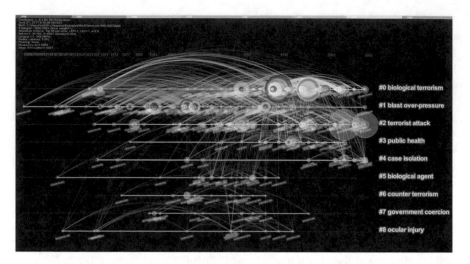

Fig. 3.11 A timeline visualization of the terrorism research (1996–2003)

the corresponding cluster have been cited. The candidate words for the labels are drawn from articles that cited the members of the cluster. For example, the largest cluster #0 biological terrorism indicates that this cluster is essentially being cited by articles relevant to the topic of biological terrorism.

Note that the citing topics and the cited topics may not be the same necessarily. The difference is the one between the intellectual base and the reference front of the corresponding specialty. In other words, each specialty has two interconnected components: the intellectual base is where the specialty draws its inspiration from and the research front is where the specialty disseminates its new contributions. A good example is the cluster #2 terrorist attack. As we will see shortly, the specialty focuses on the topic of Post Traumatic Stress Disorder (PTDS) in the context of terrorism. The difference between its intellectual base and the research front underlines the nature of the specialty. More specifically, the intellectual base is essentially on PTDS prior to the September 11 terrorist attacks and the research front is mostly produced after. A key difference is that the research front takes a new turn by recognizing the possibility that was not on the radar of PTDS research, namely, people may develop PTSD symptoms even if they have never been on a trauma site physically.

The timeline visualization makes the life cycle of a specialty visible. For example, cluster #1 blast over-pressure has the longest active time—the entire duration of the observation. In contrast, the presence of cluster #6 counter terrorism is much short lived. The timeline visualization also makes it easier to identify active specialties—clusters with many items with circles in red—they have citation bursts.

The analyst can drill down by moving from one level of granularity to another. There are several ways to drill down. CiteSpace allows the user to apply the same analytic procedure repeatedly on a cluster of the network and then on a cluster of the cluster. A research front at the top level may turn out to have finer structures at a lower level. The user can inspect the titles of those articles that cited a particular cluster and explore various aspects of the cluster. This can be done in CiteSpace with the Cluster Explorer function. Unless we note otherwise, the labels of the clusters are generated from the top 25% of the most cited citing articles for each cluster.

The largest cluster, #0 biological terrorism, has 61 cited references. It has a silhouette value of 0.658. The silhouette value measures the homogeneity of a cluster. Its value ranges between −1 and 1. The value 0.658 is strong enough to make the cluster meaningful, especially as the largest cluster, although we may expect to see higher silhouette values in some domains. The linkage between a citing article and the cluster can be measured in terms of the extent it cited members of the cluster. For cluster #0, there are six citing articles that each cited over 15% of the member references of the cluster (Table 3.1). The one with the strongest linkage, RICHARDS1999, cited 21% of them, is an article published in 1999 entitled "Emergency physicians and biological terrorism." The title of each of the six articles contains the term biological terrorism or bioterrorism, except HAIL1999. The median year of publication is 1999.

Similarly, we can inspect citing articles from the research front of the second largest cluster #1 blast over-pressure, containing 50 references. This cluster has a silhouette value of 0.862, much higher than that of the largest cluster. The median year of publication is 1985. We can also tell from the timeline view that overall this seems to be an older cluster than the largest one. The top three citing articles are shown in Table 3.2.

Two of them explicitly mentioned blast over-pressure. In fact, one mentioned blast over-pressure and the other mentioned blast overpressure-induced injury. Although the two semantically equivalent terms do not have the identical forms, they are grouped together because of the references they cited. This is an additional advantage of citation indexing, as opposed to approaches purely based on matching words or lexical patterns.

The next cluster is #2 terrorist attack, containing 47 references and an even higher silhouette value of 0.915. The median year of publication of the cluster is 1994. The first three strongest citing articles to this cluster all have the term terrorist attacks in their titles (Table 3.3). In fact, they all contain the longer phrase September 11th terrorist attacks in their titles. It is also clear that the central theme is to do with PTSD after the September 11th terrorist attacks. Each of the three articles has cited over 21% of the members.

In summary, by visualizing the citation patterns in articles published between 1996 and 2003, we have learned that the three most prominent areas of the research domain are bioterrorism, injuries caused by blast over-pressure, and PTSD caused by September 11th terrorist attacks. We have also found gateways that we could drill down as further as we like. We can check where those landmark articles are located and pinpoint articles that played critical roles in the course of the development of the complex domain. As we have seen, we can reach this macroscopic level of under-standing of a domain that we knew little about. This is largely due to the way we tap into the domain expertise of numerous researchers through their publications. This procedure is generic. It is applicable to a wide range of scientific disciplines.

Table 3.1 Major citing articles of Cluster #0

Coverage	Citing article
0.21	Richards, CF (1999) Emergency physicians and biological terrorism
0.20	Atlas, RM (1999) Combating the threat of biowarfare and bioterrorism
0.16	Inglesby, TV (1999) Anthrax as a biological weapon—medical and public health management
0.16	Relman, DA (2001) Bioterrorism preparedness: what practitioners need to know
0.15	Dhawan, B (2001) Bioterrorism: a threat for which we are ill prepared
0.15	Hail, AS (1999) Comparison of noninvasive sampling sites for early detection of bacillus anthracis spores from rhesus monkeys after aerosol exposure

Table 3.2 Major citing articles of Cluster #1

Coverage	Citing article
0.58	Elsayed, NM (1997) Toxicology of blast over-pressure
0.24	Elsayed, NM (1997) A proposed biochemical mechanism involving hemoglobin for blast overpressure-induced injury
0.04	Stein, M (1999) Medical consequences of terrorism—the conventional weapon threat

Table 3.3 Major citing articles of Cluster #2

Coverage	Citing articles
0.28	Galea, S (2002) Posttraumatic stress disorder in manhattan, New York City, after the September 11th terrorist attacks
0.23	Vlahov, D (2002) Increased use of cigarettes, alcohol, and marijuana among manhattan, New York, residents after the September 11th terrorist attacks
0.21	Galea, S (2002) Psychological sequelae of the September 11 terrorist attacks in New York City

In the following example, we will look at the terrorism research again, but this time over a wider window, especially containing articles published between 1980 and 2017. We would like to see where the three prominent topic areas are located in the broader context. We would also like to see any major topic areas emerged since 2003.

Structural Variations

The goal of a structural variation analysis is to identify two types of links added to the current network representation of a domain's knowledge, namely, incremental links and transformative links (Chen 2012). Incremental links are within the boundary of a particular cluster, whereas transformative links connect different clusters. Thus, incremental links do not change the structure of the system at the cluster level, but transformative links do.

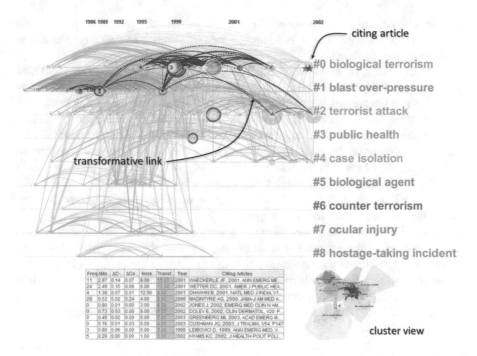

Fig. 3.12 Structural variation by transformative link count

Given a particular year Y, articles published in year Y will be examined against the structure of the network representation of the domain's knowledge over the last three years prior to Y. We refer this network as the baseline network for year Y. The analyst may choose the number of years prior to year Y to form the baseline network. The longer the baseline network extends back in time, the more accurate the structural variation measures would be because a network with a longer exposure time is likely to capture more links than a network with a shorter exposure time.

Figure 3.12 shows the footprints of top 10 articles with the largest modularity change rate. Dashed lines represent novel transformative links. Solid lines represent existing lines. Transformative links mostly connect the largest three clusters, namely, #0, #1, and #2.

Terrorism Research (1980–2017)

We used the same simplistic topic search in the Web of Science using broader terms of terrorist OR terrorism and limited to two types of publications: articles of original research and review articles. The new search found 14,656 relevant records. If one prefers to obtain additional articles that didn't use these topic search terms but may be relevant otherwise, one option is to use the citation expansion strategy to include articles that cite this set of records. We did not perform the citation expansion for this particular case because it is adequate for our purpose to focus on the scope defined by the topic search.

We used the g-index to select articles for each time slice between 1980 and 2017. In addition, these articles must have received two or more citations themselves. Since we are dealing with a timespan of 38 years, imposing this minimum citation condition can filter out many publications that do not make sufficient impact on the research domain. Admittedly, this condition is likely to be relatively harsh for recently published articles, although the use of g-index may compensate the citation distribution to an extent. One remedy is to conduct a separate study using a lower threshold on articles published within the recent few years. We limited the Look Back Years to 5, which means we will ignore citations to references that are more than five years ago.

Figure 3.13 shows the cluster view visualization of the terrorism research (1980–2017). It depicts the largest connected component of the network of 908 cited references. The largest connected component contains 694 references (76% of the entire network). Each cluster is shown with a polygon colored to indicate the median of its citing articles' publication years. In this visualization, clusters located near the top are the oldest, whereas clusters near the bottom are the most recent ones.

Fig. 3.13 A cluster-view visualization of the terrorism research (1980–2017). Node selection by g-index (k = 10)

The oldest cluster in the visualization is #9 war trauma survivor. Cluster #1 biological weapon is the second oldest, containing two articles authored by INGLESBY, which remind us the biological terrorism cluster identified in the terrorism research (1996–2003). Moving downwards, the cluster #8 blast injury is likely to be connected to the blast over-pressure cluster identified in the 1996–2003 study. The cluster #0 terrorist attack and references such as GALEA2002 indicate that this is the PTSD cluster identified before.

Moving further down, we encounter clusters such as #2 suicide bombing, #3 domestic terrorism, #7 word trade center health, and #4 islamic state. Evidently, many new topic areas emerged since 2003.

In Fig. 3.14, the network of the terrorism research in 1996–2003 is superimposed over the network in 1980–2017. This function is called a network overlay, which highlights the relationship between a subnetwork and a larger network.

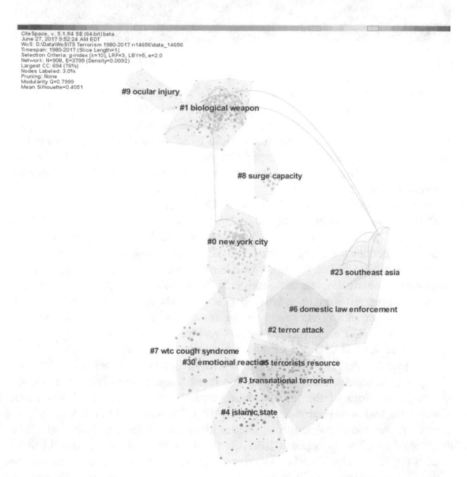

Fig. 3.14 A network overlay shows the 1996–2003 network in the context of the 1980–2017 network

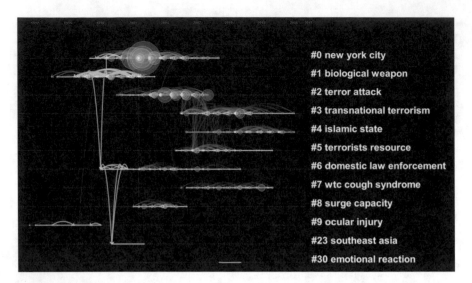

Fig. 3.15 A timeline visualization of terrorism research (1980–2017)

A new timeline visualization of the terrorism research (1980–2017) is shown in Fig. 3.15. The lines in yellow are from the earlier network (1996–2003). The timeline view shows a big picture of the terrorism research. The previously predominant topic areas such as bioterrorism and PTSD are no longer in active at this level of granularity, although there may be publications that are excluded by our selection criteria. In contrast, the two current lines of research are #3 transnational terrorism and #4 islamic state.

Using the Cluster Explorer function in CiteSpace, we can inspect the major clusters and see what the major topics are and how they may differ from their counterparts in the earlier visualization. The previously third largest cluster on PTSD now becomes the largest cluster of 127 references and an even higher silhouette value than before (0.962), which means that the specialty becomes more specialized. The median year of the cluster is 2003. As we can tell from the timeline view, the cluster remained to be active until about 2009. As shown in Table 3.4, the top five citing articles of the largest cluster are clearly related to the September 11 terrorist attacks. The major theme of PTDS and mental health in general continues the theme of the PTSD cluster identified in the previous study.

The second largest cluster #1 biological weapon has 116 member references with a very high silhouette value of 0.972. The median age of the cluster is 1999. As shown in the titles of the top five citing articles, this cluster is clearly about bioterrorism and biological weapons (Table 3.5). We notice that ATLAS1999 also appears in the biological terrorism cluster identified in the previous study. The timeline of the cluster stopped at 2003. On the other hand, there are some connections between this cluster and a few other clusters, notably #2 terror attack and #6 domestic law enforcement. It is possible that topics concerning bioterrorism may have transformed into research topics under other clusters. It is also possible, of course, the topic of bioterrorism is no longer an active line of research.

Table 3.4 Major citing articles of Cluster #0 in Terrorism Research (1980–2017)

Coverage	Citing article
0.08	Boscarino, JA (2004) Mental health service use 1-year after the world trade center disaster: implications for mental health care
0.06	Boscarino, JA (2004) Mental health service and medication use in New York City after the September 11, 2001, terrorist attack
0.06	Adams, RE (2006) Alcohol use, mental health status and psychological well-being 2 years after the world trade center attacks in New York City
0.06	Boscarino, JA (2004) Adverse reactions associated with studying persons recently exposed to mass urban disaster
0.06	Pulcino, T (2003) Posttraumatic stress in women after the September 11 terrorist attacks in New York City

Table 3.5 Major citing articles of Cluster #1 in Terrorism Research (1980–2017)

Coverage	Citing articles
0.11	Atlas, RM (1999) Combating the threat of biowarfare and bioterrorism
0.09	Fidler, DP (1999) Facing the global challenges posed by biological weapons
0.09	Greenfield, RA (2002) Bacterial pathogens as biological weapons and agents of bioterrorism
0.09	Klietmann, WF (2001) Bioterrorism: implications for the clinical microbiologist
0.08	Atlas, RM (2001) Bioterrorism before and after September 11

The third largest cluster #2 terror attack has 95 references and a silhouette value of 0.835. This cluster is relatively younger than the first two clusters. Its median age is 2004. The relatively low silhouette value is perhaps reflected on the lack of a clear consensus among the top five citing articles' titles (Table 3.6). On the other hand, the timeline view shows that this cluster has connections with #3 transnational terrorism and #5 terrorist resource.

The cluster #3 transnational terrorism is a currently active area of research. With 85 references, it has a silhouette value of 0.834 and an even younger median age of 2010. Along with cluster #4 islamic state, this cluster represents essentially the current research focus of the terrorism research community. The titles of the top citing articles suggest that research in this cluster is concerned with questions concerning the causes of terrorism (Table 3.7).

Cluster #4 islamic state is the youngest one, containing 84 references with a silhouette value of 0.891 and the median age of 2012. Four of the top five citing articles of the cluster were published in 2016 (Table 3.8). The titles of the top 5 citing articles suggest that the cluster focuses on deeper reasons of terrorism.

Figure 3.16 shows a timeline visualization that is rendered with citation bursts in red circles, which correspond to the duration of citation burst. Almost every reference in the visualization had a citation burst. We have not seen this degree of citation burst in other domains we have analyzed. We will drill down one more level deeper to characterize each cluster's theme in more detail.

Table 3.6 Major citing articles of Cluster #2 in Terrorism Research (1980–2017)

Coverage	Citing articles
0.17	Gould, ED (2010) Does terrorism work?
0.12	Rosendorff, BP (2010) Suicide terrorism and the backlash effect
0.07	Czinkota, MR (2010) Terrorism and international business: a research agenda
0.07	Fielding, D (2010) 'An eye for an eye, a tooth for a tooth': political violence and counter-insurgency in egypt
0.07	Plumper, T (2010) The friend of my enemy is my enemy: international alliances and international terrorism

Table 3.7 Major citing articles of Cluster #3 in Terrorism Research (1980–2017)

Coverage	Citing articles
0.18	Gassebner, M (2011) Lock, stock, and barrel: a comprehensive assessment of the determinants of terror
0.18	Krieger, T (2011) What causes terrorism?
0.11	Berrebi, C (2011) Earthquakes, hurricanes, and terrorism: do natural disasters incite terror?
0.11	Chenoweth, E (2013) Terrorism and democracy
0.09	Freytag, A (2011) The origins of terrorism: cross-country estimates of socio-economic determinants of terrorism

Table 3.8 Major citing articles of Cluster #4 in Terrorism Research (1980–2017)

Coverage	Citing articles
0.07	Pearson, E (2016) The case of roshonara choudhry: implications for theory on online radicalization, ISIS women, and the gendered Jihad
0.06	Horgan, J (2016) Actions speak louder than words: a behavioral analysis of 183 individuals convicted for terrorist offenses in the united states from 1995 to 2012
0.06	Schuurman, B (2016) Rationales for terrorist violence in homegrown Jihadist groups: a case study from The Netherlands
0.05	Capellan, JA (2015) Lone wolf terrorist or deranged shooter? a study of ideological active shooter events in the united states, 1970–2014
0.05	Cold, JW (2016) Extremism, religion and psychiatric morbidity in a population-based sample of young men

Semantic Structures of Clusters

Titles of leading articles that cite a cluster and summarizing labels of a cluster provide a top-level characterization of the predominant theme of the cluster. Developing a deeper understanding of each cluster's theme is possible if we can construct an ontological structure of key concepts associated with the cluster.

The procedure consists of the following steps. First, we extract terms from articles that cite members of clusters such that extracted terms are representative to

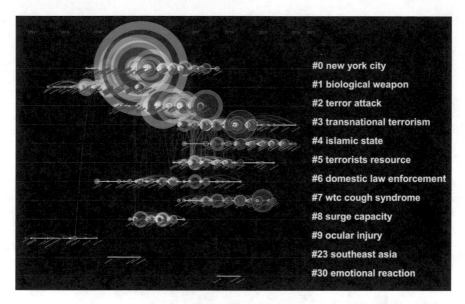

Fig. 3.16 A timeline visualization of citation bursts

individual clusters. Terms may come from titles, abstracts, and/or keywords of these articles. Next, the extracted representative terms are filtered by a given cluster so that we will only retain terms that actually appeared in the cluster. Then, co-occurrences of filtered terms within citing articles of the clusters are identified. These co-occurrences are used as the input to the construction of a hierarchical structure. Since co-occurring terms form a network, hierarchical relations between terms can be derived based on the concept of m-reachability. A term with a higher reachability is assigned to have a higher level position. The resultant hierarchical structure is finally visualized as a concept tree. A concept tree is a hierarchically organized set of concepts. Since our terms are representative of their own clusters, the resultant concept tree serves as a proxy of an ontological representation of the cluster's content.

Figure 3.17 shows a closer view of the largest cluster #0. It primarily focuses on PTSD resulted from the September 11 terrorist attacks in New York. Its member references are published between 1997 and 2009. The largest circles belong to SCHUSTER2001, GALEAS2002, and GALEAS2003. Figure 3.18 depicts the concept tree of terms extracted from the abstracts of the articles that cited the cluster. The root of the tree is on the left. The children of a term are placed on its right-hand side. Nodes that do not have any children nodes are called leave nodes. A path starting from a node to a leave node consists of all nodes along the way. The length of a path is the number of these nodes. The longest path in the concept tree of cluster #0 contains 5 nodes. In fact, three paths have the same length. They share the first four concepts: mental health treatment → terrorist attack → mental health status → New York City, then the path splits into three more specific concepts:

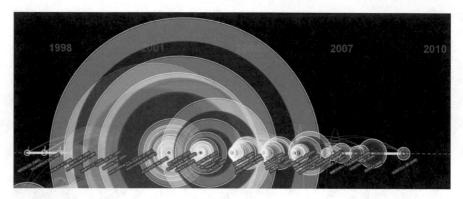

Fig. 3.17 Cluster # New York City (1997–2009)

Fig. 3.18 A concept tree of cluster #0 based on terms extracted from the abstracts of its citing articles, i.e. the research front

representative sample, probable PTSD, and mental health service. We may consider the longest path as the main path that characterizes the fundamental theme of a cluster. The term main path in network analysis has a special meaning. We address how to conduct a main path analysis in next section. Here we use the term main path for its meaning in its intuitive sense.

The mental health branch is the largest one for the cluster. It echoes what we have learned from the visualizations so far but now more specific contextual details will be very valuable for us to strengthen our understanding of the specialty. For example, the mental health dimension is particularly concerned with terrorist attack, which is further connected to terms such as world trade center. The post-traumatic stress disorder leads a branch of its own. Another branch is alcohol dependence, which is also related to mental health. The concept tree further clarifies the knowledge structure of the largest cluster in terrorism research.

Similarly, we obtained a concept tree for cluster #1 biological weapon, which has three branches (see Fig. 3.19). The largest branch starts with biological weapon, followed by civilian population, which leads to four leave nodes. Under the biological weapon node, we can see topics on bacillus anthracis, review article, growing concern, and mass destruction. The second branch consists of bioterrorist attack, bioterrorism preparedness, and public health.

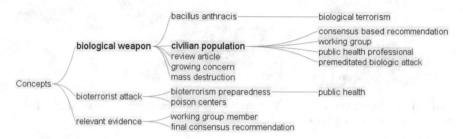

Fig. 3.19 A concept tree of Cluster #1 biological weapon

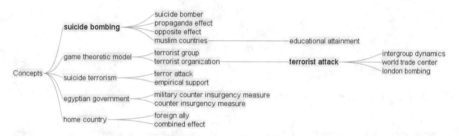

Fig. 3.20 A concept tree of Cluster #2 terror attack

The third largest cluster's concept tree contains branches on suicide bombing, game theoretical model, and three smaller branches (see Fig. 3.20). The suicide bombing branch appears to focus on individual suicide bombing, whereas the game theoretic model branch appears to focus on group dynamics and organizations of terrorist groups. Given that the overall cluster is labeled as terror attack, these branches further elaborate the research focus in these areas.

Recall that the two currently active specialties of research are associated with clusters #3 transnational terrorism and #4 islamic state. In cluster #3, transnational terrorism is a prominent concept, which leads to a few related concepts and a branch of several levels deep (Fig. 3.21). Although the information is still patchy, we can learn the most relevant vocabulary in the context of the cluster, including advanced democracies and domestic terrorism. Some contradicting terms such as democracies and nondemocratic countries appear to underline the role of democracy or the lack of it in understanding transnational terrorism. Similarly, the contrast between transnational terrorism and domestic terrorism can be observed in the concept tree as well. Closely related terms such as international terrorism and transnational terrorism invite further investigations on how these terms differ and how they are related. We will illustrate shortly how we can address these questions by exploring the actual contexts in which these concepts are discussed. We will construct a full-fledged concept tree of terms identified in the abstracts of citing articles. Furthermore, we can instantly reveal the instances of a given concept in their original contexts. The current discussion is at a higher level of granularity than

Fig. 3.21 A concept tree of Cluster #3 transnational terrorism based on terms extracted from abstracts of citing articles to the cluster

Fig. 3.22 A concept tree of Cluster #4 islamic state

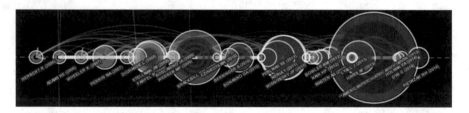

Fig. 3.23 Cluster #7—WTC cough syndrome

the full-fledged concept-to-concept exploration. We will drill down to the next level of granularity after our exploration at the current level. This is also a recommended search strategy. Instead of diving into a specific area first, seek a good understanding of the system at one level of granularity at a time.

Cluster #4 Islamic state is another cluster that is still active. The concept tree reveals some key concepts of this cluster such as religious extremism, psychiatric morbidity, and far right extremist (Fig. 3.22). We will also drill down this cluster deeper using a concept tree that is formed to reveal the concept-in-context details.

Figure 3.23 shows the timeline of cluster #7 WTC cough syndrome. Its publications range between 2006 and 2014. There are a few big and red circles, indicating that they are not only highly cited but also have strong citation bursts, namely, BRCKBILL2009, WISNIVESKY2011, and a 2013 publication of the American Psychiatric Association.

As shown in Fig. 3.24, this cluster's concept tree contains concepts concerning a particular population such as wtc disaster worker and wtc exposed firefighter, symptoms related concepts such as respiratory symptom, and PTSD related topics such as baseline PTSD symptom count.

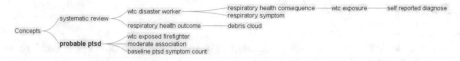

Fig. 3.24 A concept tree of cluster #7 world trade center cough syndrome

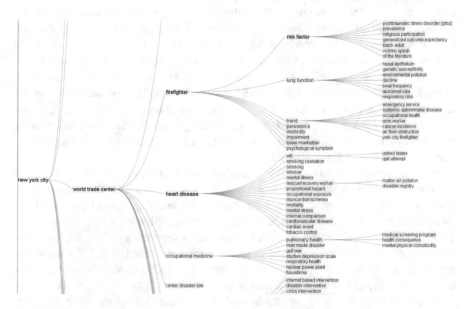

Fig. 3.25 An unfiltered concept tree of the WTC cough syndrome cluster (#7)

Figure 3.25 shows part of an unfiltered concept tree of the WTC cough syndrome cluster. It retains terms that are not deemed to be unique to the cluster, but the inclusion of such terms provides details that may be missing from the concept tree based on filtered terms. The hierarchical relations are easy to understand. For example, the New York city is the parent node of world trade center, which in turn has children nodes such as firefighter, heart disease, occupational medicine and a few other branches. Along the firefighter branch, the sub-branch of risk factor is prominent with many children nodes such as PTSD. The firefighter branch also includes a sub-branch of lung function with more specific terms such as nasal epithelium and respiratory cilia. In parallel to the firefighter branch, the heart disease branch is also prominent due to the number of its children nodes on smoking and mental illness. The contextual information provided by the concept tree is valuable as we can better plan our search strategy and explore the knowledge domain more effectively. Furthermore, the provision of such a concept tree serves the role of an organizing framework so that we can organize various concepts encountered in our search more easily, which tends to reduce the overall complexity of the task.

Concepts in Context

One way to learn about a subject domain is to explore how a concept is used in a
variety of contexts by researchers in this specialty. Although the concept tree shown
in Fig. 3.26 resembles those concept trees we have seen earlier, they differ in some
important ways. First of all, in this concept tree, children nodes represent attributes
of their parent node. For example, the concept of aid is divided into foreign aid and
military aid. In fact, children nodes are modifiers of their parent node in the original
text. If we organize the details in this way, we would be able to see the most
common features on the left and more specific features on the right, or further down
the tree structure. Secondly, the hierarchical relations differ in their semantics.
A parent-child hierarchical relation in the concept-in-context tree indicates that the
child node serves as a modifier of the parent node as in the aid-foreign example, the
term foreign modifies the term aid. In contrast, the parent-child hierarchical relation
in a concept tree in previous sections represents a broad-narrow relation as in New
York City → World Trade Center.

We can interactively explore the original text not in the conventional way to read
the text in its original sequential order; instead, we can hop over the text and read
various contexts side by side. In the foreign aid example, as we hover over the
foreign-aid node with the mouse, a list of sentences will appear in a window. All
these sentences are about the concept foreign aid in the context of the transnational
terrorism cluster. We can see various topics concerning foreign aid, for example,
using foreign aid as a counterterrorism instrument and identifying sectors that have
been particularly effective as the target of foreign aid such as education and health.

A common theme in cluster #3 is that democracies tend to experience more
terrorism than dictatorships, autocracies, or other non-democracies (see Fig. 3.27).
Much of these discussions are revolving around the democracy-autocracy divide,
for example, newly established democracies are more vulnerable to terrorism than
established democracies (1) and democracies experience more terrorism than
non-democracies (2–4).

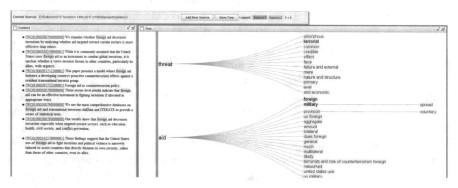

Fig. 3.26 The contexts of foreign aid in a concept-in-context tree of cluster #3 transnational
terrorism

- [WOS:000317343900006] One of the earlier empirical studies of the relationship between regime type and terrorism published in International Interactions determined that while established democracies were significantly less likely to experience terrorist attacks than were nondemocratic countries, newly established democracies were highly vulnerable to terrorism. **1**
- [WOS:000332371500009] The study finds that within-system groups are significantly less likely to emerge in democracies that have a proportional representation system and higher levels of district magnitude, while neither of these factors affects the emergence of anti-system groups.
- [WOS:000325495000011] Although empirical research has generally demonstrated that democracies experience more terrorism than autocracies, research suggests that this depends upon complex institutional differences that go beyond the democracy-autocracy divide. **2**
- [WOS:000319804100003] Research on terrorism in democracies borrows from the literature on civil war and rebellion to argue that more proportional representation decreases the likelihood of terrorist violence.
- [WOS:000284604200005] Conceiving democratic rule of law as the coexistence of effective and impartial judicial systems and citizens recognition of the law as legitimate, the author presents a causal explanation in which a high-quality rule of law is considered to dampen ordinary citizens opportunity and willingness to engage in political violence, protecting democracies from becoming victims of terrorism.
- [WOS:000321769600017] Intermediately wealthy and transitioning democracies with internally inconsistent institutions were more likely to experience domestic terrorism than advanced democracies and authoritarian regimes.
- [WOS:000341817900009] Approaching elections lead to an increase in the volume of attacks in democracies with low electoral permissiveness but not in others.
- [WOS:000332371500009] A wide range of studies find that democracies experience more terrorism than non-democracies. **3**
- [WOS:000273945700002] Why is terrorist activity more prevalent in democracies than in nondemocracies? **4**

Fig. 3.27 Sentences that mentioned the term democracies in abstracts of Cluster #3 transnational terrorism

- [WOS:000375999900005] From the outset, radicalization was conceived of as an intellectual process through which an individual would increasingly come under a spell of extremist ideas.
- [WOS:000375999900005] For more than a decade, radicalization has been a keyword in our understanding of terrorism. **1**
- [WOS:000356993800001] This Article examines the federal governments community engagement efforts with American Muslim communities as part of a larger infrastructure for policing radicalization and countering violent extremism (CVE).
- [WOS:000375999900005] Considering recent events, including the November 2015 Paris attacks, the present article sets out to reassess the above-mentioned intellectualist understanding of radicalization and come up with new suggestions as to how radicalization may be understood today.
- [WOS:000376226800002] It appears that radicalization is grounded on a vacuum of existential significance and by an ascending process, through the sharing of an ideology prone to violent actions, leads individuals to engage in terrorist goals for a collective one, greater than them.
- [WOS:000375999900005] In the second part of the article it is argued that although radicalization is often conceived of as an individual process, pathways towards terrorism are inherently social and political. **2**
- [WOS:000350893700005] In particular, little is known about the early processes and pathways to radicalization.
- [WOS:000352355500002] While the cases of Anders Behring Breivik and Mohamed Merah clearly demonstrate the impact of social networks and the role of the Internet and prison on the radicalization process, the killings in Norway and France in fact expose larger issues that exist within contemporary Europe, including profound identity crises manifesting as Islamist extremism in some others and far-right extremism in others. **3**
- [WOS:000289794000004] Transformative learning theory, developed from the sciences in education and rehabilitation, offers an interdisciplinary lens with which to study the processes of personal change associated with radicalization.
- [WOS:000375999900005] But the way we understand radicalization has specific consequences for the way we manage and fight the scourge of terrorism.
- [WOS:000281084600001] The aim is to take stock of the current state of research within this field and to answer the question: From an empirical point of view, what is known and what is not known about radicalization connected to militant Islamism in Europe?.

Fig. 3.28 Some of the contexts of radicalization in cluster # Islamic state

Radicalization is a key concept in cluster #4 (see Fig. 3.28). A common theme is the process, or the pathways, of radicalization. Radicalization is not a new topic (1). It is considered in connection with social and political influences as well as an individual process (2–3). In addition to the term radicalization, the term radicalisation, in British spelling, is used in several articles particularly from the UK's point of view.

The above examples have illustrated that one can develop an understanding of a scientific domain at multiple levels of granularity with a relatively low threshold of prior knowledge of the domain. The strategic exploration is essentially a top-down approach in the same spirit as Shneiderman's visual information search mantra advocates: overview first, zoom and filter, then details on demand. The key here is that scholarly significant patterns can be represented as prominent visual cues. Once we learn what the most common and critical patterns may look like, it usually takes little domain knowledge to recognize visually salient cues, which will lead us to the valuable information that we should concentrate on.

With little adjustments, we can transform the same methodology into a viable analytic approach to the analysis of a scientific domain at a finer level of granularity —semantic predications. Scientometric studies typically focus on structural and dynamic patterns at the level of article or higher levels of granularity such as journals and groups of journals. Analyzing a scientific domain in terms of its semantic predications and their evolving patterns across disciplinary boundaries enables us to address research questions directly.

Main Path Analysis

Representing the scientific literature of a knowledge domain as a network lends us many analytic tools and methods to identify valuable patterns and trends. The main path analysis is a method that can simplify a usually complex network to a small number of paths that would characterize the major development of the underlying domain. Early studies of main paths include (Hummon and Doreian 1989; Carley et al. 1993; Batagelj 2003; Lucio-Arias and Leydesdorff 2008). More recent examples of main path studies include (Liu and Lu 2012; Liu and Kuan 2015). Here we illustrate how to perform a main path analysis of the terrorism research using a combination of CiteSpace and Pajek. Pajek is a computer program for processing large-scale networks, including visualizing a network and analyzing a network with a wide variety of algorithms. Pajek is freely available and it is very powerful. It can handle a network with millions of nodes (Batagelj and Mrvar 1998).

The procedure starts with the bibliographic data downloaded from the Web of Science. The next step is to generate a directed citation network from the bibliographic dataset. Each node in a directed citation network is an article. The article can cite other articles in the network. The article itself can be cited by other articles in the network as well. CiteSpace provides a function that takes the Web of Science records as the input and generates a directed citation network in the Pajek's .net format. Then we can use Pajek to generate main paths, which are a sub-network. First, open the directed citation network in Pajek (Fig. 3.29).

Next, retain the largest connected component of the network so that main paths can be selected from the largest connected component. This requires two steps in Pajek. First, identify the weakly connected components (Fig. 3.30). Then, select the largest one to retain (Fig. 3.31). A strongly connected component in a directed graph is defined as a sub-graph in which every node is reachable from every other node. A weakly connected component is similarly defined, except that we will have to ignore the directed links and consider them as undirected. For example, if we are at the end of a one-way street, we can reach the beginning of the one-way street only if we ignore the one-way restriction.

Pajek identified 243,964 components from the network of 330,662 article nodes. Most of them are singleton components that contain one article only. The largest connected component, cluster 2, contains 84,770 nodes, representing about 25% of the original loosely connected network (Fig. 3.31). Cluster 2 is the largest connected component to retain (Fig. 3.32).

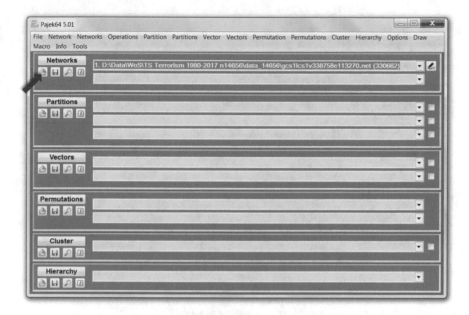

Fig. 3.29 Open the directed citation network in Pajek

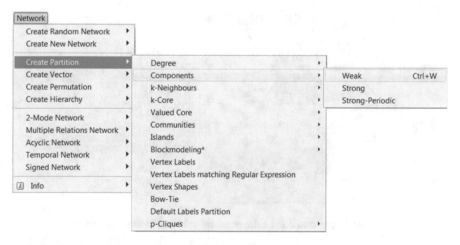

Fig. 3.30 Retain the largest connected component of the network

Main path analysis requires the target network is a directed acyclic network. If the largest connect component contains strongly connected components, such strongly connected components would violate the acyclic condition. If a few articles cite each other, they can be considered as a group. Thus the next step is to shrink strongly connected components as shown in Fig. 3.33.

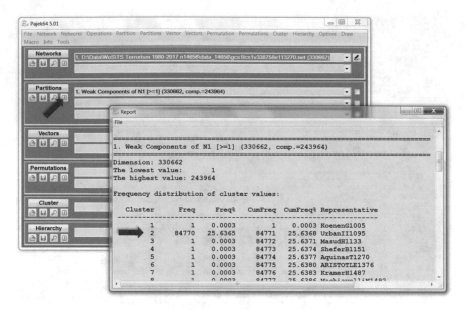

Fig. 3.31 Identify the largest connected component to retain

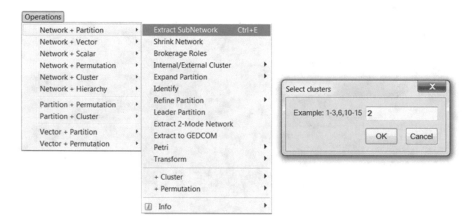

Fig. 3.32 Extract the largest connected component by selecting cluster 2 to retain

After removing loops from the remaining network (Fig. 3.34), we have obtained an acyclic network. There are several ways to compute traversal weights in order to identify main paths, including Search Path Count (SPC), Search Path Link Count (SPLC), and Search Path Node Pair (SPNP) (Fig. 3.35). The next step is to extract the main paths as a subgraph. The user has several options too. We illustrate an option called Key-Route, which presents a combination of a number of significant paths identified (Fig. 3.36).

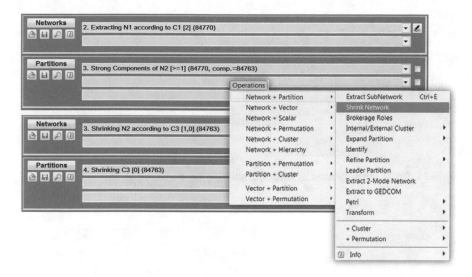

Fig. 3.33 Shrink strongly connected components

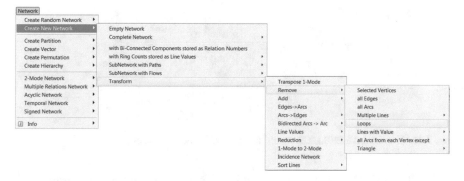

Fig. 3.34 Remove loops from the largest connected component

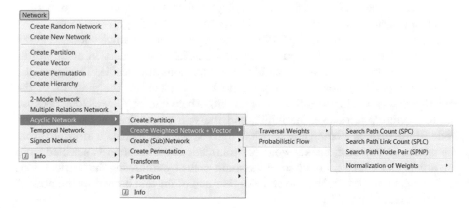

Fig. 3.35 Compute traversal weights along main paths

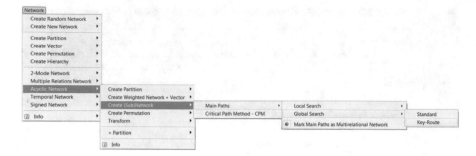

Fig. 3.36 Create the main paths by including multiple key routes, e.g. 1–10

Name	Date modified	Type	Size
Ancestors.mcr	9/28/2010 8:21 PM	MCR File	1 KB
Cognatic.mcr	9/28/2010 8:21 PM	MCR File	1 KB
Cognatic3.mcr	3/27/2006 4:40 AM	MCR File	1 KB
Descendants.mcr	9/28/2010 8:21 PM	MCR File	1 KB
LAYERS.MCR	9/28/2010 8:22 PM	MCR File	1 KB
LAYERS1.MCR	Type: MCR File 22 PM	MCR File	1 KB
LAYERS2.MCR	Size: 609 bytes 22 PM	MCR File	1 KB
LongestMatrilineage.mcr	Date modified: 9/28/2010 8:22 PM 9/28/2010 8:14 PM	MCR File	1 KB
LongestPatrilineage.mcr	9/28/2010 7:49 PM	MCR File	1 KB
NumAncestors.mcr	1/8/2015 6:13 AM	MCR File	1 KB
NumDescendants.mcr	1/8/2015 6:13 AM	MCR File	1 KB
PATH.MCR	9/28/2010 8:21 PM	MCR File	1 KB
zLayers.mcr	3/27/2006 4:14 AM	MCR File	1 KB

Fig. 3.37 Use the LAYERS.MCR macro to draw the generated main paths

Finally, the extracted main paths can be displayed by using a macro from Pajek called LAYERS.MCR (Fig. 3.37). The resultant display of the main paths is shown in Fig. 3.38. The user can refine the display further, for example, by applying a community detection algorithm and color the nodes accordingly.

The main paths shown in Fig. 3.38 are arranged such that articles located at the top of the diagram are published earlier than articles below them. In other words, the top of the diagram is where the oldest publications are located. Therefore, the group of nodes in yellow at the top would be considered as the pioneers of the terrorism research. They are all cited by one node below them, GaleaS2002. We use the first author's last name, the initial, and the year of publication to identify the node. If the node is a citing article in the dataset, we will include a keyword from the article. In this case, the keyword posttraumatic stress disorder is selected from the article GaleaS2002. Let us trace the main paths by following the vertical lines that connect two nodes: the node at the higher end of the line is cited by the node at the lower end of the line.

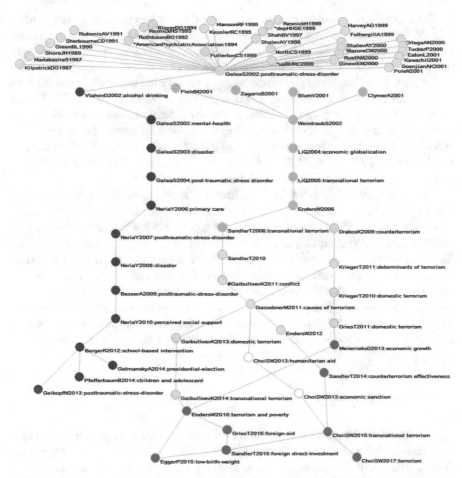

Fig. 3.38 Main paths derived from the direct citation network based on search path count (SPC)

GaleaS2002 leads to a chain of nodes in red, which means that these articles cited GaleaS2002 on PTSD. The red group is identified by a community detection algorithm. The first node of the group is VlahovD2002: alcohol drinking, followed by three more articles that appear to be authored by Galea between 2002 and 2004 on mental health, disaster, and PTSD. These articles are likely to have multiple authors, but in part due to the limitation of the data from the Web of Science, the complete authorship is not readily available. Besides our focus here is mainly on the course of evolution and less so on the authorship per se. The large node in the red group is by Neria Y. published in 2006 on primary care. As suggested by the common keywords, the red group is a line of research on mental health, especially on PTSD in relation to the September 11 terrorist attacks. Neria's 2006 article leads to a chain of nodes in blue published between 2007 and 2010. PTSD remains to be the most prominent keyword for this segment of the main path. The blue route ends with three articles that all cited BergerR2012 on school-based intervention.

The second main path starts with a group of nodes in green on the right-hand side of the diagram. More specifically, this main path begins with four articles published in 2001 followed by WeintraubS2002, which converged the multiple threads to a single-line path. The green segment of the path is characterized by keywords such as economic globalization and transnational terrorism. Then the path is split into two routes, which remained separated between 2006 and 2011. One route contains keywords such as counterterrorism, determinants of terrorism, and domestic terrorism. The other contains keywords such as transnational terrorism. However, these routes became increasingly interwoven into each other through articles such as GassebnerM2011 on causes of terrorism, EndersW2016 on terrorism and poverty, and ChoiSW2015 on transnational terrorism.

The main path analysis of the terrorism research (1980–2017) depicts a big picture that echoes what we have seen in the timeline views of co-cited references. The main path on PTSD corresponds to the largest cluster in the co-citation network, which is also clear that this line of research is no longer as active as before. The second and more complex main path on domestic and transnational terrorism corresponds to the two currently active clusters of co-cited references, which share a higher-level goal to identify the causes of terrorism from economic, social, political, and other dimensions. The questions to be answered are fundamentally significance because eliminating the environment that breeds terrorism would be more effective in a long run than focusing on dealing with aftermaths of terrorism alone.

Structural Variations

The structural variations of the structure of a domain's knowledge can be detected in terms of the changes of the modularity metric of the networks over time (Fig. 3.39). For the terrorism research (1980–2017), the domain has the lowest modularity in 2002. One interpretation would be that the overall connectivity of the network was the strongest in 2002 and a profound common theme made it hard to divide the network in 2002 into clearly separated parts. The predominant position of the PTSD specialty after the September 11 terrorist attacks was evident in both studies of the domain in 1996–2003 and 1980–2017. The September 11 terrorist attacks are likely to be the reason behind the low modularity.

The structural variation analysis of the terrorism research (1980–2017) reveals an interesting pattern: transformative links are all associated with the PTSD research (cluster #0). Since this once predominant cluster stopped its growth before 2010, the structural variation analysis did not find transformative connections in the two currently active lines of research on domestic terrorism and transnational terrorism. The lack of transformative links after the PTSD research could be explained as a sign of a period of normal science as in Kuhn's paradigmatic theory or Shneider's evolution theory. Researchers in these specialties have a clearly established conceptual framework to work with. It is less likely to observe transformative links during this period of time.

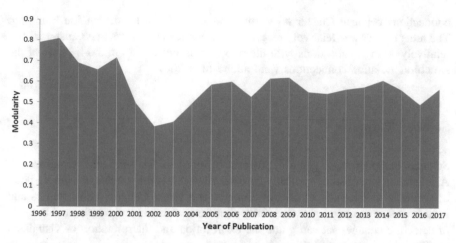

Fig. 3.39 The modularity of the baseline network changes over the years

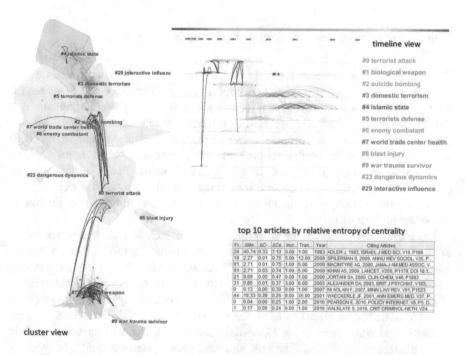

Fig. 3.40 Structural variations measured by the relative entropy of the distributions of betweenness centrality

Structural variations measured by the distribution of betweenness centrality reveal different patterns (Fig. 3.40). Earlier cross-cluster links were added between Cluster #1 biological weapon and Cluster #9 war trauma survivor and between Cluster #0 and Cluster #1. Articles identified with transformative potentials made connections between Cluster #0 terrorist attack and Cluster #2 suicide bombing. The latter cluster was relatively recent. The timeline view shows that Cluster #2 has relatively strong connections with the two current clusters #3 and #4. None of the structural variation connections were added after 2007.

Science Mapping

Science mapping is a generic process of domain analysis and visualization. A science mapping study typically consists of several components, notably a body of scientific literature, a set of scientometric and visual analytic tools, metrics and indicators that can highlight potentially significant patterns and trends, and theories of scientific change that can guide the exploration and interpretation of visualized intellectual structures and dynamic patterns.

The Interplay Between Science and Theories of Science

Science mapping approaches typically aim to represent patterns and trends of the development of science at macroscopic levels such as disciplines and fields over a long period of time. Indeed, science mapping is a very promising combination of the domain analysis method originated in information science and visual analytics from computer science. In particular, science mapping provides a unique means to verifying the validity of individual theories of scientific change. In return, each of the macroscopic theories such as Kuhn's paradigm shifts, Fuchs' mutual dependencies and competition, Shneider's four-stage evolution provides a rich and yet potentially biased perspective that may guide us to interpret how a scientific field is unfolding in front of us.

Commonly used sources of scientific literature include the Web of Science, Scopus, Google Scholar, and PubMed. Scientometric methods include author co-citation analysis (ACA) (White and McCain 1998; Chen 1999), document co-citation analysis (DCA) (Small 1973; Chen 2006), co-word analysis (Callon et al. 1983), and many other variations. Visualization techniques include graph or network visualization (Herman et al. 2000), visualizations of hierarchies or trees (Johnson and Shneiderman 1991), visualizations of temporal structures (Morris et al. 2003), geospatial visualizations, and coordinated views of multiple types of visualizations. Metrics and indicators of research impact include citation counts (Garfield 1955), the h-index (Hirsch 2005) and its numerous extensions, and a rich set of altmetrics on social media (Thelwall et al. 2013).

Theories of scientific change include the paradigmatic views of scientific revolutions, scientific advances driven by competitions, and evolutionary stages of a scientific discipline. In order to conduct a science mapping study, researchers need to develop a good understanding of each of the categories of skills and knowledge outlined above. Furthermore, each of these categories is a current and active research area in its own right, for instance, the current research on finding the optimal field normalization method and the debates over how various potentially conflicting theories of scientific change may be utilized to reveal the underlying mechanisms of how science advances.

The complexity of science mapping is shared by many research fields. We will illustrate the process of a systematic review based on a series of visual analytic functions implemented in CiteSpace (Chen 2004, 2006; Chen et al. 2010). We demonstrate the steps of preparing a representative dataset, how to generate visualizations that can guide our review, and how to identify salient patterns at various levels of granularity.

Characterizing the Field of Study

The dataset to represent the research field is collected through multiple topic search queries to the Web of Science. The rationale of the query construction is as follows. First, we would like to ensure that currently widely used science mapping tools such as VOSViewer, CiteSpace, HistCite, SciMAT, and Sci2 are covered by our topic search query. The inclusion of software tools is based on the characterization of Shneider's second evolutionary stage. Thus, publications that mention any of these software tools in their titles, abstracts, and/or keyword lists will be included. This query generates 135 records as Set #1 (Fig. 3.41).

Second, since the goal of science mapping is to identify the intellectual structure of a scientific domain, the second query focuses on the object of science mapping, including topic terms such as intellectual structure, scientific change, research front, invisible college, and domain analysis. This query is motivated by the first evolutionary stage in Shneider's evolution model. The query may also capture major paradigms because these concepts are fundamental to the research. As we will see later on, terms such as domain analysis may be ambiguous as they are also used in other contexts that are irrelevant to science mapping. In practice, one should defer the assessment of relevance until the analysis stage. This query produces 13,242 records as Set #2.

The third query focuses on scientometric and visual analytic techniques that are potentially relevant to science mapping. Topic terms include science mapping, knowledge domain visualization, information visualization, citation analysis, co-citation analysis. Some of these techniques are enabling techniques developed elsewhere in fields such as computer science. This query would capture the development and application of these techniques. This query leads to 4772 records.

The queries #4–#10 aim to retrieve bibliographic records on the common data sources for science mapping, including Scopus (6782 records), the Web of Science (15,401 records), Google Scholar (5170 records), Pubmed (46,760 records), and MEDLINE (61,405 records).

Set	Results	Save History / Create Alert	Open Saved History
# 15	350	#13 AND #2 *Indexes=SCI-EXPANDED, SSCI, A&HCI, ESCI Timespan=All years*	
# 14	17,731	#13 OR #2 *Indexes=SCI-EXPANDED, SSCI, A&HCI, ESCI Timespan=All years*	
# 13	4,839	#3 OR #1 *Indexes=SCI-EXPANDED, SSCI, A&HCI, ESCI Timespan=All years*	
# 12	111	(TS=("literature-based discovery")) AND **LANGUAGE:** (English) AND **DOCUMENT TYPES:** (Article OR Review) *Indexes=SCI-EXPANDED, SSCI, A&HCI, ESCI Timespan=All years*	
# 11	23,275	#8 OR #7 OR #6 OR #5 OR #4 *Indexes=SCI-EXPANDED, SSCI, A&HCI, ESCI Timespan=All years*	
# 10	61,405	(TS=Medline) AND **LANGUAGE:** (English) AND **DOCUMENT TYPES:** (Article OR Review) *Indexes=SCI-EXPANDED, SSCI, A&HCI, ESCI Timespan=All years*	
# 9	46,760	(TS=Pubmed) AND **LANGUAGE:** (English) AND **DOCUMENT TYPES:** (Article OR Review) *Indexes=SCI-EXPANDED, SSCI, A&HCI, ESCI Timespan=All years*	
# 8	354	(TS=("PubMed Central")) AND **LANGUAGE:** (English) AND **DOCUMENT TYPES:** (Article OR Review) *Indexes=SCI-EXPANDED, SSCI, A&HCI, ESCI Timespan=All years*	
# 7	24	(TS=("Microsoft Academic Search")) AND **LANGUAGE:** (English) AND **DOCUMENT TYPES:** (Article OR Review) *Indexes=SCI-EXPANDED, SSCI, A&HCI, ESCI Timespan=All years*	
# 6	5,170	(TS=("Google Scholar")) AND **LANGUAGE:** (English) AND **DOCUMENT TYPES:** (Article OR Review) *Indexes=SCI-EXPANDED, SSCI, A&HCI, ESCI Timespan=All years*	
# 5	15,401	(TS=("web of science" OR "web of knowledge" OR "science citation index")) AND **LANGUAGE:** (English) AND **DOCUMENT TYPES:** (Article OR Review) *Indexes=SCI-EXPANDED, SSCI, A&HCI, ESCI Timespan=All years*	
# 4	6,782	(TS=(Scopus)) AND **LANGUAGE:** (English) AND **DOCUMENT TYPES:** (Article OR Review) *Indexes=SCI-EXPANDED, SSCI, A&HCI, ESCI Timespan=All years*	
# 3	4,772	(TS=("science mapping" OR "science mapping analysis" OR "scientometric study" OR "scientometric analysis" OR "scientometric review" OR "knowledge domain visualization" OR "knowledge visualization" OR "bibliometric mapping" OR "bibliographic network visualization" OR "historiographic analysis" OR "research impact assessment" OR "information visualization" OR "domain visualization" OR "network visualization" OR "citation analysis" OR "co-citation analysis" OR "cocitation analysis" OR "co-word analysis" OR "coword analysis" OR "author co-citation" OR "document co-citation" OR "journal co-citation")) AND **LANGUAGE:** (English) AND **DOCUMENT TYPES:** (Article OR Review) *Indexes=SCI-EXPANDED, SSCI, A&HCI, ESCI Timespan=All years*	
# 2	13,242	(TS=("intellectual structure*" OR "scientific change" OR "scientific revolutions" OR "scientific progress" OR "research trend" OR "conceptual evolution" OR "domain analysis" OR "research front" OR "knowledge diffusion" OR "research programme" OR "scientific programme*" OR "invisible college" OR "scientific frontiers" OR "literature dynamics")) AND **LANGUAGE:** (English) AND **DOCUMENT TYPES:** (Article OR Review) *Indexes=SCI-EXPANDED, SSCI, A&HCI, ESCI Timespan=All years*	
# 1	135	TS=(VOSViewer OR CiteSpace OR HistCite OR SciMAT OR Sci2) *Indexes=SCI-EXPANDED, SSCI, A&HCI, ESCI Timespan=All years*	

Fig. 3.41 Topic search queries used for data collection

The final dataset is Set #14, containing 17,731 bibliographic records of the types of Article or Review in English (Fig. 3.42). This query formation strategy is generic enough to be applicable to a science mapping study unless of course one has access to the entire database.

Patents and research grants are other types of data sources one may consider, but for this particular review, we are limited to the scientific literature indexed by the Web of Science.

Visual Analysis of the Literature

We visualize and analyze the dataset with CiteSpace. CiteSpace takes a set of bibliographic records as its input and models the intellectual structure of the underlying domain in terms of a synthesized network based on a time series of networks derived from each year's publications. CiteSpace has been continuously developed for more than a decade. CiteSpace supports several types of bibliometric studies, including collaboration network analysis, co-word analysis, author co-citation analysis, document co-citation analysis, text and geospatial visualizations. In this case, we focus on the document co-citation analysis within the period of time between 1995 and 2016 (Fig. 3.43).

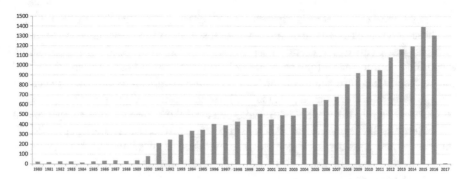

Fig. 3.42 The distribution of the bibliographic records in Set #14

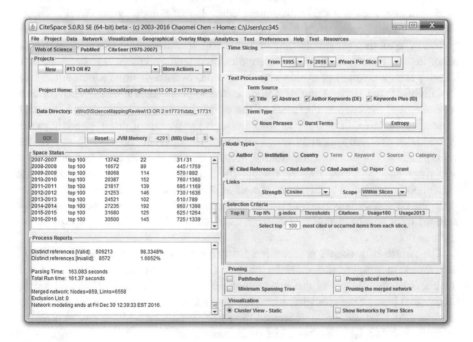

Fig. 3.43 The main user interface of CiteSpace

The Set #14 contains 16,250 records published in the range of 1980–2017. These records collectively cited 515,026 references. The document co-citation analysis function in CiteSpace constructs networks of cited references. Connections between references represent co-citation strengths. CiteSpace uses a time slicing technique to build a time series of network models over time and synthesize these individual networks to form an overview network for the systematic review of the relevant literature.

The synthesized network is divided into co-citation clusters of references. Citers to these references are considered as the research fronts associated with these clusters. Each cluster represents the intellectual base of the underlying specialty. According to Shneider's four stage model, the intellectual base of a specialty and the corresponding research fronts provide valuable insights into the current stage of the specialty as well as the intellectual milestones in the evolution of the specialty.

Our first step in the review is to make sense of the nature of major clusters and characteristics that may inform us about the stage of the underlying specialties. In this study, we consider a cluster as the embodiment of an underlying specialty. Thus, science mapping consists of multiple specialties that contribute to various aspects of the domain.

In each cluster, we focus on cluster members that are identified by structural and temporal metrics of research impact and evolutionary significance. A commonly used structural metric is the betweenness centrality of a node in a network. Studies have shown that nodes with high betweenness centrality values tend to identify boundary spanning potentials that may lead to transformative discoveries (Chen et al. 2009). Burst detection is a computational technique that has been used to identify abrupt changes of events and other types of information (Kleinberg 2002). In CiteSpace, the sigma score of a node is a composite metric of the betweenness centrality and the citation burstness of the node, i.e. the cited reference. CiteSpace represents the strength of these metrics through the design of visual encoding such that articles that are salient in terms of these metrics will be easy to see in the visualizations. For example, the citation history of a node is depicted as a number of tree rings and each tree ring represents the number of citations received in the corresponding year of publication. If a citation burst is detected for a cited reference, the corresponding tree ring will be colored in red. Otherwise, tree rings will be colored by a spectrum that ranges from cold colors such as blue to warm colors such as orange.

The nature of a cluster is identified from the following aspects: a hierarchy of key terms in articles that cite the cluster (Tibély et al. 2013), the prominent members of the cluster as the intellectual milestones in its evolution and as the intellectual base of the specialty, recurring themes in the citing articles to the cluster to reflect the interrelationship between the intellectual base and the research fronts. In particular, we will pay attention to indicators of the evolutionary stages of a specialty such as the original conceptualization, research instruments, applications, and routinization of the domain knowledge of the specialty.

In addition to the study of citation-based patterns, we will demonstrate the concept of citation trajectories in the context of distinct clusters. According to the theory of structural variation, the transformative potential of an article may be reflected by the extent to which it varies the existing intellectual structure (Chen 2012). For example, if an article adds many inter-cluster links, it may alter the overall structure. If the structural change is subsequently accepted and reinforced by other researchers, then transformative changes of the knowledge become significant in a socio-cognitive view of the domain.

Visualizing the Field

A dual-map overlay of the science mapping literature represents the entire dataset in the context of a global map of science generated from over 10,000 journals indexed in the Web of Science (Chen and Leydesdorff 2014). The dual-map overlay in Fig. 3.44 shows that science mapping papers are published in almost all major disciplines. Publications in the discipline of information science (shown in the map as curves in cyan) are built on top of at least four disciplines on the right-hand side of the map.

A hierarchical visualization of index terms, i.e. keywords, is generated to represent the coverage of the dataset (Fig. 3.45). Five semantic types of nodes are annotated in the visualized hierarchy:

What: a fundamental phenomenon of a specialty and the object of a study, for example, the intellectual structure or the dynamics of a research field.
How: methodologies, procedures, and processes of science mapping, for example, author co-citation analysis, bibliometric mapping, and co-citation analysis.
Abstraction: computational models of an underlying phenomenon identified from the bibliographic data, representations such as Pathfinder networks, metrics and indicators such as the h-index and the g-index.
Tools: computational techniques, algorithms and software tools for visualization and ranking scholarly publications.
Data: data sources used by science mapping studies, for example, Scopus and Google Scholar.

These semantic types will be also used to identify the evolutionary stage of a specialty. For example, if a cluster contains several articles that report the development of software tools, then the underlying specialty is considered as a specialty that has reached at least Stage II. If the methodologies appear in a cluster of knowledge domains external to information science, such as regenerative medicine

Fig. 3.44 A dual-map overlay of the science mapping literature

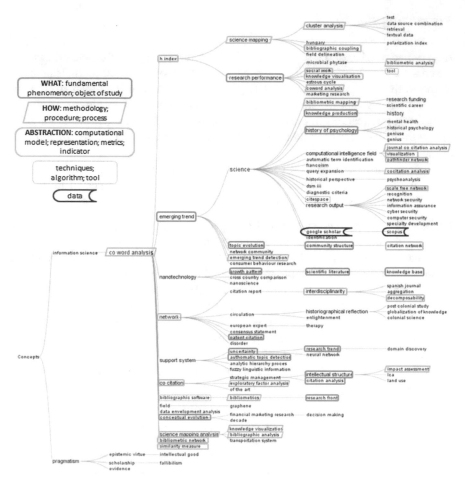

Fig. 3.45 A hierarchy of indexing terms derived from Set #14

and strategic management research, then we will consider the specialty has reached Stage III—tools developed by the specialty are applied to other subject domains. In the following analysis, we will use the terms in the hierarchy as the primary source of our vocabulary to identify the role of the contributions made by a scientific publication to a specialty.

Major milestones in the development of science mapping can be identified from the list of references that have strong citation bursts between 1995 and 2016 (Fig. 3.46). References with strong values in the Strength column tend to be significant milestones for the science mapping research. We label such references with high-level concepts. For example, the first milestone paper in the study is a landmark ACA study of information science (White and McCain 1998). The next milestone is a major collection of seminal papers in information visualization by

	References	Year	Strength	Begin	End	1995 - 2016
	MALIK M, 1992, J AM COLL CARDIOL, V20, P127	1992	10.4711	1995	1999	
	MANCINI DM, 1993, CIRCULATION, V87, P1083	1993	5.8964	1996	2000	
	CAMM AJ, 1996, CIRCULATION, V93, P1043	1996	46.4658	1997	2004	
	TAFLOVE A, 1995, COMPUTATIONAL ELECTR, V, P	1995	23.0034	1997	2002	
	CAMM AJ, 1996, EUR HEART J, V17, P354	1996	5.4362	1998	2004	
	MACROBERTS MH, 1996, SCIENTOMETRICS, V36, P435, DOI	1996	7.9414	1999	2004	
Author Co-Citation Analysis	WHITE HD, 1998, J AM SOC INFORM SCI, V49, P327, DOI	1998	40.1581	1999	2006	
	WHITE HD, 1997, ANNU REV INFORM SCI, V32, P99	1997	10.5983	1999	2003	
Information Visualization	CARD SK, 1999, READINGS INFORMATION, V, P	1999	25.7498	2000	2006	
	MAY RM, 1997, SCIENCE, V275, P793, DOI	1997	4.9544	2000	2004	
	BALDI S, 1998, AM SOCIOL REV, V63, P829, DOI	1998	4.8083	2001	2005	
	ALTSCHUL SF, 1997, NUCLEIC ACIDS RES, V25, P3389, DOI	1997	7.217	2001	2005	
	GOODRUM AA, 2001, INFORM PROCESS MANAG, V37, P661, DOI	2001	7.2483	2002	2009	
	CRONIN B, 2001, J INFORM SCI, V27, P1	2001	12.5667	2002	2008	
Citation Mapping	SMALL H, 1999, J AM SOC INFORM SCI, V50, P799, DOI	1999	15.8489	2002	2006	
Domain Analysis	HJORLAND B, 2002, J DOC, V58, P422, DOI	2002	11.7233	2003	2010	
	BORGMAN CL, 2002, ANNU REV INFORM SCI, V36, P3	2002	13.2442	2003	2010	
Graph Visualization	HERMAN I, 2000, IEEE T VIS COMPUT GR, V6, P24, DOI	2000	15.7512	2003	2008	
	TAFLOVE A, 2000, COMPUTATIONAL ELECTR, V, P	2000	11.9635	2004	2008	
	ADAM D, 2002, NATURE, V415, P726	2002	8.4895	2004	2009	
	WHITE HD, 2003, J AM SOC INF SCI TEC, V54, P423, DOI	2003	10.5968	2004	2008	
Domain Visualization	BORNER K, 2003, ANNU REV INFORM SCI, V37, P179, DOI	2003	15.4015	2004	2011	
	CHEN CM, 2002, J AM SOC INF SCI TEC, V53, P678, DOI	2002	4.5481	2004	2008	
	AKSNES DW, 2003, SCIENTOMETRICS, V56, P235, DOI	2003	3.3291	2005	2010	
	KING DA, 2004, NATURE, V430, P311, DOI	2004	6.6773	2005	2010	
	SAHA S, 2003, J MED LIBR ASSOC, V91, P42	2003	5.414	2005	2010	
	TUFTE ER, 2001, VISUAL DISPLAY QUANT, V, P	2001	4.598	2005	2009	
	KEIM DA, 2002, IEEE T VIS COMPUT GR, V8, P1, DOI	2002	6.2489	2005	2010	
Mapping Scientific Frontiers	CHEN C, 2003, MAPPING SCI FRONTIER, V, P	2003	6.6665	2005	2010	
	GLANZEL W, 2002, SCIENTOMETRICS, V53, P171, DOI	2002	7.5022	2005	2010	
	AHLGREN P, 2003, J AM SOC INF SCI TEC, V54, P550, DOI	2003	16.1924	2006	2011	
UCINET	BORGATTI S P, 2002, UCINET WINDOWS SOFTW, V, P	2002	9.3193	2006	2010	
Visual Analytics	THOMAS J J, 2005, ILLUMINATING PATH RE, V, P	2005	9.9577	2006	2011	
	GMUR M, 2003, SCIENTOMETRICS, V57, P27, DOI	2003	6.6503	2006	2010	
	RAMOS-RODRIGUEZ AR, 2004, STRATEGIC MANAGE J, V25, P981, DOI	2004	12.7007	2006	2012	
	WHITE HD, 2003, J AM SOC INF SCI TEC, V54, P1250, DOI	2003	9.6883	2006	2011	
	ALAHBABI MN, 2005, J OPT SOC AM B, V22, P1321, DOI	2005	4.291	2006	2012	
Cytoscape	SHANNON P, 2003, GENOME RES, V13, P2498, DOI	2003	16.4941	2007	2011	
h-index	HIRSCH JE, 2005, P NATL ACAD SCI USA, V102, P16569, DOI	2005	40.6435	2007	2013	
	MOED H F, 2005, CITATION ANAL RES EV, V, P	2005	17.9615	2007	2012	
	MEHO LI, 2007, J AM SOC INF SCI TEC, V58, P2105, DOI	2007	3.315	2008	2013	
CiteSpace I	CHEN CM, 2004, P NATL ACAD SCI USA, V101, P5303, DOI	2004	6.8456	2008	2012	
	BOYACK KW, 2005, SCIENTOMETRICS, V64, P351, DOI	2005	11.8903	2009	2013	
CiteSpace II	CHEN CM, 2006, J AM SOC INF SCI TEC, V57, P359, DOI	2006	6.7057	2009	2014	
	SINGH J, 2005, MANAGE SCI, V51, P756, DOI	2005	4.6233	2009	2013	
	GARFIELD E, 2006, JAMA-J AM MED ASSOC, V295, P90, DOI	2006	7.1782	2009	2013	
	CHIU WT, 2007, SCIENTOMETRICS, V73, P3, DOI	2007	4.6127	2010	2016	
	WUCHTY S, 2007, SCIENCE, V316, P1036, DOI	2007	7.1104	2011	2016	
	XIE SD, 2008, SCIENTOMETRICS, V77, P113, DOI	2008	4.1387	2011	2016	

Fig. 3.46 49 references with citation bursts of at least 5 years

Card et al. (1999). Other major milestones include visual analytics (Thomas and Cook 2005), and the h-index (Hirsch 2005).

Landscape View

The landscape view in Fig. 3.47 is generated based on publications between 1995 and 2016. Top 100 most cited publications in each year are used to construct a network of references cited in that year. Then individual networks are synthesized. The synthesized network contains 3145 references. The network contains 603

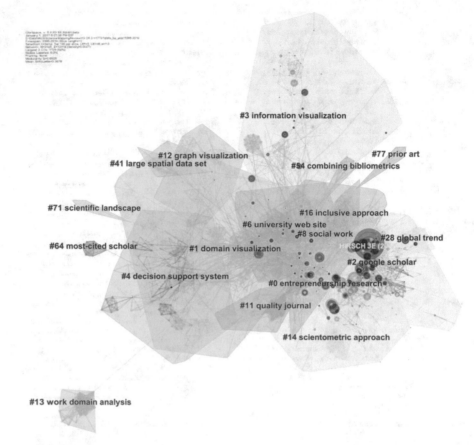

Fig. 3.47 A landscape view of the co-citation network, generated by top 100 per slice between 1995 and 2016 (LRF = 3, LBY = 8, and e = 1.0)

co-citation clusters. The three largest connected components include 1729 nodes, which account for 54% of the entire network. The network has a modularity of 0.8925, which is considered as very high, suggesting that the specialties in science mapping are clearly defined in terms of co-citation clusters. The average silhouette score of 0.3678 is relative low mainly because of the numerous small clusters. The major clusters that we will focus on in the review are sufficiently high.

The areas of different colors indicate the time when co-citation links in those areas appeared for the first time. Areas in blue were generated earlier than areas in green. Areas in yellow were generated after the green areas and so on. Each cluster can be labeled by title terms, keywords, and abstract terms of citing articles to the cluster. For example, the yellow-colored area at the upper right quadrant is labeled as #3 information visualization, indicating that Cluster #3 is cited by articles on information visualization. The largest node is the paper that introduces the h-index. Other nodes with tree rings in red are references with citation bursts.

Timeline View

A timeline visualization in CiteSpace depicts clusters along horizontal timelines (Fig. 3.48). Each cluster is displayed from left to right. The legend of the publication time is shown on top of the view. The clusters are arranged vertically in the descending order of their size. The largest cluster is shown at the top of the view. The colored curves represent co-citation links added in the year of the corresponding color. Large-sized nodes or nodes with red tree rings are of particular interest because they are either highly cited or have citation bursts or both. Below each timeline the three most cited references in a particular year are displayed. The label of the most cited reference is placed at the lowest position. References published in the same year are placed so that the less cited references are shifted to the left. The new version of CiteSpace supports the function to generate labels of a cluster year by year based on terms identified by Latent Semantic Indexing (LSI) (Deerwester et al. 1990). The year-by-year labels can be displayed in a table or above the corresponding timeline. Users may control the displays interactively.

Clusters are numbered from 0, i.e. Cluster #0 is the largest cluster and Cluster #1 is the second largest one. As shown in the timeline overview, the sustainability of a specialty varies. Some clusters sustain a period over 20 years, whereas some clusters are relatively short-lived. Some clusters remain active until the 2015, the most recent year of publication for a cited reference in this study.

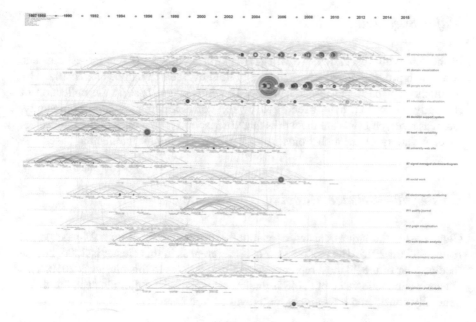

Fig. 3.48 A timeline visualization of the largest clusters of the total of 603 clusters

Table 3.9 The five largest clusters of co-cited references of the network of 3145 references

Cluster	Size	Mean (Year)	Silhouette	% of the network	Accumulated % of network	% of top 3 LCCs	Accumulated % of LCCs
0	214	2006	0.748	4.5	4.5	8.1	8.1
1	209	1997	0.765	2.3	6.7	4.1	12.2
2	190	2009	0.845	3.3	10.0	6.0	18.2
3	160	2005	0.954	2.9	12.9	5.3	23.5
4	152	1992	0.890	1.7	14.6	3.0	26.5

The largest three connected components include 1729 of the references

As shown in Table 3.9, each of the largest five clusters has over 150 members. The largest cluster's homogeneity in terms of the silhouette score is slightly lower than that of the smaller clusters. The largest cluster represents 4.5% of the references from the entire network and 8.1% of the largest three connected components of the network (LCCs). In this study, our review will primarily focus on the largest five clusters.

The duration of a cluster is particularly interesting (see Table 3.10). The largest cluster lasts 21 years and it is still active. Cluster #3 spans a 19-year period and also remains to be active. In contrast, Cluster #6 on webometrics ends by 2006, but as we will see, relevant research finds its way in new specialties, notably in the form of altmetrics.

Major Specialties

In the following discussion, we will particularly focus on the five largest clusters. A research programme, or a paradigm, in a field of research can be characterized by its intellectual base and research fronts. The intellectual base is the collection of scholarly works that have been cited by the corresponding research community, whereas research fronts are the works that are inspired by the ones of the intellectual base. A variety of research fronts may rise from a common intellectual base.

Cluster #0—Science Mapping

Cluster #0 is the largest cluster, containing 214 references across a 21-year period from 1995 till 2015. The median year of all references in this cluster is 2006, but the median year of the 20 most representative citing articles to this cluster is 2010. This cluster's silhouette value of 0.748 is the lowest among the major clusters, but this is generally considered a relatively high level of homogeneity.

Table 3.10 Temporal properties of major clusters

Cluster ID	Size	Silhouette	From	To	Duration	Median	Sustainability	Activeness	Theme
0	214	0.748	1995	2015	21	2006	+++++	Active	Science mapping
1	209	0.765	1990	2006	17	1997	++	Inactive	Domain analysis
2	190	0.845	2000	2015	16	2009		Active	Research evaluation
3	160	0.954	1996	2014	19	2005	++++	Active	Information visualization/visual analytics
4	152	0.890	1988	1999	12	1993		Inactive	Applications of ACA
6	125	0.925	1995	2006	12	2001		Inactive	Webometrics
8	93	0.882	1994	2010	17	2002	++	Inactive	Bibliometric studies of social work in health
11	48	0.965	1994	2006	13	2000		Inactive	Bibliometric studies of management research
12	44	0.966	1990	1999	10	1996		Inactive	Graph visualization
16	29	0.977	1999	2007	9	2003		Inactive	Bibliometric studies of information systems
28	15	0.995	2004	2013	10	2008		Inactive	Global trend; water resources

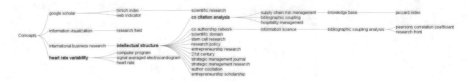

Fig. 3.49 A hierarchy of key concepts selected from citing articles of Cluster #0 by log-likelihood ratio test

The primary focus of the large and currently active cluster is on the intellectual structure of a scientific discipline, a field of research, or any sufficiently self-contained domain of scientific inquiry. Key concepts identified from the titles of citing articles to this cluster can be algorithmically organized according to hierarchical relations derived from co-occurring concepts (Fig. 3.49). The largest branch of a such hierarchy typically reflects the core concepts of scholarly publications produced by the specialty behind the cluster. For example, concepts such as intellectual structure, co-citation analysis, co-authorship network underline the primary interest of this specialty.

We can use a simple method to classify various terms into two broad categories: domain-intrinsic or domain-extrinsic. Domain-intrinsic terms belong to the research field that aims to advance the conceptual and methodological capabilities of science mapping, for example, intellectual structure and co-citation analysis. In contrast, domain-extrinsic terms belong to the domain to which science mapping techniques are applied. In other words, they belong to the domain that is the object of a science mapping study. For example, stem cell research per se may not directly influence the advance of a specialty that is mainly concerned with how to identify the intellectual structure of a research field from scientific literature. Information science has a unique position. On the one hand, it is the discipline that hosts a considerable number of fields relevant to science mapping. On the other hand, it is the most frequent choice of a knowledge domain to test drive newly developed techniques and methods.

The timeline visualization reveals three periods of its development (Fig. 3.50). The first period is from 1995 to 2002. This period is relatively uneventful without high-profile references in terms of citation counts or bursts. Two visualization-centric domain analysis articles, Boyack2002 and Chen2002, preluded the subsequent wave of high-impact studies appeared in the second period. This period also features a social network analysis tool UCINET Borgatti2002.

The second period is from 2003 to 2010. Unlike the first period, the second period is full of high impact contributions—large citation tree rings and periods of citation bursts colored in red. Several types of high impact contributions appeared in this period, notably

- literature reviews—Börner 2003
- software tools—CiteSpace (Chen 2004), CiteSpace II (Chen 2006), CiteSpace III (Chen et al. 2010), VOSViewer (Van Eck and Waltman 2010)

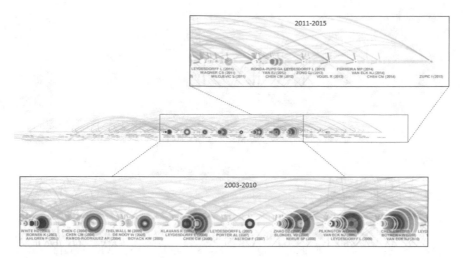

Fig. 3.50 High impact members of Cluster #0

- science mapping applications—visualization of information science White 2003, mapping the backbone of science Boyack et al. 2005 and a global map of science based on ISI subject categories Leydesdorff and Rafols 2009
- metrics and indicators—a critique on the use of Pearson's correlation coefficients as co-citation similarities—a previously common practice in ACA studies Ahlgren 2003
- applications to other domains—a bibliometric study of strategic management research Ramos-Rodriguez 2004 and another ACA of strategic management research Nerur 2008

The third period is from 2010 to 2015. Although no citation bursts were detected so far in this period, the themes of this period sheds additional insights into the more recent developmental status of the specialty. Most cited publications in this period include a study of the cognitive structure of library and information science —Milojević 2011 and a few studies that focus on domains with no apparent overlaps with computer and information science, for example regenerative medicine (Chen et al. 2012, 2014a, b) and strategic management—Vogel 2013.

A specialty may experience the initial conceptualization stage, the growth of research capabilities through the flourish of research tools, the expansion stage when researchers apply their methods to subject domains beyond the original research problems, and the final stage of decay (Shneider 2009). The largest cluster is dominated by an overwhelming number of tool-related references. As shown in Fig. 3.51, the top 20 most cited members of the cluster include several software tools such as CiteSpace, UCINET, VOSviewer, and global maps of science. If we follow Shneider's four-stage evolution model, the high concentration of software tools seems to suggest that the specialty behind this cluster evidently reached the second stage of its evolution by 2010. However, the several types of high-impact articles in this cluster, especially in the second period, suggest a far more complex picture.

Freq	Burst	Centrality	Σ	PageRank	...	Author	Year	Tit.	Source	Vol	Page	HalfLife	Cluster
80	24.90	0.02	1.81	0.00	...	Chen CM	2006	...	J AM SOC I...	V57	P359	6	0
39	12.90	0.04	1.69	0.00		White HD	2003	...	J AM SOC I...	V54	P423	5	0
67	19.84	0.01	1.25	0.00		Leydesdorff L	2009	...	J AM SOC I...	V60	P348	3	0
63	22.61	0.01	1.20	0.00		Boyack KW	2005	...	SCIENTOM...	V64	P351	6	0
73	19.33	0.01	1.17	0.00		Nerur SP	2008	...	STRATEG...	V29	P319	6	0
53	18.11	0.01	1.16	0.00		Ahlgren P	2003	...	J AM SOC I...	V54	P550	6	0
38		0.01	1.16	0.00		Van Eck NJ	2009	...	J AM SOC I...	V60	P1635	5	0
32	10.78	0.01	1.13	0.00		White HD	2003	...	J AM SOC I...	V54	P1250	5	0
82		0.00	1.12	0.00		Van Eck NJ	2010	...	SCIENTOM...	V84	P523	5	0
27	11.59	0.01	1.11	0.00		Borgatti S P	2002	...	UCINET WI...	V	P	6	0
16	6.06	0.02	1.10	0.00		Boyack KW	2002	...	J AM SOC I...	V53	P764	6	0
51	16.88	0.00	1.08	0.00		Borner K	2003	...	ANNU REV...	V37	P179	7	0
16	5.54	0.01	1.08	0.00		White HD	2001	...	J AM SOC I...	V52	P87	7	0
25		0.01	1.07	0.00		De Nooy W	2005	...	EXPLORAT...	V	P	4	0
58	20.32	0.00	1.06	0.00		Chen CM	2010	...	J AM SOC I...	V61	P1386	4	0
54	18.09	0.00	1.06	0.00		Ramos-rodriguez...	2004	...	STRATEGI...	V25	P981	7	0
20	9.13	0.01	1.06	0.00		Gmur M	2003	...	SCIENTOM...	V57	P27	6	0
10	4.22	0.01	1.06	0.00		Ding Y	2000	...	SCIENTOM...	V47	P55	6	0
30	10.77	0.00	1.05	0.00		Rafols I	2009	...	J AM SOC I...	V60	P1823	4	0
18	6.82	0.01	1.05	0.00		Morris SA	2003	...	J AM SOC I...	V54	P413	6	0

Fig. 3.51 Top 20 most cited references in the largest cluster

The cluster includes several author co-citation studies of disciplines and research areas such as information science and strategic management. White 2003 revisits the intellectual structure of information science. Instead of using multidimensional scaling technique as they did in a previous study of the domain, the new study applied the Pathfinder network scaling technique and demonstrated the advantages of the technique. Pathfinder network scaling was first introduced to author co-citation analysis in (Chen 1999). The studies of strategic management research can be seen as applications outside the original specialty of author co-citation analysis. Furthermore, as we can see here, the application of ACA to a new target domain was made by researchers from the target domain several years after the analytic procedure was developed in information science. The techniques evidently spread to domains beyond information science. Fuchs's theory explains the speed of such diffusion in terms of the density of scientists' social network. Information travels faster in tightly coupled networks than loosely connected ones.

According to Shneider's evolution model, the application of tools to a new target should mark the beginning of the third stage. However, it seems we are seeing a considerable overlap between the second stage and the third stage. On the one hand, the development of new tools appears to be strengthening. There is no obvious sign that this trend would slow down anytime soon. On the other hand, the application of science mapping techniques to subject domains beyond information science appears to be a gradual process. As new tools have been developed, their applications are likely to follow. This particular example seems to suggest that techniques may be transferred in waves and that the speed of transfer is influenced by the structure of the networks of the researchers at the providing and the receiving ends.

Articles that cited members of the cluster convey additional information for us to understand the dynamics of the specialty (Fig. 3.52). The top 20 citing articles ranked by the bibliographic overlap with the cluster reveal similar types of contributions, namely software tools and techniques (1, 2, 5, 8, 14), new methods (9, 11, 16, 19, 20), surveys and reviews (3, 10, 13), and applications of bibliometric studies (6, 12, 17).

1. 0.11 van Eck, NJ (2010) software survey: vosviewer. a computer program for bibliometric mapping
2. 0.09 Rafols, I (2010) science overlay maps: a new tool for research policy and library management
3. 0.09 Bar-Ilan, J (2008) informetrics at the beginning of the 21st century - a review
4. 0.09 van Eck, NJ (2010) a comparison of two techniques for bibliometric mapping: multidimensional scaling and vos
5. 0.08 Vargas-Quesada, B (2010) showing the essential science structure of a scientific domain and its evolution
6. 0.07 Dolfsma, W (2010) the citation field of evolutionary economics
7. 0.07 Leydesdorff, L (2010) mapping the geography of science: distribution patterns and networks of relations among cities and institutes
8. 0.07 Quirin, A (2010) graph-based data mining: a new tool for the analysis and comparison of scientific domains represented as scientograms
9. 0.07 Waltman, L (2010) a unified approach to mapping and clustering of bibliometric networks
10. 0.06 Morris, SA (2008) mapping research specialties
11. 0.06 Takeda, Y (2010) tracking modularity in citation networks
12. 0.06 Uysal, OO (2010) business ethics research with an accounting focus: a bibliometric analysis from 1988 to 2007
13. 0.05 Borner, K (2003) visualizing knowledge domains
14. 0.05 Chen, CM (2010) the structure and dynamics of cocitation clusters: a multiple-perspective cocitation analysis
15. 0.05 Hicks, CC (2010) interdisciplinarity in the environmental sciences: barriers and frontiers
16. 0.05 Minguillo, D (2010) toward a new way of mapping scientific fields: authors' competence for publishing in scholarly journals
17. 0.04 Hofer, KM (2010) conference proceedings as a matter of bibliometric studies: the academy of international business 2006-2008
18. 0.04 Kurtz, MJ (2010) usage bibliometrics
19. 0.04 Leydesdorff, L (2013) interactive overlays of journals and the measurement of interdisciplinarity on the basis of aggregated journal-journal citations
20. 0.04 Zhao, DZ (2008) information science during the first decade of the web: an enriched author cocitation analysis

Fig. 3.52 Major citing articles to the largest cluster

The timeline visualization suggests that the specialty represented by the largest cluster has cumulated sufficient research techniques and tools by the end of the third period. It is likely that the specialty is ready for a larger scale of applications to subject domains rather than information science. According to Shneider's four-stage model, this is also the stage in while researchers may encounter anomalies that could lead to new discoveries and even the emergence of a new field.

At a more pragmatic level, one may monitor the further development of the specialty by tracking research fronts that are building on the early stages of the specialty. One can monitor emerging trends and patterns in terms of the major dimensions in the latent semantic space spanned by each year's publications connected to this particular cluster. For example, the growing number of domain-extrinsic terms such as nanotechnology, case study, and solar cell, suggest an expansion of the research scope—a hallmark of a third-stage specialty.

In summary, taken all the characteristics into account, the specialty seems to have a sustained second stage while clearly showing characteristics of the third stage in terms of Shneider's evolutionary model. Fuchs' theory provides a framework that one may pursue the diffusion of techniques from the origin of their developers to their users. In particular, one may trace the paths of the diffusion in the context of social networks of the researchers involved. Shneider's theory provides the most concrete account of how a specialty develops. Fuchs' theory provides the mid-range framework to embed the development of techniques in the context of social networks. Kuhn's theory seems to capture the dynamics at the highest level of abstraction. It is more likely that one would find evidence of a paradigm shift between distinct clusters than within the same cluster.

Cluster #1—Domain Analysis

Cluster #1 is the second largest cluster, containing 209 references that range a 17-year duration from 1990 to 2006. The cluster, or its underlying specialty, is

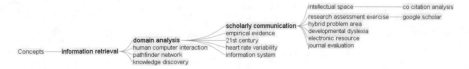

Fig. 3.53 A hierarchy of key concepts in Cluster #1

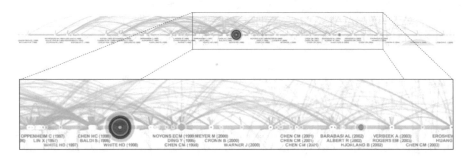

Fig. 3.54 Key members of Cluster #1

largely inactive with reference to the resolution of this study. This cluster is dominated by representative terms such as information retrieval, domain analysis, scholarly communication, and intellectual space (Fig. 3.53). Although information retrieval is the root node in the hierarchy of key terms in this cluster, domain analysis underlines the conceptual foundation of this cluster, as we will see shortly.

Two outstanding references from the timeline visualization of this cluster have strong citation burstness (Fig. 3.54). One is a domain analysis of information science (White and McCain 1998), in which the multidimensional scaling of an author co-citation space was utilized to visualize the intellectual structure of the domain. The other is a study of major approaches to domain analysis—Hjørland 2002. In early 1990s, Hjørland developed a domain-analytic approach, also known as sociological-epistemological approach or a socio-cognitive view, as a methodological alternative to the then methodological individualism and cognitive perspective towards information science that largely marginalized the social, historical, and cultural roles in understanding a domain of scientific knowledge. Hjørland's another article published in 1997 on domain analysis is also a member of the cluster.

The sigma score of a cited reference reflects its structural and temporal significance. In addition to the author co-citation analysis of information science (White and McCain 1998), two more author co-citation studies are ranked highly by their sigma scores, namely an author co-citation study of information retrieval—Ding 1999, and an author co-citation study of hypertext—Chen 1999 (Fig. 3.55).

The review article by White and McCain (1997) on visualization of literatures is an important member of the cluster, whereas Tabah's (1999) review of the study of literature dynamics is a citing article to the cluster. Although the term domain

Freq	Burst	Centrality	Σ	PageRank	...	Author	Year	Tit.	Source	Vol	Page	HalfLife	Cluster
71	26.53	0.02	1.57	0.00	...	White HD	1998	...	J AM SOC I...	V49	P327	5	1
18	6.70	0.02	1.17	0.00	...	Ding Y	1999	...	J INF SCI	V25	P67	4	1
14	5.94	0.03	1.17	0.00	...	Noyons ECM	1999	...	J AM SOC I...	V50	P115	3	1
26	10.32	0.01	1.08	0.00	...	White HD	1997	...	ANNU REV ...	V32	P99	5	1
18		0.01	1.08	0.00	...	Chen CM	1999	...	INFORM P...	V35	P401	4	1
14	5.50	0.01	1.07	0.00	...	Lin X	1997	...	J AM SOC I...	V48	P40	5	1
12	4.76	0.01	1.07	0.00	...	Baldi S	1998	...	AM SOCIOL...	V63	P829	6	1
10	3.93	0.01	1.05	0.00	...	Luukkonen T	1997	...	SCIENTOM...	V38	P27	5	1
12	4.76	0.01	1.04	0.00	...	Oppenheim C	1997	...	J DOC	V53	P477	6	1
13		0.01	1.03	0.00	...	Barabasi AL	2002	...	PHYSICA A	V311	P590	7	1
11		0.01	1.03	0.00	...	Chen C	1999	...	INFORMATI...	V	P	4	1
9	3.57	0.01	1.03	0.00	...	Kostoff RN	1998	...	SCIENTOM...	V43	P27	4	1
7	3.32	0.01	1.03	0.00	...	Garfield E	1998	...	LIBRI	V48	P67	2	1
15	6.50	0.00	1.02	0.00	...	Albert R	2002	...	REV MOD P...	V74	P47	6	1
10	3.93	0.00	1.02	0.00	...	Small H	1997	...	SCIENTOM...	V38	P275	5	1
9	3.82	0.01	1.02	0.00	...	Noyons ECM	1998	...	J AM SOC I...	V49	P68	2	1
8		0.01	1.02	0.00	...	Chen CM	2001	...	J AM SOC I...	V52	P315	2	1
7	3.60	0.00	1.02	0.00	...	Borgman CL	1992	...	J AM SOC I...	V43	P397	6	1
10		0.00	1.01	0.00	...	Hjorland B	1997	...	INFORMATI...	V	P	6	1
8	3.73	0.00	1.01	0.00	...	Ingwersen P	1997	...	J AM SOC I...	V48	P205	4	1

Fig. 3.55 Key members of Cluster #1, sorted by sigma

1. 0.65 Wilson, CS (1999) informetrics
2. 0.23 Ingwersen, P (1999) cognitive information retrieval
3. 0.18 Ding, Y (2000) bibliometric information retrieval system (birs): a web search interface utilizing bibliometric research results
4. 0.15 Boyack, KW (2002) domain visualization using vxinsight (r) for science and technology management
5. 0.15 Ding, Y (2000) incorporating the results of co-word analyses to increase search variety for information retrieval
6. 0.14 Borner, K (2003) visualizing knowledge domains
7. 0.14 Boyack, KW (2002) information visualization, human-computer interaction, and cognitive psychology: domain visualizations
8. 0.13 Cointet, JP (2007) how realistic should knowledge diffusion models be?
9. 0.13 Tabah, AN (1999) literature dynamics: studies on growth, diffusion, and epidemics
10. 0.11 Roth, C (2007) empiricism for descriptive social network models
11. 0.11 Borgman, CL (2002) scholarly communication and bibliometrics
12. 0.1 Ellis, D (1999) information science and information systems: conjunct subjects disjunct disciplines
13. 0.1 Ingwersen, P (2000) applying diachronic citation analysis to research program evaluations
14. 0.1 Rowlands, I (1999) patterns of scholarly communication in information policy: a bibliometric study
15. 0.08 Chen, CM (2002) visualizing and tracking the growth of competing paradigms: two case studies
16. 0.08 Jarvelin, K (2000) a user-oriented interface for generalised informetric analysis based on applying advanced data modelling techniques
17. 0.08 White, HD (2001) authors as citers over time
18. 0.07 Li, X (2007) patent citation network in nanotechnology (1976-2004)
19. 0.07 Song, M (2000) visualization in information retrieval: a three-level analysis
20. 0.06 Cronin, B (2000) semiotics and evaluative bibliometrics

Fig. 3.56 Citing articles to Cluster #1

analysis was not used consistently during the period of this cluster, the contributions consistently focus on holistic views of a knowledge domain. As Hjørland argued, domain analysis serves a fundamental role in information science because its goal is to understand the subject matter from a holistic view of sociological, cognitive, historical, and epistemological dimensions.

Citing articles to Cluster #1 include some of the earliest attempts to integrate information visualization techniques to the methodology of a domain analysis— Börner 2003, Boyack 2002, Chen 2002 (Fig. 3.56). Interestingly, some of these citing articles appear as cited references in Cluster #0. In other words, the downturn of Cluster #1 does not mean that researchers lost their interest in the domain analysis approaches. Rather, they shifted their focus to explore a new generation of domain analysis with the support of a variety of computational and visualization techniques. As a result, the specialty underline Cluster #0 continues the vision conceived in the works of Cluster #1. The citers of Cluster #1 identify the group of researchers who would be the core members of the specialty of the new generation of domain analysis.

Author co-citation analysis (ACA) plays an instrumental role in the development of the domain analysis specialty embodied in Cluster #1. It is not only a bibliometric method that has been adopted by researchers beyond information science, but also a research instrument that helps to reveal challenges that the next generation of domain analysis must deal with.

In their 1998 ACA study of information science, White and McCain masterfully demonstrated the power and the potential of what one may learn from a holistic view of the intellectual landscape of a discipline. They utilized the multidimensional scaling technique as a vehicle for visualization and tapped into their encyclopedic knowledge of the information science discipline in an intellectually rich guided tour across the literature. In an attempt to enrich and enhance the conventional methodology of ACA, Chen (1999) introduced the Pathfinder network scaling technique. Using Pathfinder networks brings several advantages to the methodology of ACA, including the ability to identify and preserve salient structural patterns and algorithmically derived visual cues to assist the navigation and interpretation of resultant visualizations. White (2003) revisited the ACA study of information science with Pathfinder network scaling. A fast algorithm to compute Pathfinder networks is published in 2008 (Quirin et al. 2008).

The re-introduction of the network thinking opens up a wider variety of computational techniques to an ACA study, notably network modeling and visualization. Furthermore, technical advances resulted from the improvement of ACA have been applied to a broader range of bibliometric studies, notably document co-citation analysis (DCA). As we will see shortly, the adaptation of network modeling and information visualization techniques in general results from a Stage III specialty of information visualization and visual analytics.

Cluster #2—Research Evaluation

Cluster #2 is the third largest cluster with 190 cited references and a silhouette value of 0.845, which is slightly higher than the previous two larger clusters #0 and #1, suggesting a higher homogeneity. In other words, one would consider this specialty a more specialized than the previously identified specialties. This cluster is active over a 16-year period from 2000 till 2015. It represents an active specialty.

The overarching theme of the cluster is suggested by the two major branches shown in the hierarchy of key terms of this cluster: the *information visualization* branch and the much larger branch of *research evaluation* (Fig. 3.57). The information visualization branch highlights the recurring themes of intellectual structure and co-citation analysis. The research evaluation branch highlights numerous concepts that are central to measuring scholarly impact, notably h-index, bibliometric ranking, bibliometric indicator, sub-field normalization, web indicator, citation distribution, social media metrics, and alternative metrics.

The 6-year period from 2005 through 2010 is a highly active period of the cluster (Fig. 3.58). The most prominent contributions in this period include the

Fig. 3.57 A hierarchy of key concepts in Cluster #2

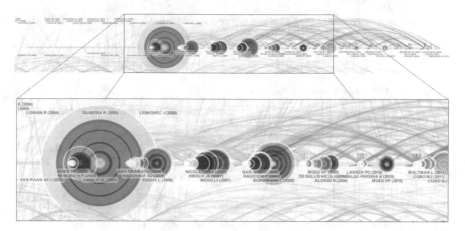

Fig. 3.58 High impact members of Cluster #2

original article that introduces the now widely known h-index (Hirsch 2005), the subsequent introduction of g-index as a refinement by taking citations into account (Egghe 2006), a 2007 study that compares the impact of using the Web of Science, Scopus, and Google Scholar on citation-based ranking—Meho 2007, a 2008 review entitled "What do citation counts measure?"—Bornmann 2008, and a study of the universality of citation distributions (Radicchi et al. 2008). These papers are also among the top sigma ranked members of this cluster because of their structural centrality as well as the strength of their citation burstness (Fig. 3.59).

The top 20 citing articles of the cluster reveal a considerable level of thematic consistency (Fig. 3.60). The overarching theme of research evaluation is evidently behind all these articles with popular title terms identified by latent semantic indexing such as citation impact, scientific impact, impact measures, bibliometric indicators, research evaluation, and web indicators.

Some of the more recent and highly cited members in Cluster #2 include a comparative study of 11 altmetrics and counterpart articles matched in the Web of Science (Thelwall et al. 2013) and the Leiden manifesto for research metrics (Hicks et al. 2015).

Freq	Burst	Centrality	Σ	PageRank	...	Author	Year	Tit.	Source	Vol	Page	HalfLife	Cluster
221	82.97	0.00	1.34	0.00	...	Hirsch JE	2005	...	P NATL AC...	V102	P16569	6	2
89	22.91	0.01	1.16	0.00	...	Bornmann L	2008	...	J DOC	V64	P45	5	2
68	17.76	0.01	1.13	0.00	...	Meho LI	2007	...	J AM SOC I...	V58	P2105	5	2
54	16.13	0.01	1.12	0.00	...	Radicchi F	2008	...	P NATL AC...	V105	P17268	5	2
17		0.01	1.12	0.00	...	Batista PD	2006	...	SCIENTOM...	V68	P179	4	2
49	12.52	0.01	1.11	0.00	...	Bar-ilan J	2008	...	SCIENTOM...	V74	P257	4	2
49	17.16	0.01	1.09	0.00	...	Moed HF	2010	...	J INFORME...	V4	P265	3	2
74	23.20	0.00	1.07	0.00	...	Egghe L	2006	...	SCIENTOM...	V69	P131	5	2
39	12.38	0.01	1.07	0.00	...	Nederhof AJ	2006	...	SCIENTOM...	V66	P81	6	2
48	12.54	0.01	1.07	0.00	...	Nicolaisen J	2007	...	ANNU REV...	V41	P609	5	2
50		0.00	1.05	0.00	...	Cobo MJ	2011	...	J AM SOC I...	V62	P1382	4	2
32		0.00	1.05	0.00	...	Moed HF	2009	...	ARCH IMM...	V57	P13	5	2
42	14.70	0.00	1.04	0.00	...	Alonso S	2009	...	J INFORME...	V3	P273	5	2
68	25.47	0.00	1.04	0.00	...	Moed H F	2005	...	CITATION A...	V	P	5	2
32	10.20	0.00	1.04	0.00	...	Harzing A W K	2008	...	ETHICS SC...	V8	P61	5	2
19		0.00	1.03	0.00	...	Kousha K	2007	...	J AM SOC I...	V58	P1055	2	2
48	12.52	0.00	1.03	0.00	...	Hirsch JE	2007	...	P NATL AC...	V104	P19193	6	2
10	4.34	0.01	1.02	0.00	...	Egghe L	2006	...	SCIENTOM...	V69	P121	7	2
30	11.30	0.00	1.02	0.00	...	Van Raan AFJ	2006	...	SCIENTOM...	V67	P491	5	2
10		0.00	1.02	0.00	...	Bauer K	2005	...	D LIB MAGA...	V11	P	3	2

Fig. 3.59 High impact members of Cluster #2

1. 0.16 Egghe, L (2010) the hirsch index and related impact measures
2. 0.12 Zhang, L (2011) the diffusion of h-related literature
3. 0.12 Franceschet, M (2010) a comparison of bibliometric indicators for computer science scholars and journals on web of science and google scholar
4. 0.12 Haddow, G (2010) citation analysis and peer ranking of australian social science journals
5. 0.12 Lacasse, JR (2011) evaluating the productivity of social work scholars using the h-index
6. 0.1 Wu, H (2010) scientific impact at the topic level: a case study in computational linguistics
7. 0.09 Bornmann, L (2011) a multilevel modelling approach to investigating the predictive validity of editorial decisions: do the editors of a high profile journal select manuscripts that are highly cited after publication?
8. 0.09 Mas-Bleda, A (2014) do highly cited researchers successfully use the social web?
9. 0.09 Thelwall, M (2015) web indicators for research evaluation. part 2: social media metrics
10. 0.09 Waltman, L (2016) a review of the literature on citation impact indicators
11. 0.08 Thelwall, M (2016) are there too many uncited articles? zero inflated variants of the discretised lognormal and hooked power law distributions
12. 0.08 Thelwall, M (2016) the precision of the arithmetic mean, geometric mean and percentiles for citation data: an experimental simulation modelling approach
13. 0.07 Boell, SK (2010) journal impact factors for evaluating scientific performance: use of h-like indicators
14. 0.07 Mingers, J (2015) a review of theory and practice in scientometrics
15. 0.07 Saad, G (2010) applying the h-index in exploring bibliometric properties of elite marketing scholars
16. 0.07 Serenko, A (2010) the development of an ai journal ranking based on the revealed preference approach
17. 0.07 Thelwall, M (2015) web indicators for research evaluation. part 1: citations and links to academic articles from the web
18. 0.07 Thelwall, M (2016) interpreting correlations between citation counts and other indicators
19. 0.07 Thelwall, M (2016) the discretised lognormal and hooked power law distributions for complete citation data: best options for modelling and regression
20. 0.06 Bornmann, L (2010) from black box to white box at open access journals: predictive validity of manuscript reviewing and editorial decisions at atmospheric chemistry and physics

Fig. 3.60 Citing articles of Cluster #2

Cluster #3—Information Visualization and Visual Analytics

Cluster #3 is the fourth largest cluster. Its duration ranges from 2004 through 2014. The topic hierarchy has two branches: information visualization and heart rate variability (Fig. 3.61). The heart rate variability does not belong to the domain analysis in the context of information science. In fact, its inclusion in the original results of the topic search was due to the ambiguity of the term domain analysis across multiple disciplines. Pragmatically it is easier and more efficient to simply skip an irrelevant branch than keep refining the original topic search query until all noticeable irrelevant topics are eliminated. This is one of the fundamental challenges for information retrieval and this is where domain analysis has an instrumental role to play (Hjørland 2002).

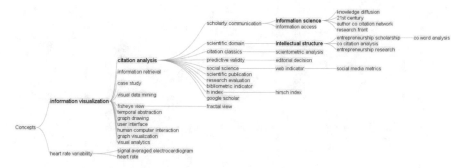

Fig. 3.61 A hierarchy of key concepts in Cluster #3

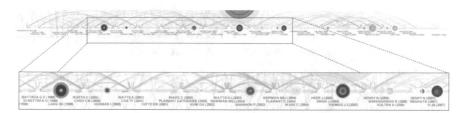

Fig. 3.62 High impact members of Cluster #3

The information visualization branch includes a mixture of information visualization techniques such as fisheye view, group drawing, graph visualization, and visual analytics and topics that are center to information science such as citation analysis, information retrieval. The mixture is a sign of attempts to apply information visualization and visual analytic techniques to bibliometric approaches to the study of intellectual structure of a research domain. The vision of information visualization is to identify insightful patterns from abstract information (Card et al. 1999). The subsequently emerged visual analytics emphasizes the critical and more specific role of sense-making and analytic reasoning in accomplishing such goals (Thomas and Cook 2005) (See Fig. 3.62).

High-impact contributions in Cluster #3 include the collection of seminal works in information visualization—Card 1999, a survey of graph visualization techniques —Herman 2000, Cytoscape—a widely used software tool for visualizing biomolecular interaction networks—Shannon 2003, the ground breaking work of visual analytics (Thomas and Cook 2005), Many Eyes—the popular web-based visualization platform—Viégas 2007, and a framework of seven types of interaction techniques in information visualization—Yi 2007 (Fig. 3.63).

In addition to the above high-impact contributions, this cluster features information visualization tools such as the InfoVis toolkit—Fekete 2004, NodeTrix— Henry 2007, Jigsaw—a visual analytic tool—Stasko 2008, and D3—Bostock 2011. The most widely used information visualization tools such as Many Eyes and D3 became available between 2007 and 2011. Figure 3.64 shows a list of citing articles of Cluster #3.

Freq	Burst	Centrality	Σ	PageRank	...	Author	Year	Tit.	Source	Vol	Page	HalfLife	Cluster
51	14.89	0.03	1.45	0.00	...	Thomas J J	2005	...	ILLUMINATI...	V	P	4	3
46		0.01	1.18	0.00	...	Bostock M	2011	...	IEEE T VIS ...	V17	P2301	4	3
46	18.80	0.01	1.14	0.00	...	Shannon P	2003	...	GENOME R...	V13	P2498	6	3
48	19.93	0.01	1.12	0.00	...	Card SK	1999	...	READINGS ...	V	P	4	3
19	7.26	0.01	1.07	0.00	...	Tufte ER	2001	...	VISUAL DIS...	V	P	5	3
40		0.00	1.07	0.00	...	Ware C	2004	...	INFORM VI...	V	P	6	3
31		0.00	1.07	0.00	...	Lam H	2012	...	IEEE T VIS ...	V18	P1520	2	3
44	11.47	0.00	1.05	0.00	...	Yi JS	2007	...	IEEE T VIS ...	V13	P1224	5	3
9		0.01	1.05	0.00	...	Henry N	2006	...	IEEE T VIS ...	V12	P677	2	3
31	11.71	0.00	1.04	0.00	...	Herman I	2000	...	IEEE T VIS ...	V6	P24	6	3
19	7.80	0.01	1.04	0.00	...	Henry N	2007	...	IEEE T VIS ...	V13	P1302	5	3
10		0.01	1.04	0.00	...	Watts DJ	1998	...	NATURE	V393	P440	7	3
17		0.01	1.03	0.00	...	Shneiderman B	2006	...	IEEE T VIS ...	V13	P733	3	3
12		0.00	1.03	0.00	...	Fekete J	2004	...	P IEEE S IN...	V,	P	5	3
14		0.00	1.02	0.00	...	Plaisant C	2004	...	P WORK C ...	V,	P	5	3
24	10.30	0.00	1.01	0.00	...	Keim DA	2002	...	IEEE T VIS ...	V8	P1	6	3
19	8.26	0.00	1.01	0.00	...	Singh J	2005	...	MANAGE SCI	V51	P756	7	3
9	3.33	0.00	1.01	0.00	...	Stasko J	2008	...	INFORM VI...	V7	P118	6	3
14		0.00	1.01	0.00	...	Keim D A	2010	...	MASTERIN...	V	P	4	3
11		0.00	1.01	0.00	...	Liu ZC	2010	...	IEEE T VIS ...	V16	P999	6	3

Fig. 3.63 Key members of Cluster #3

1. 0.14 Liu, SX (2014) a survey on information visualization: recent advances and challenges
2. 0.14 Morone, P (2004) knowledge diffusion dynamics and network properties of face-to-face interactions
3. 0.14 Morone, P (2004) small world dynamics and the process of knowledge diffusion: the case of the metropolitan area of greater santiago de chile
4. 0.12 Blanch, R (2007) browsing zoomable treemaps: structure-aware multi-scale navigation techniques
5. 0.1 Sedlmair, M (2014) visual parameter space analysis: a conceptual framework
6. 0.08 Cowan, R (2004) network structure and the diffusion of knowledge
7. 0.08 Gleicher, M (2011) visual comparison for information visualization
8. 0.08 von Landesberger, T (2011) visual analysis of large graphs: state-of-the-art and future research challenges
9. 0.07 Elmqvist, N (2010) hierarchical aggregation for information visualization: overview, techniques, and design guidelines
10. 0.07 Elmqvist, N (2011) fluid interaction for information visualization
11. 0.07 Lee, B (2006) treeplus: interactive exploration of networks with enhanced tree layouts
12. 0.05 Gelernter, J (2007) visual classification with information visualization (infoviz) for digital library collections
13. 0.05 Heer, J (2007) animated transitions in statistical data graphics
14. 0.05 Henry, N (2007) nodetrix: a hybrid visualization of social networks
15. 0.05 Hornbaek, K (2007) untangling the usability of fisheye menus
16. 0.05 Hornbaek, K (2011) the notion of overview in information visualization
17. 0.05 Liu, ZC (2014) ploceus: modeling, visualizing, and analyzing tabular data as networks
18. 0.05 Wong, PC (2006) generating graphs for visual analytics through interactive sketching
19. 0.04 Brandes, U (2011) asymmetric relations in longitudinal social networks
20. 0.04 Chen, M (2014) visual multiplexing

Fig. 3.64 Citing articles of Cluster #3

According to Shneider's four stage model, the information visualization and visual analytics specialty in the context of domain analysis and literature visualization has demonstrated properties of a Stage IV specialty. For example, in the most recent few years of the cluster, researchers reflect on empirical evaluations of information visualization in various scenarios—Lam 2012, revisit taxonomic organizations of abstract visualization tasks—Brehmer 2013, and synthesize and codify domain knowledge in the forms of textbooks—Munzner 2014.

Trajectories of Citations Across Cluster Boundaries

Cluster analysis helps us to understand the major specialties associated with science mapping. Now we turn our attention to the trajectories of several leading contributors in the landscape of these clusters. We are interested in what we may learn from citation links made in publications of a scholar, especially those links bridging distinct clusters.

Trajectories of Prolific Authors

The first example is the citation trajectory of Howard White (Fig. 3.65 left). He is the author of several seminal papers featured in several clusters. His citation trajectories move across the citation landscape from the left to the center, ranging from #4 decision support system (applications of ACA), #1 domain visualization (domain analysis), and #8 social work (another cluster of bibliometric studies).

The second example is the citation trajectory of Mike Thelwall (Fig. 3.65 right). He is a prolific researcher who contributed to webometrics and altmetrics among other areas of bibliometrics. An overlay of his citation trajectories on a citation landscape view shows that his trajectories spanning clusters such as #6 university websites (webometrics) and # google scholar (research evaluation).

In both examples of citation trajectories, we have observed that their citation trajectories span across a wide area over the citation landscape. Monitoring the movement of citation trajectories in such a way provides an intuitive insight into the evolution of the underlying specialties and the context in which high-impact researchers make their contributions.

Articles with Transformative Potentials

It is widely known that a major limitation of any citation-based indicators is their reliance on citations accumulated over time. Thus, citation-based indicators are likely to overlook newly published articles. An alternative method is to focus on the extent to which a newly published article brings to the conceptual structure of the knowledge domain of interest (Chen 2012). The idea is to identify the potential of an article to make extraordinary or unexpected connections across distinct clusters.

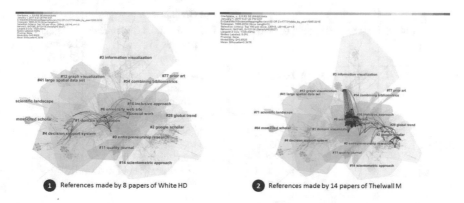

Fig. 3.65 Novel co-citations made by 8 papers of White HD (left) and by 14 papers of Thelwall M (right)

According to theories of scientific discovery, many significant contributions are resulted from boundary spanning ideas.

Table 3.11 lists three articles each year for the last five years. These articles have the highest geometric mean of three structural variation variables generated by CiteSpace. For example, in 2016, the highest score goes to the review of citation impact indicators—Waltman 2016, followed by two bibliometric analyses—one contrasts two closely related but distinct domains and the other studies the research over a 20-year span. In 2015, two bibliometric studies followed by a review of theory and practice in scientometrics (Mingers and Leydesdorff 2015).

Table 3.11 Potentially transformative papers published in recent years (2012–2016)

Year	ΔM	ΔCL_w	C_{KL}	Geometric mean	GC	Title	References
2016	6.0541	0.0152	0.0251	0.1322	5	A review of the literature on citation impact indicators	Waltman (2016)
2016	0.9235	0.0019	0.3407	0.0842	0	How are they different? A quantitative domain comparison of information visualization and data visualization (2000–2014)	Kim et al. (2016)
2016	0.8207	0.0017	0.0640	0.0447	2	A bibliometric analysis of 20 years of research on software product lines	Heradio et al. (2016)
2015	1.7498	0.0073	0.0380	0.0786	0	Global ontology research progress: A bibliometric analysis	Zhu et al. (2015)
2015	1.9873	0.0052	0.0397	0.0743	9	Bibliometric Methods in Management and Organization	Zupic (2015)
2015	1.9906	0.0029	0.0238	0.0516	13	A review of theory and practice in scientometrics	Mingers and Leydesdorff (2015)
2014	1.6240	0.0087	0.0434	0.0850	3	Research dynamics: Measuring the continuity and popularity of research topics	Yan (2014)
2014	1.1837	0.0031	0.0463	0.0554	1	Making a Mark: A computational and visual analysis of one researcher's intellectual domain	Skupin (2014)

(continued)

Table 3.11 (continued)

Year	ΔM	ΔCL$_w$	C$_{KL}$	Geometric mean	GC	Title	References
2014	0.4462	0.0024	0.0270	0.0307	12	The Knowledge Base and Research Front of Information Science 2006–2010: An Author Co citation and Bibliographic Coupling Analysis	Zhao and Strotmann (2014)
2013	2.5398	0.0112	0.0643	0.1223	13	Analysis of bibliometric indicators for individual scholars in a large data set	Radicchi and Castellano (2013)
2013	1.0781	0.0065	0.2180	0.1152	6	A visual analytic study of retracted articles in scientific literature	Chen et al. (2013)
2013	1.7978	0.0064	0.0542	0.0854	24	Quantitative evaluation of alternative field normalization procedures	Li et al. (2013)
2012	3.6274	0.0107	0.0811	0.1466	29	SciMAT: A new science mapping analysis software tool	Cobo et al. (2011)
2012	3.4380	0.0248	0.0259	0.1302	15	A forward diversity index	Carley and Porter (2012)
2012	1.0719	0.0032	0.0321	0.0479	11	Visualizing and mapping the intellectual structure of information retrieval	Rorissa and Yuan (2012)

These highly ranked articles represent a few types of studies that may serve as predictive indicators, namely review papers (Mingers and Leydesdorff 2015; Waltman 2016), applications of bibliometric studies to specific domains, software tools for science mapping (Cobo et al. 2011), new metrics and indicators (Li et al. 2013), and visual analytic studies of unconventional topics—retractions (Chen et al. 2013). Figure 3.66 shows the trajectories of three articles with high modularity change rates.

The Emergence of a Specialty

The emergence of a specialty is determined by two factors: the intellectual base and the research fronts associated with the intellectual base. The intellectual base is what the specialty cites, whereas the research fronts are what the specialty is

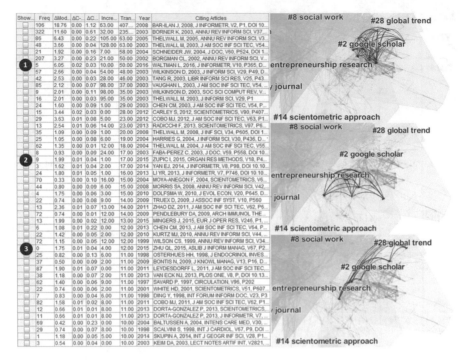

Fig. 3.66 Three examples of articles with high modularity change rates: (1) Waltman (2016), (2) Zupic (2015), and (3) Zhu et al. (2015)

currently addressing. As we have seen, on the one hand, a research front may remain in the same co-citation cluster as in the case of Cluster #2 Research Evaluation. On the other hand, a research front may belong to a different specialty and become the intellectual base of a new specialty as in the case of Cluster # Domain Analysis and Cluster #0 Bibliometric Mapping.

The citation trajectories of a researcher's publications and the positions of these publications as cited references can be simultaneously shown by overlaying trajectories (dashed lines for novel links or solid lines for existing links) and citing papers as stars if they also appear in a co-citation cluster as cited references. For example, the series of stars in the visualization shown in Fig. 3.67 tell us two things: First, the author is connecting topics in two clusters (Cluster #0 Science Mapping and Cluster 2 Research Evaluation) and second, the author belongs to the specialty of science mapping.

The example in Fig. 3.68 illustrates the citation trajectories of Howard White's publications and their own positions in the timelines of clusters. His publications appear in the early stage of the science mapping cluster (#0) and make novel connections between science mapping and domain analysis (Cluster #1), domain analysis (Cluster #1) and applications of ACA (Cluster #4), domain analysis (Cluster #1) and webometrics (Cluster #6).

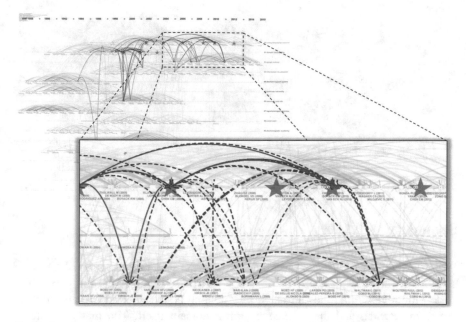

Fig. 3.67 Stars indicate articles that are both cited and citing articles. Dashed lines indicate novel co-citation links. Illustrated based on 15 papers of the author's own publications

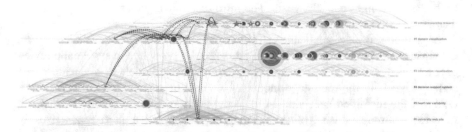

Fig. 3.68 Citation trajectories of Howard White's publications and their own locations

The next example in Fig. 3.69 depicts the novel co-citation links made by a review paper of informetrics (Bar-Ilan 2008). These novel links include within-cluster links as well as between-cluster links. It should be easy to tell that the scope of the review is essentially limited to research papers published about 6–7 years prior to the time of the review. Furthermore, we can see that the review systematically emphasizes the diversity of topics instead of tracing to the origin of any particular specialty.

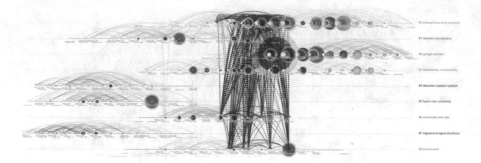

Fig. 3.69 Novel links made by a review paper of informetrics (Bar-Ilan 2008)

Summary

We present three examples of visually exploring the scientific literature of a field of study. Our intention is twofold. First, our goal is to demonstrate the depth of a systematic review that one can reach by applying a science mapping approach to terrorism research and the science mapping domain itself. The first example of terrorism research is based on publications between 1996 and 2003. The second example of terrorism research is based on a much longer timespan between 1980 and 2017, with particular interests in how the visual analytic approach is sensitive to the latent changes over the years. The third example is the science mapping field itself.

In addition to the application of computational functions available in the CiteSpace software, we also enrich the procedure of producing a systematic review of a knowledge domain by incorporating evolutionary models of a scientific specialty—especially the four-stage model of a scientific discipline into the interpretation of the identified specialties. Our interpretation not only identifies thematic milestones of major streams of science mapping research, but also characterizes the developmental stages of the underlying specialties and the dynamics of transitions from one specialty to another.

Second, our goal is to provide a reliable historiographic survey of the science mapping research. The survey identifies the major clusters in terms of their high-impact members and citing articles that form new research fronts. We also demonstrate new insights that one can intuitively obtain through an inspection of citation trajectories and the positions of citing papers. The enhanced science mapping procedure introduced in this article is applicable to the analysis of other domains of interest. Researchers can utilize these visual analytic tools to perform timely surveys of the literature as frequently as they wish and find relevant publications more effectively.

References

Abt HA (1998) Why some papers have long citation lifetimes. Nature 395:756–757
Bar-Ilan J (2008) Informetrics at the beginning of the 21st century—a review. J Informetrics 2 (1):1–52
Batagelj V (2003) Efficient algorithms for citation network analysis. arXiv:cs/0309023v1
Batagelj V, Mrvar A (1998) Pajek—program for large network analysis. Connections 21(2):47–57
Boyack KW, Klavans R (2010) Co-citation analysis, bibliographic coupling, and direct citation: which citation approach represents the research front most accurately? J Am Soc Inform Sci Technol 61(12):2389–2404. doi:10.1002/asi.21419
Boyack KW, Klavans R, Börner K (2005) Mapping the backbone of science. Scientometrics 64 (3):351–374
Burt RS (1992) Structural holes: the social structure of competition. Harvard University Press, Cambridge, MA
Bush V (1945) As we may think. Atlantic Monthly 176(1):101–108
Callon M, Courtial JP, Turner WA, Bauin S (1983) From translations to problematic networks—an introduction to co-word analysis. Soc Sci Inf Sci Soc 22(2):191–235
Card S, Mackinlay DJ, Shneiderman B (1999) Readings in information visualization: using vision to think. Interactive technologies. Morgan Kaufmann Publisher, San Francisco, CA USA
Carley KM, Hummon NP, Harty M (1993) Scientific influence—an analysis of the main path structure in the journal of conflict-resolution. Knowledge-Creation Diffus Utilization 14 (4):417–447
Carley S, Porter AL (2012) A forward diversity index. SCIENTOMETRICS 90(2):407–427. doi:10.1007/s11192-011-0528-1
Chen C (2004) Searching for intellectual turning points: progressive knowledge domain visualization. Proc Natl Acad Sci U S A 101 (Suppl):5303–5310
Chen C (1999) Visualising semantic spaces and author co-citation networks in digital libraries. Inf Process Manage 35(2):401–420
Chen C (2006) CiteSpace II: detecting and visualizing emerging trends and transient patterns in scientific literature. J Am Soc Inform Sci Technol 57(3):359–377
Chen C (2012) Predictive effects of structural variation on citation counts. J Am Soc Inform Sci Technol 63(3):431–449. doi:10.1002/asi.21694
Chen C (2014) The fitness of information: quantitative assessments of critical evidence. Wiley
Chen C (2017) Science mapping: a systematic review of the literature. J Data Inform Sci 2(2):1–40
Chen C, Leydesdorff L (2014) Patterns of connections and movements in dual-map overlays: a new method of publication portfolio analysis. J Assoc Inform Sci Technol 65(2):334–351
Chen C, Chen Y, Horowitz M, Hou H, Liu Z, Pellegrino D (2009) Towards an explanatory and computational theory of scientific discovery. J Informetrics 3(3):191–209
Chen C, Ibekwe-SanJuan F, Hou J (2010) The structure and dynamics of co-citation clusters: a multiple-perspective co-citation analysis. J Am Soc Inform Sci Technol 61(7):1386–1409
Chen C, Hu Z, Liu S, Tseng H (2012) Emerging trends in regenerative medicine: a scientometric analysis in CiteSpace. Exp Opin Biol Ther 12(5):593–608
Chen C, Hu Z, Milbank J, Schultz T (2013) A visual analytic study of retracted articles in scientific literature. Journal of the American Society for Information Science and Technology 64 (2):234–253
Chen C, Dubin R, Kim MC (2014a) Emerging trends and new developments in regenerative medicine: a scientometric update (2000–2014). Expert Opin Biol Ther 14(9):1295–1317
Chen C, Dubin R, Kim MC (2014b) Orphan drugs and rare diseases: a scientometric review (2000–2014). Exp Opin Orphan Drugs 2(7):709–724
Cobo MJ, Lopez-Herrera AG, Herrera-Viedma E, Herrera F (2011) Science mapping software tools: review, analysis, and cooperative study among tools. J Am Soc Inform Sci Technol 62 (7):1382–1402. doi:10.1002/asi.21525

Deerwester S, Dumais ST, Landauer TK, Furnas GW, Harshman RA (1990) Indexing by latent semantic analysis. J Am Soc Inform Sci 41(6):391–407

Egghe L (2006) Theory and practise of the g-index. Scientometrics 69(1):131–152. doi:10.1007/s11192-006-0144-7

Garfield E (1955) Citation indexes for science: a new dimension in documentation through association of ideas. Science 122(3159):108–111

Heer J (2007) The prefuse visualization toolkit. http://prefuse.org/

Heradio R, Perez-Moragoa H, Fernandez-Amorosa D, Cabrerizoa FJ, Herrera-Viedmab E (2016) A bibliometric analysis of 20 years of research on software product lines. Information and Software Technology 72:1–15. doi:10.1016/j.infsof.2015.11.004

Herman I, Melançon G, Marshall MS (2000) Graph visualization and navigation in information visualization: a survey. IEEE Trans Visual Comput Graph 6(1):24–44

Hicks D, Wouters P, Waltman L, Rijcke Sd, Rafols I (2015) Bibliometrics: the Leiden Manifesto for research metrics. Nature 520(7548):429–431. doi:10.1038/520429a

Hirsch JE (2005) An index to quantify an individual's scientific research output. Proc Natl Acad Sci U S A 102(46):16569–16572. doi:10.1073/pnas.0507655102

Hjørland B (2002) Epistemology and the socio-cognitive perspective in information science. J Am Soc Inform Sci Technol 53(4):257–270

Hummon NP, Doreian P (1989) Connectivity in a citation network—the Development of DNA Theory. Soc Netw 11(1):39–63

Johnson B, Shneiderman B (1991) Tree-maps: a space filling approach to the visualization of hierarchical information structures. IEEE Vis 91(Oct 1991):284–291

Kim MC, Zhu Y, Chen C (2016) How are they different? A quantitative domain comparison of information visualization and data visualization (2000-2014). SCIENTOMETRICS 107:123

Kleinberg J (2002) Bursty and hierarchical structure in streams. In: Proceedings of the 8th ACM SIGKDD international conference on knowledge discovery and data mining, pp 91–101

Leydesdorff L, Rafols I (2009) A global map of science based on the ISI subject categories. J Am Soc Inform Sci Technol 60(2):348–362

Li Y, Radicchi F, Castellano C, Ruiz-Castillo J (2013) Quantitative evaluation of alternative field normalization procedures. J Informetrics 7(3):746–755

Liu JS, Kuan C-H (2015) A new approach for main path analysis: Decay in knowledge diffusion. J Assoc Inform Sci Technol. doi:10.1002/asi.23384

Liu JS, Lu LYY (2012) An integrated approach for main path analysis: Development of the Hirsch Index as an example. J Am Soc Inform Sci Technol 63(3):528–542

Lucio-Arias D, Leydesdorff L (2008) Main-path analysis and path-dependent transitions in HistCite-based historiograms. J Am Soc Inform Sci Technol 59(12):1948–1962

Mingers J, Leydesdorff L (2015) A review of theory and practice in scientometrics. Eur J Oper Res 246(1):1–19

Morris SA, Yen G, Wu Z, Asnake B (2003) Timeline visualization of research fronts. J Am Soc Inform Sci Technol 55(5):413–422

Price DD (1965) Networks of scientific papers. Science 149:510–515

Quirin A, Cordon O, Santamaria J, Vargas-Quesada B, Moya-Anegon F (2008) A new variant of the Pathfinder algorithm to generate large visual science maps in cubic time. Inform Process Manag 44(4):1611–1623. doi:10.1016/j.ipm.2007.09.005

Radicchi F, Castellano C (2013) Analysis of bibliometric indicators for individual scholars in a large data set. SCIENTOMETRICS 97(3):627–637. doi:10.1007/s11192-013-1027-3

Radicchi F, Fortunato S, Castellano C (2008) Universality of citation distributions: toward an objective measure of scientific impact. PNAS 105(45):17268–17272

Rorissa A, Yuan X (2012) Visualizing and mapping the intellectual structure of information retrieval. Information Processing & Management 48(1):120–135. doi:10.1016/j.ipm.2011.03.004

Schvaneveldt RW (ed) (1990) Pathfinder associative networks: studies in knowledge organization. Ablex series in computational sciences. Ablex Publishing Corporations, Norwood, New Jersey

Shneider AM (2009) Four stages of a scientific discipline: four types of scientists. Trends Biochem Sci 34(5):217–223

Shneiderman B (1996) The eyes have it: a task by data type taxonomy for information visualization. In: IEEE workshop on visual language, Boulder, CO, Sept 3–6, 1996. IEEE Computer Society Press, pp 336–343

Skupin A (2014) Making a Mark: a computational and visual analysis of one researcher's intellectual domain. International Journal of Geographical Information Science 28(6):1209–1232. doi:10.1080/13658816.2014.906040

Smalheiser NR, Swanson DR (1998) Using ARROWSMITH: a computer-assisted approach to formulating and assessing scientific hypotheses. Comput Meth Programs Biomed 57(3):149–153

Small H (1973) Co-citation in the scientific literature: a new measure of the relationship between two documents. J Am Soc Inform Sci 24:265–269

Tabah AN (1999) Literature dynamics: studies on growth, diffusion, and epidemics. Annu Rev Inform Sci Technol 34:249–286

Thelwall M, Haustein S, Larivière V, Sugimoto CR (2013) Do altmetrics work? Twitter and ten other social web services. PLoS ONE 8(5):e64841

Thomas JJ, Cook AK (2005) Illuminating the path: the research and development agenda for visual analytics. IEEE Press

Tibély G, Pollner P, Vicsek T, Palla G (2013) Extracting tag-hierarchies. PLoS ONE 8(12):e84133

Van Eck NJ, Waltman L (2010) Software survey: VOSviewer, a computer program for bibliometric mapping. Scientometrics 84(2):523–538

Van Eck NJ, Waltman L (2014) CitNetExplorer: a new software tool for analyzing and visualizing citation networks. J Informetrics 8(4):802–823

van Raan A (2000) On growth, ageing, and fractal differentiation of science. Scientometrics 47(2):347–362

Waltman L (2016) A review of the literature on citation impact indicators. J Informetr 10(2):365–391. doi:10.1016/j.joi.2016.02.007

Watts DJ, Strogatz SH (1998) Collective dynamics of 'small-world' networks. Nature 393(6684):440–442

White HD, McCain KW (1997) Visualization of Literatures. Annu Rev Inform Sci Technol 32:99–168

White HD, McCain KW (1998) Visualizing a discipline: an author co-citation analysis of information science, 1972–1995. J Am Soc Inform Sci 49(4):327–356

Yan E (2014) Research dynamics: Measuring the continuity and popularity of research topics. Journal of Informetrics 8(1):98–110. doi:10.1016/j.joi.2013.10.010

Zhu Q, Kong X, Hong S, Li J, He Z (2015) Global ontology research progress: a bibliometric analysis. Aslib Journal of Information Management 67(1):27–54. doi:10.1108/AJIM-05-2014-0061

Zupic I (2015) Bibliometric methods in management and organization. Organizational Research Methods 18(3):429–472

Chapter 4
Measuring Scholarly Impact

Abstract The ability to measure scholarly impact, ranging from individual scientists to an institution of researchers, is crucial to both research assessment and the advance of science itself. In this chapter, we summarize an array of fundamental and widely used concepts and computational methods for measuring scholarly impact as well as identifying more generic properties such as semantic relatedness, burstness, clumping, and centrality. Most of these common ideas are applicable to a wide variety of needs as long as we can identify the profound issues that are in common across distinct phenomena. Normalizations of metrics across scientific fields and the year of publication are discussed with concrete examples.

Introduction

Quantitative measures of scholarly impact are rooted in the measurement of information, uncertainty, proximity, novelty, rarity, connectivity, and other numerous indicators of significance. Some of these indicators and domain independent, whereas others are domain specific (Piffer 2012, Shwed and Bearman 2010).

The pragmatic question to many of these diverse metrics is whether and to what extent we may learn something useful or something new from the input or signals we receive, including text and other types of messages. The value of information is that it brings changes to our knowledge or our belief. This property can be seen as the fitness of information (Chen 2014). Information entropy (Shannon 1948) can be seen as a measure of the potential of what we may learn. Equivalently, it can be seen as a measure of the amount of uncertainty that can be resolved. For example, a dialogue between a physician and a patient reduces the initial entropy as various uncertainties are progressively narrowed down. An assumption that has been commonly seen in the reasoning of many information metrics is that we are more likely to learn something from a relatively rare event or word than from a common one. We expect to find creative ideas in areas that have not been well studied. We expect that boundary spanning may inspire extraordinary ideas.

© Springer International Publishing AG 2017

C. Chen and M. Song, *Representing Scientific Knowledge*,
https://doi.org/10.1007/978-3-319-62543-0_4

Another strategy is to measure the importance or saliency of something by comparing it to a baseline. The strategy has been used in novelty detection, intrusion detection, burst detection, measuring rarity, and identifying outliers. The importance can be also measured in terms of connectivity, such as degree centrality, betweenness centrality, or eigenvector centrality.

Semantic similarities are often measured with reference to an existing ontological structure or a taxonomy. Ontology-based semantic similarity measures include path-based such as Wu and Palmer (1994), information-content-based such as Resnik (1995), feature-based such as Tversky (1977), and other types. WordNet is one of the most popular resources of choice in defining semantic similarity measures.

The ultimate utility of an indicator is to make easy and simple comparisons. Normalization is essential when we need to assure different measurements are comparable. The examples included in this chapter are representative and influential because they are designed based on some of the most fundamental principles that have been used in the design of a wide variety of indicators.

Information Metrics

Information Content

The concept of information content (IC) is used in a wide variety of many information metrics as well as on its own. More importantly, the principles behind the quantitative measure are applicable to a broad range of scenarios. The idea is to measure how much we can learn from a source of information. When we receive a message, the message may tell us nothing that we don't already know. On the other hand, a message may turn what we believe or what we think we know upside down!

Given a transmitted message m of information, its information content IC(m) is defined as the negative of the log likelihood of the message.

$$\mathbf{IC(m)} = -\mathbf{log_2}\, p(\mathbf{m})$$

As shown in Fig. 4.1, as the probability of an event increases, the value of IC decreases. In particular, the IC value is the lowest for very common events, whereas the IC values are larger for rare events.

Shannon entropy quantifies the information in a message as something that would be new to the recipient of the message. If a message brings nothing new to the recipient, then the message does not carry any information as far as the recipient is concerned. Shannon entropy, or information entropy, is defined in terms of information content across all the possible events of a random variable X:

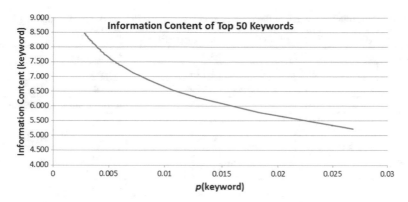

Fig. 4.1 Information content of top 50 most common keywords in 17,731 science mapping articles

$$H(X) = -\sum_{i=1}^{n} p(x_i)\log(p(x_i)) = \sum_{i=1}^{n} p(x_i)IC(x_i)$$

The value of the term $p(x_i) \times IC(x_i)$ amplifies the small probability of a rare event with a large IC value but suppresses the large probability of a common event with a small IC value.

In general, we expect to learn a lot from rare events than a common event. Since we probably haven't experienced a rare event, it is likely that information associated with the rare event is new to our cognitive or belief system. Measuring interestingness or a degree of surprise often adopts similar principles.

Consider a dataset of science mapping publications we used in a systematic review (Chen 2017). The dataset contains 17,731 publications. These publications are indexed by 56,159 distinct keywords. From the relative frequency of a keyword, the information content of the keyword with respect to this particular dataset is calculated as $-log_2(f_k/f_N)$. We can use the following MySQL query to generate frequencies, relative frequencies, and information content of top 50 keywords to illustrate the concept of IC.

```
SELECT count(*), count(*)/56159, -log2(count(*)/56159), keyword
FROM keywords
WHERE project='sciencemapping17731' AND type!='sc'
GROUP BY keyword
ORDER BY count(*) DESC
LIMIT 50;
```

Table 4.1 list top 10 most common keywords. Keywords in this group have the lowest IC values because they occurred most frequently. Indeed, in the context of science mapping, keywords such as science, model, system, and impact do not tell us anything new, in part because they are field-independent words and in part they are almost applicable to any science mapping articles. Although keywords such as

Table 4.1 Information content of the most common keywords in a set of science mapping articles

Frequency (F)	Relative F (RF)	IC-log2(RF)	Keyword
1506	0.0268	5.221	Citation analysis
1026	0.0183	5.774	Science
724	0.0129	6.277	Model
716	0.0127	6.293	Information visualization
714	0.0127	6.297	System
603	0.0107	6.541	Time-domain analysis
477	0.0085	6.879	Impact
475	0.0085	6.885	Network
471	0.0084	6.898	Bibliometrics
425	0.0076	7.046	Journal

Table 4.2 Information contents of low-frequency keywords

Frequency	Relative frequency	Information content	Keyword
10	0.0002	12.4553	Latent semantic analysis
10	0.0002	12.4553	Explanation
10	0.0002	12.4553	Health policy
10	0.0002	12.4553	Randomized controlled trial
10	0.0002	12.4553	Nonlinear-system
5	0.0001	13.4553	Citation classic
5	0.0001	13.4553	Cross-section
5	0.0001	13.4553	Circuit modeling
5	0.0001	13.4553	Semantic network
5	0.0001	13.4553	Xylanase
1	0.0000	15.7773	Dysplastic nevus
1	0.0000	15.7773	Saturation time
1	0.0000	15.7773	Fiber-optics sensor
1	0.0000	15.7773	Ale metaanalysis
1	0.0000	15.7773	Terrorist

citation analysis and information visualization are field-dependent, their frequent occurrences serve little more than reinforce what we already know.

Table 4.2, generated by the MySQL query below, illustrates the information content scores of low-frequency keywords. In this dataset, keywords appear 10 times have a relevant frequency of 0.0002 and the information content of 12.4553. The ICs of keywords appeared for 5 times have even higher ICs of 13.4553. The ICs of 15.7773 are the highest possible for this particular dataset for keywords that appeared only once. The highest possible value depends on the total number of distinct keywords in the set.

```
SELECT *
FROM (
  SELECT
    count(*) AS c1,
    count(*)/56159 AS c2,
    -log2(count(*)/56159) AS c3,
    keyword AS c4
  FROM keywords
  WHERE project='sciencemapping17731' AND type!='sc'
  GROUP BY keyword
  ORDER BY count(*)
) AS a
WHERE c1=10
LIMIT 10;
```

Keywords such as latent semantic analysis and randomized controlled trial are less informative as keywords such as citation classic and semantic network, which in turn have lower information contents than dyplastic nevus, ale metaanalysis, and terrorist (Table 4.2).

Year-by-Year Labels of a Cluster

The evolution of a cluster may demonstrate various subthemes over time. CiteSpace supports a function to extract terms from each year's publications to characterize the nature of a cluster on a year-by-year basis (Fig. 4.2). The extraction is based on the LSI technique. We can select extracted terms from multiple dimensions of the latent semantic space so as to develop a good understanding of the major subthemes.

Selecting Noun Phrases with LSI

Figure 4.3 reveals further details of the biological terrorism cluster by extracting title terms from articles published in each year. Terms from the first two dimensions of the LSI latent semantic space are inspected here. Changes in these terms over time may give us additional insights into the evolution of the cluster.

The more detailed year-by-year terms are shown in Table 4.3. Top five terms for the largest three dimensions of the latent semantic space are listed for each year between 1999 and 2003, indicating that the cluster's research fronts started in 1999. The terms bioterrorism and biological terrorism appeared persistently in the first four years of the 5-year period. It seems that it reached its peak in 2001 because both the first and second dimensions are led by the semantically equivalent terms.

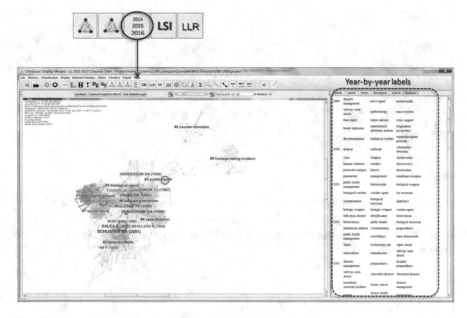

Fig. 4.2 Generating year-by-year labels of a cluster in CiteSpace

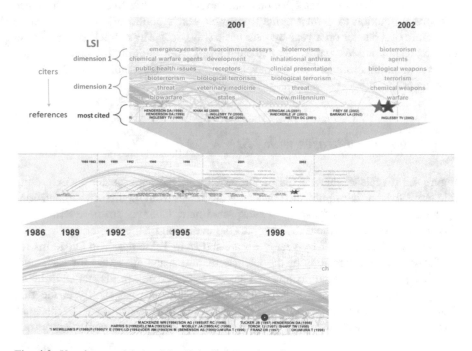

Fig. 4.3 Year-by-year labels of the biological terrorism cluster

Table 4.3 Year-by-year label terms of the biological terrorism cluster

Year	Dimension 1	Dimension 2	Dimension 3
1999	Emergency	**Bioterrorism**	Chemical
	Chemical warfare agents	Threat	Biological agents
	Public health issues	Biowarfare	Psychiatric aspects
	Hazmat	Emergency physicians	Domestic terrorism
	Anthrax	Reason	Emergency physicians
2000	Sensitive fluoroimmunoassays	**Biological terrorism**	Tetanus toxin
	Development	Veterinary medicine	Bind
	Receptors	States	Identification
	Using ganglioside-bearing liposomes	**Bioterrorism**	Novel
	Gangliosides	Tetanus toxin	Small molecule
2001	**Bioterrorism**	**Biological terrorism**	Emergency
	Inhalational anthrax	Threat	Ethics
	Clinical presentation	New millennium	Medical care
	Following **bioterrorism** exposure	Short-term safety experience	Chemical
	Surviving patients	Public health	Victims
2002	**Bioterrorism**	Terrorism	Food
	Agents	Chemical weapons	Thought
	Biological weapons	Warfare	Deployment locations
	Bacterial pathogens	Public health law	Vulnerability
	Terrorist attacks	Common goods	Terrorist attack
2003	Health care facility decontamination	Medical emergency	Report
	Protective equipment	Chemical terrorist attack	Drexel university emergency department
	Recommendations	Evaluations	Terrorism preparedness consensus panel
	Personnel	Teams	Radiation disasters
	Evaluations	Terrorism preparedness consensus panel	Children

Selecting Indexing Terms with LSI

Table 4.4 shows indexing terms extracted as the year-by-year labels for the biological terrorism cluster. Semantically equivalent terms such as biological warfare,
bioterrorism, and biological terrorism appeared in the 1st, 3rd, 4th, and 5th years of
the cluster. Terms such as disaster management, public health management, and
management clearly identified the primary motivation of the research behind this
cluster. Terms such as subway sarin attack, Tokyo subway, and Chernobyl disaster

Table 4.4 Year-by-year cluster labels extracted from indexing terms of the biological terrorism cluster

Year	Dimension 1	Dimension 2	Dimension 3
1999	Disaster management	Nerve agent	Mental health
	Subway sarin attack	Epidemiology	Mass hysteria
	Blast injury	Tokyo subway	Crisis support
	Bomb explosion	Experimental inhalation anthrax	Longitudinal perspective
	Decontamination	**Biological warfare**	Organophosphate pesticide
2000	Anthrax	Outbreak	Colorimetric detection
	Virus	Weapon	Cholera toxin
	Human volunteer	Warfare	Fluorescence
	Protective antigen	History	Deactivation
	Pneumonic	Management	Membrane receptor
2001	Public health management	**Bioterrorism**	Biological weapon
	Biological warfare	Warfare agent	**Bio terrorism**
	Contamination	**Biological terrorism**	Epidemics
	Biologic weapon	Biologic weapon	Warfare agent
	Infectious disease	Identification	**Bioterrorism**
2002	**Bioterrorism**	Public health	**Biological terrorism**
	Inhalational anthrax	Contamination	Preparedness
	Public health management	Surveillance	Mass destruction
	States	*Escherichia coli*	Septic shock
	Tuberculosis	Transmission	Subway sarin attack
2003	Disaster management	Preparedness	Hospital preparedness
	Subway sarin attack	Chernobyl disaster	Chernobyl disaster
	Hazardous materials incident	Breast cancer	Disaster management
	Patient	Atomic bomb survivor	**Bioterrorism**
	Breast cancer	Risk factor	Recommendation

indicate the influence of these attacks or disasters on research in bioterrorism over multiple years.

Semantic Relatedness

Two concepts are related if there is an incident or an event that involves both of them. A bank and a robber can be related by a bank robbery instance. Relatedness is a relation that connects two entities or abstract concepts. Association is commonly used to describe a relationship. A semantic relation is defined between two entities. In natural language, a semantic relation is typically represented by a triple, namely, the subject, the object, and the relation. In the statement JOHN TEACHES CALCULUS, JOHN is the subject, CALCULUS is the object, and the verb TEACHES established the connection. JOHN is a teacher and CALCULUS is a course. At a higher level of abstraction, a teacher TEACHES a course. The two concepts of teacher and course are semantically related.

Scientific articles routinely include a section on related work. Authors often discuss previous studies that addressed the same problem in some ways, but they are not considered as similar studies. They are related to each other because they more or less addressed the same problem.

The similarity between two concepts implies that we are comparing the two concepts in terms one or more attributes. Two smartphones may be similar because of their appearance such as size or color or internal design such as apps or controls.

The concept of semantic similarity is typically defined based on an underlying ontology or taxonomy, where concepts are organized to reflect their semantic relations. Notable sources such as WordNet are widely used in related research.

Semantic relatedness between two concepts can be established in a given domain ontology. If the two concepts can be connected with a path in the ontological representation, then the semantic relatedness is evident.

Semantic similarity is a special case of semantic relatedness. Two semantically related concepts may not be semantically similar, whereas two semantically similar concepts must be semantically related. In the earlier example, a bank and a robber are semantically related, but it does not make much sense if we say that they are similar in terms of some attributes or aspects.

Resnik's Semantic Similarity

The most influential work on measuring semantic similarities is the work by Resnik (1995). His approach makes use of the IS-A semantic links in a taxonomy, namely the WordNet, and measure the semantic similarity based on the information content over the most relevant semantic structure. The results were very encouraging, with a correlation of 0.79 to the upper bound of 0.90 of human subjects.

Given a taxonomy of concepts, the semantic similarity between two nodes in the taxonomy can be estimated in many ways. Here we consider IS-A links only in the taxonomy. The most straightforward way is to measure the distance between the two concepts. The shorter the connecting path between them, the more similar the two concepts are. If there are multiple paths, the length of the shortest path should be used to represent the semantic similarity. In fact, this edge-counting approach was proposed by Rada and Bicknell (1989). However, each link in a taxonomy is usually considered to have a length of 1 unit. All the links have this property regardless which part of the taxonomy they belong to. In a taxonomy like the WordNet, the semantic strength of a link near to the top, i.e. the broadest possible term may differ considerably from the semantic strength of a link near to the bottom of the taxonomy, where concepts are much more concrete and specific.

Intuitively, the edge-counting similarity is sensitive to the concepts' positions in the taxonomy. Such a sensitivity is not desirable because a similarity measure should not depend on additional factors. Resnik offered an alternative method to measure semantic similarity over a taxonomy of IS-A relations. His solution is based on the notion of information content. His approach also makes uses corpus-based statistics to estimate the probability of a concept. Connecting to the underlying data source makes it possible to use the same taxonomy with multiple contexts.

The semantic similarity between two concepts should reflect the extent to which they share information. In the context of an IS-A taxonomy, concepts are linked by IS-A relations. The extent to which two concepts share information is equivalent to finding a concept that subsumes both concepts. In WordNet, COIN subsumes both NICKEL and DIME. The semantic similarity between NICKEL and DIME is therefore reflected by the concept of COIN. Since CASH subsumes COIN, CASH indirectly subsumes both NICKEL and DIME as well. Both COIN and CASH are called subsumers of NICKEL and DIME. Which subsumer, CASH or COIN, makes the best candidate to represent the shared information content?

COIN is more specific than CASH. COIN has less irrelevant information than CASH. For example, CASH subsumes BILL as well as COIN. The information about BILL is irrelevant to the similarity between two COINS. Thus, the shared information content should be represented by the subsumer that has the lowest position on the taxonomy. The lower a concept on the taxonomy, the more specific it is.

The criteria discussed so far are applicable to the edge-counting method as well. The edge-counting method selects the shortest path that connects two concepts in question, for example, NICKEL—COIN—DIME. If there is a longer path connecting the two concepts, then the longer path includes broader concepts rather than narrower concepts than the shortest path, for example, NICKEL—COIN—CASH—COIN—DIME.

In order to avoid the unreliability issues with the edge-counting method, Resnik introduced probabilities of concepts in measuring semantic similarities. For each concept c in the underlying taxonomy, p(c) is the probability of encountering an instance of the concept. A concept positioned higher up in the taxonomy should have a higher probability than a concept positioned below it. If c1 IS-A c2 in the taxonomy, e.g. DIME IS-A COIN, then $p(c1) \leq p(c2)$. Thus, $p(\text{DIME}) \leq p(\text{COIN})$. The root concept r of the taxonomy should have $p(r) = 1$.

The information content IC of a concept c is: $-log_2 p(c)$. Since the probability of the broadest concept is 1, the lowest value of information content is 0. All other values of information content would be positive. Theoretically, there is no upper limit.

The semantic similarity between concepts c1 and c2 is the information content shared by the two concepts, which is in turn represented by the information content of the concepts that subsume the two concepts in the taxonomy

$$sim(c_1, c_2) = \max_{c \in S(c1,c2)} (-\log p(c))$$

where S(c1, c2) is the set of concepts that subsume both c1 and c2. Since the probability of a concept on the taxonomy is a monotonic along the IS-A links, the information content of a parent concept is less than the information content of its child concept, e.g. I(CASH) = −log p(CASH) ≤ log p(COIN) = I(COIN). Thus, the concept that reaches the maximum information content must be the subsumer that has the lowest position in S(c1, c2), or equivalently, the most specific concept that subsumes c1 and c2.

Resnik (1995) estimated the probability of a concept based on the Brown Corpus of American English, which is a collection of 1 million words of various genres of text, including news articles and scientific fictions. The occurrences of a word are counted towards all its parent concepts as well as its own concept in the taxonomy because an occurrence of DIME is also an occurrence of COIN and that of CASH. The probability of a concept is then defined as the relative frequency of the corresponding noun to the total number of nouns in the corpus.

Resnik validated his information content-based semantic similarity measure based on the assumption that a good similarity measure should agree with similarity ratings made by human subjects. Computational similarity measures should be consistent with similarity ratings based on our intuitions. He replicated an experiment designed by Miller and Charles. In Miller and Charles' original experiment, 30 pairs of nouns were given to 38 undergraduate subjects to rate "similarity of meaning" on a scale from 0, which means no similarity, to 4, which means perfect synonymy. These nouns were selected based on a previous study so that various degrees of similarity are covered by the set. Resnik gave the same 30 pairs of nouns to 10 computer science students or postdocs at the University of Pennsylvania and used exactly the same instructions. The average rating for each pair provides an estimate of the semantic similarity of the pair as judged by human.

Resnik found a correlation of 0.96 between the mean ratings in his experiment and in Miller and Charles' one. In terms of correlations with human judgements in Miller and Charles' experiments, the new human ratings are the nearest (r = 0.9015), followed by the information content (r = 0.7911), then by probability (r = 0.6671), with the edge counting the lowest (r = 0.6645).

Resnik's work is influential. Researchers have developed a number of variations based on Resnik's original work.

Other Measures of Semantic Similarity

WordNet Similarity for Java (WS4J)[1] is a Java library developed by Hideki Shim when he was a doctoral student at Carnegie Mellon University. It implements

---

[1] http://code.google.com/p/ws4j/.

several algorithms to compute semantic relatedness or similarity algorithms based on semantic relations in WordNet. An online demo is available at http://ws4jdemo. appspot.com. It appears that the demo version is somewhat better than the Java library. The examples below are based on the online version.

One can enter two words to the WS4J Demo and if they are found in WordNet, then the demo will report eight types of similarity measures for the pair of words. For instance, we can enter dime and nickel to the WS4J demo interface (Fig. 4.4). Note that nickel has multiple meanings, or senses in WordNet. Its meaning as a coin is the second sense.

WS4J Demo reports the structural details for each similarity, including common subsumers of concepts in WordNet. Figure 4.5 illustrates information that can be reconstructed from WS4J's outputs. Information of the local structure is useful for understanding basic concepts used in this group of algorithms. For instance, the Lowest Common Subsumers (LCS) of dime and nickel is currency. The shortest path connecting dime and nickel has a length of 3. Both dime and nickel have the depths of 11. IC(c) is the information content of the concept c. Thus, the subsumer coin has a lower information content, IC(coin) of 9.0577, than that of dime, which has IC(dime) of 11.0726. In this example, using the third sense of the word nickel, nickle3 in WordNet, WUP(dime, nickel2) = 0.9091 and RES(dime, nickel2) = 9.0577.

Fig. 4.4 WS4J Demo at http://ws4jdemo.appspot.com

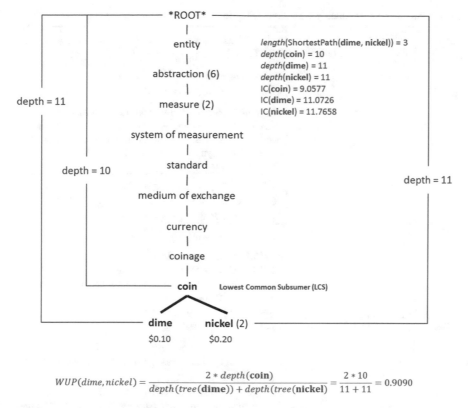

$$WUP(dime, nickel) = \frac{2 * depth(\textbf{coin})}{depth(tree(\textbf{dime})) + depth(tree(\textbf{nickel}))} = \frac{2 * 10}{11 + 11} = 0.9090$$

Fig. 4.5 The local structure of dime, nickel, and their LCS in WordNet and intermediate measures used in semantic similarity algorithms

Table 4.5 summarizes the algorithms for computing semantic similarities on WordNet. When applicable, we give sim(dime, nickel) as a concrete example to illustrate each algorithm.

Table 4.6 shows how various semantic similarity algorithms measure the semantic relatedness of 30 pairs of words and how they are correlated with ratings made by human subjects. The 30 pairs of words are the same set used by Miller and Charles (1991) in their experiment. They obtained similarity ratings from 38 human subjects on these pairs. Resnik duplicated the experiment in 1995 with 10 subjects. We consider the average rating from the Miller and Charles' experiment as the gold standard for the comparison. The comparison simply aims to see which algorithm behaves most like human raters.

Not surprisingly, human ratings in Resnik's experiment in 1995 have a strong correlation (r = 0.79) with human ratings obtained in Miller and Charles's experiment. This correlation is stronger than that from any of the computational algorithms. The algorithm that is the nearest to human ratings in Miller and Charles' experiment is the Resnik's similarity (RES), with a correlation of 0.61. RES is

Table 4.5 Semantic similarity algorithms with sim (dime, nickel) as an illustrative example

GS	Reference	Description
2904	Wu and Palmer (1994)	$WUP(s_1, s_2) = \dfrac{2*dLCS.d}{\min\limits_{dlcs \in dLCS}(s1.d - dlcs.d) + \min\limits_{dlcs \in dLCS}(s2.d - dlcs.d)}$ $WUP(dime, nickel) = \frac{2*10}{11+11} = 0.9090$ $where dLCS(s_1, s_2) = \underset{lcs \in LCS}{\text{argmax}}(lcs.d)$ The Wu-Palmer similarity measures the semantic relatedness of two synsets s1 and s2 in WordNet with respect to the LCS—the least common subsumer of s1 and s2. For a synset s, s.d is its depth in WordNet The range of the WUP is [0, 1]
3146	Jiang and Conrath (1997)	$JCN(s_1, s_2) = \frac{1}{IC(s1) + IC(s2) - 2*IC(LCS(s_1, s_2))}$ $JCN(dime, nickel) = \frac{1}{11.0726 + 11.7658 - 2*9.0577} = 0.2117$ The range of JCN is [0, +∞)
1891	Leacock and Chodorow (1998)	$LCH(s_1, s_2) = -\ln\left(\frac{length(LCS(s_1, s_2))}{2(MaxDepth(n))}\right)$ $LCH(dime, nickel) = -\ln\left(\frac{3}{2*20}\right) = 2.5903$ LCH is defined based on the shortest path between the two synsets and scale the path length by the maximum depth of the taxonomy The range of JCN is [0, +∞)
4312	Lin (1998)	$LIN(s_1, s_2) = \frac{2*IC(LCS(s_1, s_2))}{IC(s1) + IC(s2)}$ $LIN(dime, nickel) = \frac{2*9.0577}{11.0726 + 11.7658} = 0.7932$ Similar to JCN, but the range of LIN is scaled to [0, 1]
3602	Resnik (1995)	$RES(s_1, s_2) = IC(LCS(s_1, s_2))$ $RES(dime, nickel) = IC(coin) = 9.0577$ RES defined the similarity between two synsets to be the information content of their lowest super-ordinate (most specific common subsumer) The range of RES is [0, +∞)
PATH	Rada and Bicknell (1989)	$PATH(s_1, s_2) = \frac{1}{length(shortestpath(s_1, s_2))}$ $PATH(dime, nickel) = \frac{1}{3} = 0.3333$ PATH counts the number of nodes along the shortest path between the senses in the IS-A hierarchies of WordNet The range of Path is [0, +∞)
854	Banerjee and Pedersen (2002), Lesk (1986)	$LESK(s_1, s_2) = sum(dictionary\ definition\ overlaps)$ $LESK(dime, nickel) = 149.0$ LESK computes the relatedness of two words in terms of the extent to which their dictionary definitions overlap. Banerjee and Pedersen (2002) extended this notion to use WordNet as the dictionary for the word definitions The range of Path is [0, +∞)
1087	Hirst and St-Onge (1998)	$HSO(s_1, s_2) = 8 - distance - change\ Of\ Direction$ $HSO(dime, nickel) = 8 - 2 - 1 = 5.0$ HSO(s1, s2) = c − length(path(s1, s2)) − k * changes of directions (s1, s2) Links to be considered include 2 horizontal links, upward links, downward links The range of RES is [0,16]

The GS column is the citation count on Google Scholar as of July 21, 2017

Table 4.6 Comparing algorithms with Miller and Charles' (1991) experiment and Resnik's 1995 experiment

Pairs of words	Miller and Charles means	Resnik 1995	RES	WUP	LCH	LIN	PATH	HSO	JCN	LESK
Car, Automobile	3.92	8.04	7.00	1.00	3.69	1.00	1.00	16.00	12876699.50	9519.00
Gem, Jewel	3.84	14.93	2.49	0.63	1.61	0.24	0.13	0.00	0.06	8.00
Journey, Voyage	3.84	6.75	6.80	0.87	2.30	0.83	0.25	4.00	0.35	35.00
Boy, Lad	3.76	8.42	4.65	0.90	2.59	0.64	0.33	5.00	0.19	152.00
Coast, Shore	3.70	10.81	8.10	0.92	3.00	0.96	0.50	4.00	1.62	330.00
Asylum, Madhouse	3.61	15.67	3.94	0.67	1.61	0.00	0.13	0.00	0.00	6.00
Magician, Wizard	3.50	13.67	1.90	0.76	1.90	0.20	0.17	2.00	0.06	25.00
Midday, Noon	3.42	12.39	9.57	1.00	3.69	1.00	1.00	16.00	12876699.50	152.00
Furnace, Stove	3.11	1.71	2.49	0.50	1.12	0.23	0.08	5.00	0.06	190.00
Food, Fruit	3.08	5.01	0.61	0.38	1.29	0.10	0.09	0.00	0.09	130.00
Bird, Cock	3.05	9.31	0.61	0.29	0.92	0.00	0.06	2.00	0.00	17.00
Bird, Crane	2.97	9.31	1.82	0.64	1.49	0.00	0.11	0.00	0.00	12.00
Tool, Implement	2.95	6.08	6.31	0.94	3.00	0.91	0.50	4.00	0.85	509.00
Brother, Monk	2.82	2.97	1.90	0.70	1.61	0.21	0.13	0.00	0.07	29.00
Crane, Implement	1.68	2.97	1.37	0.53	1.39	0.00	0.10	0.00	0.00	7.00
Lad, Brother	1.66	2.94	1.90	0.73	1.74	0.24	0.14	0.00	0.08	14.00
Journey, Car	1.16	0.00	0.00	0.17	0.69	0.00	0.05	0.00	0.07	179.00
Monk, Oracle	1.10	2.97	1.90	0.70	1.61	0.18	0.13	0.00	0.06	30.00
Food, Rooster	0.89	1.01	0.61	0.29	0.92	0.08	0.06	0.00	0.07	18.00
Coast, Hill	0.87	6.23	6.14	0.71	2.08	0.73	0.20	3.00	0.22	123.00
Forest, Graveyard	0.84	0.00	0.00	0.24	1.05	0.00	0.07	0.00	0.06	20.00
Monk, Slave	0.55	2.97	1.90	0.80	2.08	0.20	0.20	3.00	0.07	74.00
Coast, Forest	0.42	0.00	0.00	0.29	1.29	0.00	0.09	0.00	0.06	59.00
Lad, Wizard	0.42	2.97	1.90	0.80	2.08	0.22	0.20	3.00	0.08	13.00

(continued)

Table 4.6 (continued)

Pairs of words	Miller and Charles means	Resnik 1995	RES	WUP	LCH	LIN	PATH	HSO	JCN	LESK
Chord, Smile	0.13	2.35	0.78	0.38	1.29	0.08	0.09	0.00	0.06	11.00
Glass, Magician	0.11	1.01	0.61	0.43	1.20	0.07	0.11	0.00	0.06	25.00
Noon, String	0.08	0.00	0.00	0.20	0.86	0.00	0.06	0.00	0.05	14.00
Rooster, Voyage	0.08	0.00	0.00	0.15	0.51	0.00	0.04	0.00	0.05	2.00
	1.00		**0.61**	**0.59**	**0.55**	**0.52**	**0.47**	**0.45**	**0.32**	**0.27**
		0.79	0.58	0.58	0.54	0.40	0.44	0.37	0.29	0.11

closely followed by WUP ($r = 0.59$). At the other end of the scale, JCN and LESK yielded the lowest correlations with Miller and Charles. If we use human ratings in Resnik's 1995 experiment, RES and WUP would have a tie ($r = 0.58$).

To our knowledge the largest gold standard of human ratings of similarity is the RG-65 test collection, containing similarity ratings of 65 pairs of words by 51 subjects on a scale of 0–4. The study was published in 1965 by Rubenstein and Goodenough.[2] Good enough!

The ACL wiki page lists a series of algorithms tested with the RG-65. Algorithms are ranked by Spearman and Pearson correlation coefficients. The highest correlation is achieved by an algorithm of Pilehvar and Navigli in 2015 with the Spearson correlation of 0.92 and Pearson correlation of 0.91. Several algorithms we have discussed earlier are included in the list, including HSO (0.813/0.732 by Spearson/Pearson correlations), JCN (0.804/0.731), LIN (0.788/0.834), and RES (0.731/0.800), and LSI (0.609/0.644).

In summary, as computational linguistics advances and a wide variety of resources become accessible, measuring semantic similarities has become increasingly powerful and reliable. For instance, estimating the probability of a word with the large pool of documents on Google is much more reliable than estimating it using a smaller collection of documents. The basic principles for estimating the semantic relatedness of a pair of words have fostered a large number of algorithms. Each of them has unique strengths.

Concentration

Burstness

The burstness of a variable X measures abrupt increases of the value of X over a specific period of time. Although the majority of research on burstness has focused on X as a scalar variable, the concept is intuitive enough to be expanded to a variable of multiple dimensions. In the real world, tsunamis would be a good example of a burst in a three dimensional space.

An Automaton

Kleinberg (2002) proposed a burst detection approach at the 8th ACM international conference on Knowledge Discovery and Data Mining (KDD). He models bursts in streams of text such as streams of email, publications, and speeches. The gap between the consecutive arrivals of items or events in time measures the frequency

---

[2]https://aclweb.org/aclwiki/RG-65_Test_Collection_(State_of_the_art).

of the events. A burst in a stream of email would be a period of time in which one receives many emails with small gaps. In contrast, during a period without any burst one would receive emails with much larger gaps. Such changes of frequencies are common in everyday life, for instance, distances between cars in rush hours.

Kleinberg's approach is to model the stream using an automaton that has an infinite number of states. In each state of the automaton, events take place at a particular rate. The automaton has states that characterize slow and fast rates of emission, a signal, an email, or an event. Streams with different rates can exist in the same system through state transitions. For instance, a slow-moving stream may be interwoven with a fast-moving stream by transiting from the corresponding slow-moving state to the state with a faster rate.

More formally, each stream is generated by an exponential distribution. Items in a stream are emitted probabilistically based on the exponential distribution so that the gap between one item and the next item follows the exponential density function $f(x) = \alpha e^{-\alpha x}$, where α is the rate of the arrival of the next item. If the automaton has two states that are responsible for emitting items at two different rates, low and high, then each state is modeled by its own exponential density function with α_{low} and α_{high}, respectively. The state transition probability in the automaton is p and it will remain in the same state with the probability of $1 - p$. Modeling the sequences with such an automaton is equivalent to determining the conditional probability of a state sequence based on the exponential density functions. The optimal sequence tends to minimize the number of state transitions; plus, the sequences would conform well to the corresponding gaps. Transitions to a high-frequency state will cost in proportional to a parameter gamma, but moving to a low-frequency will incur no cost.

Kleinberg demonstrated a hierarchical structure of the emails he received. The hierarchical structure revealed some bursts related to some intensive periods of emails due to proposal writing activities. His 2002 paper also included an example of 30 bursts detected from titles of all papers from two conferences between 1975 and 2001, namely SIGMOD and VLDB.

Burst Detection in CiteSpace

CiteSpace supports burst detection of several types of events, including citations to references and occurrences of keywords and noun phrases. The user may fine-tune the automaton by adjusting a few parameters of the automaton, including the minimum duration of a burst episode, state transition costs (gamma), and the ratio of the emission rates between states (Fig. 4.6).

Table 4.7 illustrates the burst durations of top 48 title terms with the strongest bursts in terrorism research between 1990 and 2017. The term biological terrorism has the strongest burst between 1996 and 2004. In terms of the automaton model, the term belongs to the state that emits articles at the fastest rate. A group of burst title terms are apparently related to the September 11 terrorist attacks in New York

Fig. 4.6 The user can modify
the automaton by adjusting a
few parameters

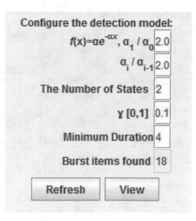

Table 4.7 The burst durations of 48 title terms between 1990 and 2017 in terrorism research

Terms	Strength	Begin	End	1990–2017
Biological weapons	9.0056	**1990**	2003	
Terrorist bombing	5.0276	**1990**	2000	
Biological terrorism	12.6472	**1996**	2004	
Nuclear terrorism	6.0997	**2001**	2006	
World trade center attack	4.742	**2001**	2001	
Public health	4.0079	**2001**	2003	
New york city	9.3846	**2002**	2007	
Islamic terrorism	8.0579	**2002**	2005	
World trade center	5.1459	**2002**	2004	
Mass destruction	4.7395	**2002**	2006	
Military commissions	4.6649	**2002**	2003	
Terrorist attack	4.2836	**2002**	2005	
New York	4.0418	**2002**	2006	
11th terrorist attacks	3.9182	**2002**	2006	
Suicide terrorism	4.9632	**2003**	2010	
Mental health	5.4509	**2006**	2010	
Hurricane katrina	4.4913	**2006**	2010	
Global war	4.8608	**2007**	2009	
Southeast Asia	4.5094	**2007**	2009	
World trade center disaster	4.2206	**2008**	2011	
Northern Ireland	3.4748	**2009**	2013	
Intimate partner violence	6.4243	**2010**	2017	

(continued)

Table 4.7 (continued)

Terms	Strength	Begin	End	1990–2017
Economic growth	4.3858	**2010**	2013	
State terrorism	4.0271	**2010**	2011	
Systematic review	6.4222	**2011**	2017	
Comparative analysis	4.3312	**2011**	2014	
Terrorist threats	4.1981	**2011**	2013	
Public opinion	3.6887	**2011**	2015	
Domestic terrorism	3.5188	**2011**	2017	
Terrorist organization	6.2394	**2012**	2017	
Political violence	5.6434	**2012**	2014	
Terrorist group	5.2063	**2012**	2015	
Civil war	5.1036	**2012**	2017	
Posttraumatic stress symptoms	4.4322	**2013**	2017	
Risk perception	4.2900	**2013**	2015	
Social media	5.9780	**2014**	2017	
Terrorism research	3.8617	**2014**	2014	
Empirical analysis	3.5993	**2014**	2017	
Lone wolf	3.4993	**2014**	2015	
Islamic state	9.7410	**2015**	2017	
Armed conflict	7.5456	**2015**	2017	
Boko haram	7.1621	**2015**	2017	
European Union	5.1440	**2015**	2017	
Boston marathon bombing	4.5943	**2015**	2017	
National security	3.8458	**2015**	2017	
Violent extremism	5.1387	**2016**	2017	
Risk factor	4.6470	**2016**	2017	
Terror attacks	3.6236	**2016**	2017	

City, notably world trade center attack, new york city, and world trade center. The term world trade center disaster also has a burst between 2008 and 2011. Among terms with a period of burst within the last three years, three of them with the strongest bursts are associated with radical violent extremism, including islamic state, armed conflict, and boko haram. Boko Haram, for instance, is Nigeria's militant Islamist group responsible for a series of bombings, assassinations and abductions.

Figure 4.7 depicts the distributions of three title terms with the strongest bursts between 1990 and 2017. The term biological terrorism has the strongest burst between 1996 and 2004. The term Islamic state has the second strongest burst

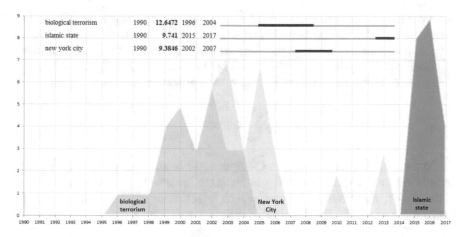

Fig. 4.7 The distributions of three title terms with the strongest bursts

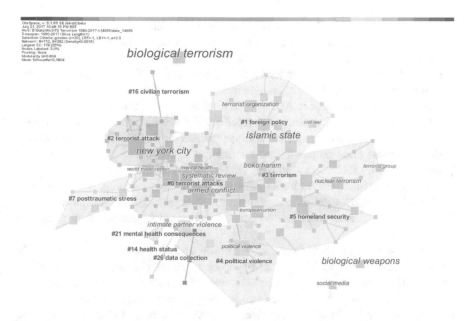

Fig. 4.8 A cluster view of title terms in terrorism research (1990–2017). Term labels are proportional to the strength of their burst. Labels starting with # are cluster labels, e.g. #0 terrorist attacks

between 2015 and 2017. The term New York City has the third strongest burst between 2002 and 2007. The term New York City appeared later on in titles in 2010 and 2013, but they are not bursts.

Figure 4.8 shows a network visualization of the title terms in terrorism research between 1990 and 2017. Publications on terrorism research in each year are selected to form the network. The selection is based on the g-index, which is an index that quantifies scientific productivity. In fact, the g-index is an extension of the h-index (Hirsch 2005) such that, unlike the h-index, it takes into account citations of these publications. The g-index was proposed by Leo Egghe (2006). The g-index is defined such that the most cited g articles have at least g^2 citations.

$$\sum_{i=1}^{g} c_i \geq g^2$$

As a node selection criterion in CiteSpace, we modify the g-index with a constant k. When k = 1, the modified g-index is the same as Egghe's original g-index. When k > 1, the modified g-index would select more articles than the g-index because the actual citation count of an article is raised by k times.

$$k \sum_{i=1}^{g} c_i = \sum_{i=1}^{g} k \times c_i \geq g^2$$

Table 4.8 shows the selection process using the modified g-index. For instance, in 1999, there are eight articles in our dataset on terrorism research. The g-index for this group is 3, which means the three most cited articles together have 9 or more citations. If we set k as 1, then the most cited three papers will be selected out of the total of eight. If we would like to include more articles and set k to 20, then all eight articles meet the condition, i.e. 20 times the total citations of the eight articles are no less than 9 citations. As another example, our dataset includes 106 articles published in 1999. The citations of these articles yielded a g-index of 4, i.e. the subtotal of the four most cited articles is greater than or equal to 16. 25 articles become qualified based on k of 20 instead of 1. As the third example, our dataset has 2957 articles published in 2016 on topics relevant to terrorism research. The g-index is 11. By using k of 20, CiteSpace selected title terms from 79 articles instead of 11. Thus, using a k greater than 1 allows us to include more articles than using the original g-index.

Figure 4.8 shows a cluster view visualization of a network of co-occurring title terms between 1990 and 2017. The top level aggregates in the visualization are clusters. The label of each cluster starts with the character #, for example, #0 terrorist attacks. The size of a title term is proportional to the strength of a burst detected. The larger the node label size, the stronger a burst it has. Thus the one with the strongest burst is the term with the largest font size—biological terrorism. The second strongest burst is with Islamic state. The third one is with New York City.

Burst detection is a very valuable technique. It helps us to focus on the important development dynamically. It is also applicable to many types of events. In addition to detect bursts in title words, we can also apply the technique to identify bursts in

Table 4.8 The number of articles selected by the g-index each year to construct the network of title terms

Time slice	g-index	Articles	Selected articles	Links/all
1990–1990	g = 3, k = 20	8	8	3/3
1991–1991	g = 2, k = 20	12	12	8/8
1992–1992	g = 2, k = 20	15	15	14/14
1993–1993	g = 2, k = 20	11	11	5/5
1994–1994	g = 2, k = 20	21	21	23/23
1995–1995	g = 2, k = 20	21	21	11/11
1996–1996	g = 2, k = 20	42	21	6/6
1997–1997	g = 3, k = 20	68	22	4/4
1998–1998	g = 3, k = 20	49	22	7/7
1999–1999	g = 4, k = 20	106	25	4/4
2000–2000	g = 4, k = 20	95	27	5/5
2001–2001	g = 5, k = 20	163	35	5/5
2002–2002	g = 10, k = 20	895	65	37/37
2003–2003	g = 8, k = 20	1009	58	22/22
2004–2004	g = 9, k = 20	1197	64	35/35
2005–2005	g = 7, k = 20	1488	58	17/17
2006–2006	g = 9, k = 20	1595	66	18/18
2007–2007	g = 7, k = 20	1682	60	16/16
2008–2008	g = 9, k = 20	1717	66	20/20
2009–2009	g = 9, k = 20	1796	62	12/12
2010–2010	g = 7, k = 20	1830	57	17/17
2011–2011	g = 10, k = 20	2065	70	28/28
2012–2012	g = 7, k = 20	1848	59	7/7
2013–2013	g = 7, k = 20	1797	57	6/6
2014–2014	g = 8, k = 20	1800	59	21/21
2015–2015	g = 8, k = 20	2191	64	22/22
2016–2016	g = 11, k = 20	2957	79	28/28
2017–2017	g = 6, k = 20	931	46	10/10

citations and bursts in institutions and individuals that are particularly active on specific topics. Unlike many popular indices of scientific productivities such as the h-index and the g-index, burst detection can tell us much more about the dynamics of the underlying process so that one can better understand how the process pans out. Burst detection can help us answer many specific questions: does an individual researcher have a burst in terms of the number of articles he/she published? If so, when did the most recent episode of burst begin? How long did the period of burst last? Is the researcher still at a state with a high productivity?

Log-Likelihood Ratio

Many commonly used statistical methods such as z-standard scores assume that the data is normally distributed. When dealing with text analysis, however, it is most likely that the normal distribution assumption is no longer valid, especially when we focus on terms that represent emerging topics or novel concepts. Researchers have shown that statistics based on the normal distribution assumption in such cases often overestimate the occurrences of rare words and that much of the content bearing words, technical jargons, and domain-specific terminologies in scientific publications are rare in the pool of English words in general.

Ted Dunning is currently the Chief Application Architect at MapR. Nearly 25 years ago, in 1993, he wrote an influential paper on text analysis (Dunning 1993). In the paper, he demonstrated the advantages of log-likelihood ratio tests for identifying relatively rare but significant patterns in text, for example surprising and unexpected combinations of words. His article now has 2773 citations on Google Scholar. With this amount of citations, the paper would be very close to the peak of the Mount Kilimanjaro. As a reference the 2006 JASIST paper on CiteSpace (Chen 2006) now has 1716 citations on Google Scholar.

Likelihood Ratio

Parametric and nonparametric are two board classifications of statistical procedures. One way to differentiate one from another is whether a statistical procedure relies on any assumptions about a probability distribution from which the data were drawn. The bottom line is whether a statistic procedure makes any use of such an assumption. For instance, to calculate z-scores, or standard scores, we need to know the mean and standard deviations of the underlying distribution of the data. The mean and standard deviations only make sense if the data were normally distributed. Therefore, the statistical procedure regarding the z-scores is parametric. In contrast, nonparametric tests are also called distribution free because they do not rely on any assumptions about the underlying distributions.

Given outcomes k as a point in the space of observations K, a set of model parameters ω as a point in the parameter space Ω, the likelihood $H(\omega; k)$ is the probability $P(k|\omega)$ that the outcome k would be observed given those parameter values at ω. $H(\omega; k)$ is the notation used by Dunning in his 1993 article.

$$H(\omega; k) = P(k|\omega)$$

For example, the likelihood function for repeated Bernoulli trials can be defined as follows:

$$H(\omega; k) = H(p; n, m) = p^m (1 - p)^{n-m} \binom{n}{m}$$

In this case, the parameter space Ω is the set of all the probabilities p, i.e. $[0, 1]$, whereas the subspace Ω_H for the hypothesis that $p = p_H$ is a singleton set $\{p_H\}$, which is a subset of $[0, 1]$.

The likelihood ratio λ for a hypothesis is the ratio of two maxima of the likelihood function. One is the maximum value of the likelihood function over a subspace Ω_H on which the hypothesis applies. The other is the maximum value of the likelihood function over the entire parameter space Ω.

$$\lambda = \frac{\max_{\omega \in \Omega_H} H(\omega; k)}{\max_{\omega \in \Omega} H(\omega; k)}$$

For two binomial processes that are characterized by p_i, m_i, and n_i for $i = 1$ and 2, the maxima are reached when $p_1 = \frac{m_1}{n_1}$, $p_2 = \frac{m_2}{n_2}$, and $p = \frac{m_1 + m_2}{n_1 + n_2}$. Let

$$L(p, m, n) = p^m (1 - p)^{n-m}$$

The log-likelihood ratio can be computed as follows:

$$-2log\lambda = 2(\log L(p_1, m_1, n_1) + \log L(p_2, m_2, n_2) - \log L(p, m_1, n_1) - \log L(p, m_1, n_1))$$

The value $-2\log \lambda$ is asymptotically distributed as χ^2 with the difference between the dimensions of Ω and Ω_H as the degree of freedom. Thus the log-likelihood ratio value is associated with a p-level, which indicates the statistical significance of the observed event. The 'oddness' measures how special the observation is.

Characterizing a Cluster

A major advantage of a likelihood ratio test helps us to identify events that are particularly more common in a subspace of the parameter space than the entire parameter space. A term that is particularly unique in one cluster but not in other clusters would have a very high likelihood ratio on the subspace associated with the matching cluster. For instance, the term post-traumatic stress disorder would stand out in terms of its likelihood ratio to differentiate a cluster on this topic from other topics in terrorism research.

Table 4.9 lists two sets of title terms selected from the three largest clusters. One set was selected by Latent Semantic Indexing (LSI) (Deerwester et al. 1990). The other was selected by Log-Likelihood Ratio (LLR). Two numbers are shown next to each term selected by LLR. The first number is the $-2\log \lambda$ value of the

Table 4.9 Representative terms selected by LSI and Log-Likelihood Ratio Tests for the largest three clusters in Project Demo 1 on terrorism research (1996–2003)

Cluster	Label (LSI)	Label (LLR)
0	Bioterrorism	Biological terrorism (8082.39, 1.0E-4)
	Reason	Front line (5684.68, 1.0E-4)
	Small molecule	New york city (5658.81, 1.0E-4)
	Family physicians	Emergency physician (5400.81, 1.0E-4)
	Thought	Blast over-pressure (4767.67, 1.0E-4)
	Nation	Terrorist attack (4541.68, 1.0E-4)
	Collaborative literature	11th terrorist attack (4210.69, 1.0E-4)
	Cure	Posttraumatic stress disorder (3605.89, 1.0E-4)
	Intentional poisoning	Chemical terrorism (3438.99, 1.0E-4)
	Bind|terrorism	Biological weapon (3269.72, 1.0E-4)
	Community-based model	Bioterrorism preparedness (3220.3, 1.0E-4)
	Protecting rural communities	Public health management (2887.06, 1.0E-4)
	Large-scale quarantine	Overpressure-induced injury (2811.21, 1.0E-4)
	Following biological terrorism	Involving hemoglobin (2811.21, 1.0E-4)
	Possible consequences	Biochemical mechanism (2811.21, 1.0E-4)
	Medical technicians	Oklahoma city bombing (2792.14, 1.0E-4)
	Panic	World trade center (2789.9, 1.0E-4)
	Common goods	Hospital preparedness (2768.94, 1.0E-4)
	Predictions	Medical response (2510.49, 1.0E-4)
		Psychological sequelae (2417.99, 1.0E-4)
1	Terrorism	Blast over-pressure (18729.92, 1.0E-4)
	Mental health responses	Overpressure-induced injury (11125.44, 1.0E-4)
	UCH experience	
	Bomb blast	Involving hemoglobin (11125.44, 1.0E-4)
	Biochemical mechanism	Biochemical mechanism (11125.44, 1.0E-4)
	Blast lung injury	Conventional weapon threat (3893.35, 1.0E-4)
	Oklahoma city bombing	Medical consequence (3893.35, 1.0E-4)
	Pulmonary blast injury	Blast injury (3456.56, 1.0E-4)
	Explosion survivors	Exercise performance (3281.15, 1.0E-4)
	Sublethal blast overpressure| major	Sublethal blast overpressure (3281.15, 1.0E-4)
	incidents	Food intake (3281.15, 1.0E-4)
	Proposal	Social consequence (2223.28, 1.0E-4)
	Dissemination	Physical injury (1588, 1.0E-4)
	Manchester bombing	Soho nail bomb (1575.04, 1.0E-4)
	Casualty profiles	UCH experience (1575.04, 1.0E-4)
	Casualty profile	Terrorist bombing (1354.2, 1.0E-4)
	Construction	Evolving threat (1329.12, 1.0E-4)
	Hazmat	Biological terrorism (1316.57, 1.0E-4)
	Suicidal deaths	Terrorist attack (1076.2, 1.0E-4)
	Pathologic features	Open-air bombing (1074.82, 1.0E-4)

(continued)

Table 4.9 (continued)

Cluster	Label (LSI)	Label (LLR)
		Confined-space explosion (1074.82, 1.0E-4)
2	September	Terrorist attack (9718.45, 1.0E-4)
	Terrorist attacks	New York city (8811.82, 1.0E-4)
	Negative changes	11th terrorist attack (7733.12, 1.0E-4)
	Following vicarious exposure	Biological terrorism (7130.79, 1.0E-4)
	Exposure	Posttraumatic stress disorder (6505.87, 1.0E-4)
	New York city children	World trade center (5195.7, 1.0E-4)
	Posttraumatic stress reactions	Psychological sequelae (4499.33, 1.0E-4)
	Stress-related mental health	Blast over-pressure (3850.16, 1.0E-4)
	Israel	Biological weapon (3797.09, 1.0E-4)
	Coping behaviors|terrorism	New York (3233.08, 1.0E-4)
	OPM-sang experience	Front line (2935.05, 1.0E-4)
	Risk assessment	Emergency physician (2882.3, 1.0E-4)
	Functional impairment	Vulnerable population (2430.25, 1.0E-4)
	Supporting children	Drug user (2430.25, 1.0E-4)
	Youth	Bioterrorism preparedness (2339.45, 1.0E-4)
	Television exposure	Overpressure-induced injury (2270.61, 1.0E-4)
	International relations	Involving hemoglobin (2270.61, 1.0E-4)
	Oklahoma city	Biochemical mechanism (2270.61, 1.0E-4)
	Warfare	Prior trauma (2247.83, 1.0E-4)
		Posttraumatic stress symptom (2247.83, 1.0E-4)

Up to top 20 terms are selected for each cluster

log-likelihood ratio. The larger this number is, the more special the term for the current cluster. The second number is the statistical significance of the $-2\log \lambda$ value from a χ^2 distribution. It is the p-level of the term.

The strongest LLR terms in the largest cluster #0 include biological terrorism (8082.39, 1.0E-4), front line (5684.68, 1.0E-4). The value 1.0E-4 is the statistical significance of the LLR value 8082.39 and 5684.68 according to a χ^2 distribution. The term biological terrorism is specific enough to give us a clear idea what the cluster is about. In contrast, the term front line is more ambiguous. Similarly, on the LSI list, terms such as reason, thought, and nature are usually too broad to be useful even within the specific context of a co-citation cluster.

Table 4.9 shows terms short selected by LSI and LLR as candidates for cluster labels. If the two lists match, then the decision would be easy. If they differ substantially, we need to investigate further. For the largest cluster (#0), it is relatively easy because bioterrorism and biological terrorism are semantically equivalent. Terms such as reason, thought, and front line are common on Google, suggesting that they are not good candidates for cluster labels because they are too broad and ambiguous to be informative. Although the LLR list includes 11th terrorist attack (LLR = 4210.69) and Oklahoma City bombing (LLR = 2792.14), their log-likelihood ratios are much lower than that of biological terrorism (LLR = 8082).

For the second largest cluster (#1), LSI identifies terms such as terrorism, mental health responses and uch experience, whereas LLR identifies blast over-pressure (LLR = 18,729.92), overpressure-induced injury (LLR = 11,125.44), and involving hemoglobin (LLR = 11,125.44). The LLR terms seem to suggest a theme on physical injuries, but the LSI list includes mental health responses. Further down the LSI list, there are terms related to physical injuries such as blast lung injury, pulmonary blast injury, and sublethal blast overpressure.

For the third largest cluster (#2), the top three LLR terms are terrorist attack (LLR = 9718.45), New York city (LLR = 8811.82), and 11th terrorist attack (LLR = 7733.12). The LSI list is topped by terms such as September and terrorist attacks. These terms strongly suggest that this cluster is about the September 11th terrorist attacks in 2001 at the World Trade Center in New York.

More generally, it is useful to differentiate two kinds of words in text, depending on their role in a sentence: function words or content-bearing words. Function words organize different parts of a sentence together but they don't mean anything on their own. In contrast, content-bearing words are the ones that carry the meaning of a sentence. A sentence would be meaningless without such contents. Content-bearing words to a sentence would be similar to the wine to a bottle. An earlier description of the distinction can be found in Charles Carpenter Fries' work (Fries 1952). Function words are also called structure words, whereas content words are also called lexical words.

Function words include prepositions, pronouns, auxiliary verbs, conjunctions, grammatical articles or particles. For instance, commonly seen function words include the, a, her, however, and otherwise. Content words are those that are not function words. Nouns, verbs, adjectives, and most adverbs are examples of content words. 99.9% of words in English are content words.

The following example is based on Project Demo 1: Terrorism Research (1996–2003) in CiteSpace. We first generated a network of co-cited references and then divided the network into several clusters. Each cluster is resulted from the citations made by a group of published articles. In order to understand what a cluster is about, one may inspect whether there are common reasons for these articles to cite the member references of the cluster together. CiteSpace implements a few functions to label a cluster based on terms selected from citing articles' titles, keywords, abstracts, or any combinations of terms from these fields. Figure 4.9 shows several clusters with automatically generated labels.

In addition to rank terms based on log-likelihood ratio tests with respect to their roles in a subspace of the underlying model, log-likelihood ratio tests can also measure associations between two terms so that one can generate an associative network of concepts or terms extracted from text. CiteSpace supports a function to compute the strengths of associations based on log-likelihood ratio tests (Fig. 4.10).

Similarities between terms can be measured in terms of how often they appear together, i.e. co-occurrences, and how likely they appear given the fact that they are published in the same journal. Figure 4.11 illustrates some of the interrelationships between title terms from publications in the journal Scientometrics. The strength of a link is based on a log-likelihood ratio test that compares the probability of co-occurrences with probabilities of the entire parameter space, including other scenarios in which only one of them appears or none of them appears. The log-likelihood ratio (LLR) between terms publications and papers is 0.8390, which

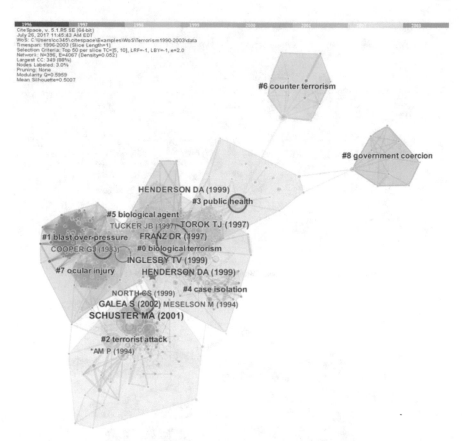

Fig. 4.9 Project Demo 1 in CiteSpace. Cluster labels are selected by LLR

is relatively low because the two terms are semantically equivalent so they are less likely to appear together. In contrast, the LLR between citation and impact is 8.1686 and the LLR between happiness and feelings is 10.6193. These relatively higher LLRs suggest some special connections between citation and impact and between happiness and feelings.

Entropy

Table 4.10 illustrates an approximate number of documents on Google that contain a term. The higher the number of instances on Google, the higher the probability of the term and the lower its information entropy is. For example, words such as 'the' and 'a' appeared most often on Google. Both are estimated to have appeared in approximately 25,270,000,000 documents on Google. In contrast, terms such as small molecule, bioterrorism, and posttraumatic stress disorder have much fewer

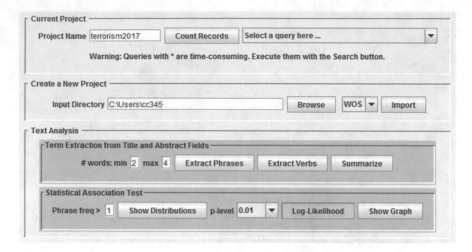

Fig. 4.10   Compute statistical associations with log-likelihood ratio tests in CiteSpace

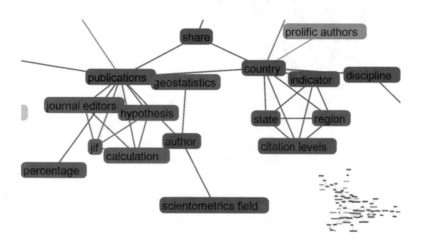

Fig. 4.11   Associations between title terms articles published in the Scientometrics

appearances, namely 22,100,000, 7,120,000, and 1,130,000, respectively. By the same method, Gone with the Wind has 90,200,000 hits. CiteSpace has 47,600 hits.

The information entropy of a term can be seen as a measure of its associated uncertainty. If we consider the appearance of a term as an event that transmits a message, then observing a rare event taking place is more information than observing a common event. Entropy is zero when we have nothing to learn from the occurrence of an event. The entropy reaches its maximum when the uncertainty is the highest, or, the occurrences of an event are completely random.

Figure 4.12 shows a plot of the information entropy of terms extracted from articles on terrorism each year. As new vocabularies are introduced into the latent

Term	Instances on Google
The	25,270,000,000
a	25,270,000,000
It	19,730,000,000
Thought	1,770,000,000
Reason	1,540,000,000
Front line	291,000,000
Terrorism	147,000,000
Gone with the Wind	90,200,000
Small molecule	22,100,000
Bioterrorism	7,120,000
Posttraumatic stress disorder	1,130,000
UCH[a] experience	627,000
Blast over-pressure	278,000
CiteSpace	47,600

Table 4.10 The popularity of a few terms on Google as of July 26, 2017

[a]UCH = University City Hospital

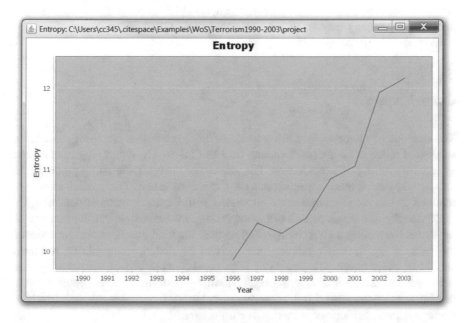

Fig. 4.12 The information entropy of terms extracted from articles on terrorism each year

semantic space, the information entropy would be higher in that year than before. The uncertainty of the latent semantic space increases due to the appearance of the additional terms. As shown in the plat, the largest increase is between 2001 and 2002 due to the September 11 terrorist attacks.

Table 4.11 illustrates top 20 terms based on their information entropy (Shannon 1948). The entropy of a term is calculated based on its distribution over the years between 1996 and 2003. The term explosion has the highest entropy. In other words, it is the least informative term in the context of terrorism. The standard deviation of the occurrences of a term can be used to measure the stability of a term. For instance, the term casualties has a standard deviation of 3.204, reflecting an uneven distribution of the term over the years. In comparison, the term explosion has a standard deviation of 1.309, reflecting a relatively stable distribution over this period of time.

Clumping Properties of Content-Bearing Words

Clumping was introduced in by Bookstein et al. (1998). They also introduced four ways to measure and identify clumping terms. Clumping metrics can be applied to an arbitrarily long text document or a chain of documents. The key assumption here is that content-bearing terms are more likely to clump than non-content-bearing ones. If non-content-bearing terms are randomly distributed throughout a text document, then one may focus on terms that their distributions deviate from the random distributions.

Condensation

The concept of clumping is similar to clustering except that clumping assumes a sequential order as an internal structure between items. Thus clumping can be seen as serial clustering of terms in text. The interest of studying clumping properties is to see whether a term appears unusually close together. The spatial closeness reminds us the temporal closeness associated with the topic of burst detection.

If we take sentences as units of observation, we would expect the number of sentences containing a given term to be less than the total number of occurrences of the term. If the term is clumping, the number of sentences containing the term should be even fewer. The degree of such condensation can be measured by the ratio of the actual number of sentences containing the term to the expected number of sentences of the term if it is randomly distributed. In practice, the unit can be a single sentence, a block of sentences, or a paragraph.

Given a term t, suppose the document to be analyzed has D units, N of them contain t, and t occurs T times in total. The number of ways to have T occurrences in D units is D^T, i.e. for each of the T instances, select a unit from D. The next step is to calculate the probability that exactly N units contain one or more instances of t. There are $\binom{D}{N}$ ways to select the units with at least one hit. A Stirling number of

Table 4.11 The information entropy of a term in articles of terrorism research (1996–2003)

TERM	Entropy	1996	1997	1998	1999	2000	2001	2002	2003	Subtotal	S.D.
Explosion	**2.914**	2	2	4	6	3	3	4	4	28	1.309
Blast	**2.901**	3	6	2	5	3	3	3	6	31	1.553
Tokyo	**2.877**	1	5	2	2	4	3	3	3	23	1.246
Injuries	**2.872**	4	2	3	4	4	6	6	9	38	2.188
Injury	**2.869**	4	5	1	4	5	4	5	8	36	1.927
Nerve	**2.840**	1	1	2	2	1	2	2	4	15	0.991
Bombing	**2.807**	3	3	4	6	6	1	9	4	36	2.449
Mortality	**2.790**	1	1	2	4	1	2	3	1	15	1.126
Israel	**2.780**	3	2	4	2	4	3	4	10	32	2.563
Gas	**2.786**	1	2	1	2	3	5	4	5	23	1.642
Sarin	**2.780**	0	3	2	2	3	2	3	3	18	1.035
Casualties	**2.765**	3	1	3	5	5	7	8	11	43	3.204
Oklahoma	**2.757**	2	2	1	5	6	1	4	3	24	1.852
Hospital	**2.745**	1	3	1	6	2	4	6	6	29	2.200
Children	**2.738**	6	2	2	3	5	2	6	11	37	3.114
Physicians	**2.733**	2	1	2	2	2	7	6	4	26	2.188
France	**2.725**	1	0	1	1	2	1	2	1	9	0.641
Disruption	**2.722**	2	0	2	1	1	1	2	1	10	0.707
Terrorist bombings	**2.722**	1	1	1	2	0	2	2	1	10	0.707
Biological terrorism	**2.720**	1	1	3	5	5	9	8	5	37	2.925

the second kind is a partition of T terms into N classes. There are N! ways to order the components of the partition. The probability p(N, T) is therefore:

$$p(N,T) = \frac{N! \binom{D}{N} \left\{ \begin{matrix} T \\ N \end{matrix} \right\}}{D^T}$$

The expected number of units containing t is as follows. See Bookstein et al. (1998) for details.

$$E_{C1} = D\left[1 - \left(1 - \frac{1}{D}\right)^T\right]$$

If N units contain term t, then N/E_{c1} measures the strength of the condensation. For a clumping term, this ratio would be less than 1. As we will see shortly, this is the clumping measure implemented in CiteSpace.

The second condensation-based measurement is based on specific distributions of terms over units. The probability that m occurrences of a term appear in any given unit can be modeled by the binomial distribution:

$$p(m) = \binom{T}{m}\left(\frac{1}{D}\right)^m\left(1 - \frac{1}{D}\right)^{T-m}$$

Thus one can expect $p(m)*D$ units to contain m occurrences.

The third measurement is the number of clumps. Here a clump is defined as a consecutive chain of units containing the term. The probability of K clumps of the term t is defined as follows. Again see Bookstein et al. (1998) for detailed reasoning.

$$p_K = \frac{\binom{N-1}{K-1}\binom{D-N+1}{K}}{\binom{D}{N}}$$

The expected number of clumps is defined by the following formula:

$$E_{L1} = (D - N + 1)\frac{N}{D} = N\left(1 - \frac{N-1}{D}\right)$$

The ratio K/E_{L1} measures linear-clustering clumping. Content-bearing terms are terms the ratio of which is substantially less than one.

Finally, the fourth measure of clumping is based on gap length between marked units. If N marked units are randomly distributed over D units of text, then the probability that a randomly chosen unit not be marked is $\gamma = 1 - N/D$.

The probability of r blank units between two marked units is given approximately by the geometric distribution.

Clumping Versus TF*IDF

Table 4.12 illustrates terms with strong condensation strengths (clumping) and top terms identified by term frequency (TF) by inverse document frequency (IDF). Perhaps more interestingly, these terms are associated with the largest co-citation cluster—the one on biological terrorism. Terms that appear on both lists are highlighted in the table.

The top clumping terms include radiation, virus, vaccine, plague, toxins, hemorrhagic, spores and toxin. These terms are clearly related to the central theme of biological terrorism. These terms are domain-dependent terms. Removing or ignoring the role of these terms will undermine the adequacy of a study. The clumping list also includes some terms that are not as tightly connected to biological terrorism as the first type of terms. The second type of terms include food, water, and protective. Yet another group of terms on the clumping list are domain independent terms. One can expect to see these terms in publications on any research topic, namely, evaluations, consensus, model, task, and final. The first three types are domain dependent. The fourth, sixth, are eighth and domain independent.

By applying the same classification heuristics to the list of terms ranked by their TF*IDF scores, the TF*IDF list has fewer Type 1 terms than the clumping list (4 vs. 10), more Type 2 terms (12 vs. 4), and about the same number of Type 3 terms (4 vs. 6).

Importance and Impact

Among the many types of importance metrics, two are particularly relevant to our understand how scientific knowledge is organized and diffused: eigenvector centrality and betweenness centrality. The eigenvector centrality is also called eigencentrality. Both of them measure the importance of a node in a network.

Degree Centrality and Eigenvector Centrality

The degree centrality of a node in a network is a simple measure of the node's importance in terms of how many nodes it connects to (Freeman 1977). Within the same network, a person with a lot of friends will have a higher degree centrality

Table 4.12 Top 20 content-bearing terms from the largest cluster of terrorism research (1990–2003) along with top terms identified by TF*IDF

TF	IDF	TF*IDF	Clumping	Term	Type	TF	IDF	TF*IDF	Clumping	Term	Type
21	3.53	74.05	0.16	*Radiation*	1	67	1.39	**92.88**	0.50	Chemical	2
15	3.26	48.87	0.29	Virus	1	127	0.69	**88.03**	0.67	Health	2
14	3.26	45.61	0.30	Vaccine	1	73	1.1	**80.2**	0.61	Bioterrorism	2
10	3.53	35.26	0.31	Evaluations	3	21	3.53	**74.05**	0.16	*Radiation*	1
28	2.56	71.82	0.32	*Food*	2	67	1.1	**73.61**	0.59	Emergency	2
13	3.26	42.36	0.33	Consensus	3	53	1.39	**73.47**	0.58	Response	3
20	2.83	56.66	0.33	*Anthracis*	1	28	2.56	**71.82**	0.32	*Food*	2
16	3	47.93	0.34	Model	3	49	1.39	**67.93**	0.59	Attack	2
16	3	47.93	0.34	Water	2	59	1.1	**64.82**	0.64	Care	2
9	3.53	31.74	0.35	Regional	2	25	2.56	**64.12**	0.36	*Training*	3
15	3	44.94	0.36	Plague	1	29	2.2	**63.72**	0.43	Physicians	2
25	2.56	64.12	0.36	*Training*	3	27	2.3	**62.17**	0.42	Smallpox	1
11	3.26	35.84	0.38	Protective	2	88	0.69	**61.00**	0.65	Medical	2
11	3.26	35.84	0.38	Task	3	23	2.56	**58.99**	0.39	*Toxins*	1
23	2.56	58.99	0.39	*Toxins*	1	36	1.61	**57.94**	0.62	Preparedness	2
8	3.53	28.21	0.39	Botulinum	1	83	0.69	**57.53**	0.66	Public	3
8	3.53	28.21	0.39	Final	3	20	2.83	**56.66**	0.33	*Anthracis*	1
8	3.53	28.21	0.39	Hemorrhagic	1	80	0.69	**55.45**	0.64	Agents	2
16	2.83	45.33	0.40	Spores	1	28	1.95	**54.49**	0.57	Detection	3
16	2.83	45.33	0.40	Toxin	1	49	1.1	**53.83**	0.71	Weapons	2

The cluster is labeled as bioterrorism. Terms appear on both lists are emphasized than that of someone with fewer friends. Two people with the same number of friends will have the same degree centrality.

More realistically, friends may have their own friends. If person A has a friend who has many friends and person B has a friend who has no more friends, should A and B have the same centrality? Unlike degree centrality, the eigenvector centrality treats friends differently. Connecting to an important friend will increase your own importance. As many have put it, it is about who you know, at least sometimes. In a social network, having many friends is generally a good idea unless all your friends are antisocial except with you.

The most famous member of the eigenvector centrality family is probably Google's PageRank. Recent research in neuroscience found that the eigenvector centrality of a neuron in a neural network is correlated with its relative firing rate.[3] The eigenvector centrality has been used to measure the prestige of a scientific journal, notably, the SJR indicator developed by a group of researchers in Spain.

The original idea can be traced to the works of Leontief (1941) and that of Seeley (1949) on reciprocal influence in social metric networks in 1949. The work of Phillip Bonacich (1972) is also widely known in relevant literature. Here we use Bonacich's notation. Given a network, the e_i centrality of node n_i in a network reflects the centralities of its neighboring nodes.

$$\lambda e_i = \sum_j R_{ij} e_j$$

Or, equivalently,

$$\lambda e = Re$$

where R is a matrix representation of the network. The diagonal values of R are zeros, i.e. $r_{ij} = 0$. By definition, e is the an eigenvector of R and λ is the corresponding eigenvalue.

As illustrated in Fig. 4.13, the visualization on the right shows that Cluster #1 at the top level (Level 0) has a concentration of nodes with high eigenvector centrality scores. In the context of a co-citation network, a high eigenvector centrality node means that it is co-cited with some well-connected references. The density of Cluster #1 is considerably higher than the density of the network overall.

The visualization on the left is generated based on articles that cited references in Cluster #1. Articles that did not cite any members of Cluster #1 are omitted from this procedure. As a result, the new network not only preserves the essential

---

[3]Fletcher, Jack McKay and Wennekers, Thomas (2017). From Structure to Activity: Using Centrality Measures to Predict Neuronal Activity. *International Journal of Neural Systems.* **0** (0): 1750013. doi:10.1142/S0129065717500137.

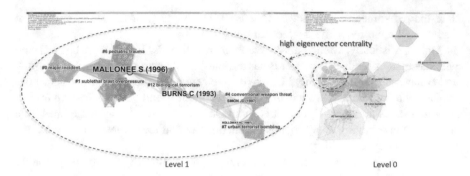

Fig. 4.13 High eigenvector centrality nodes are concentrated in Cluster #1 blast over-pressure. Zooming into #1 at the next level reveals high betweenness centrality nodes such as Mallonee1996 and Burns1993

structure of Cluster #1, but also reveals additional details. For instance, the new network reveals that it does not find any references published in 1996, 2000, and 2002. Instead, it contains publications in 1997–1999, 2001, and 2003. Cluster #1 is further divided into several clusters, including Level-1 clusters such as #0 major incident, #1 sublethal blast overpressure, #12 biological terrorism, and #6 pediatric trauma.

Two prominent nodes have strong betweenness centrality scores: MALLONEE1996 and BURNS1993. The betweenness centrality of a node measures the extent to which the node is in the middle of two or more dense areas. Suppose node v is connecting two sub-networks A and B. If the only way to reach one of the sub-networks from the other one is to go through node v, then the betweenness centrality of the node will reach the maximum possible level. The more alternative paths there are to bypass the node, the lower its betweenness centrality value will be.

The two nodes with strong betweenness centrality scores nicely illustrate the meaning of betweenness centrality in the visualization. MALLONEE1996 plays a central role in connecting at least three Level-1 clusters in three different colors. Removing MALLONEE1996 from the network will effectively disconnect these clusters because MALLONEE1996 is the only common node they share. Furthermore, MALLONEE1996 connects a 2003 cluster—#6 pediatric trauma (brown)—with clusters formed a few years ago (in blue and green years between 1997 and 1999), suggesting that in 2003 researchers revisited issues that had been addressed in 1997–1999. Such visits and revisits to the same research topics may explain the concentration of high eigenvector centrality nodes.

Figure 4.14 shows three displays of different metrics, namely, betweenness centrality, PageRank, and eigenvector centrality. The distributions of these metrics are different because they are designed to highlight different properties.

Betweenness centrality is effective in identifying critical information for understanding interrelationship between two or more clusters. Eigenvector centrality generalizes degree centrality by incorporating the importance of the

Betweenness Centrality   PageRank   Eigenvector Centrality

Fig. 4.14 The size of a node represents its betweenness centrality (left), PageRank (middle), and eigenvector centrality (right)

neighbors. Eigenvector centrality implemented in CiteSpace follows Zafarani et al. (2014).

Hirsch Index

Extrinsic factors are more common in the literature because of their relatively longer history. The most widely known examples include the Hirsch-index (Hirsch 2005), or the h-index, and the journal impact factor. Both of them have been extensively used and both have been subject to a wide variety of criticisms and modifications.

The *h*-index was introduced as an indicator of the productivity of a scientist in terms of all his/her N publications $\{a_i\}$ and corresponding citations $\{c(a_i)\}$, where $i = 1, 2, \ldots, h, \ldots, N$. For simplicity, assume the publications are sorted by their citations in descending order. The magic number h is the largest number of top cited h publications that have at least h citations, $c(a_i) \geq$ h, for the scientist.

$$h = \max_i(i|c(a_i) \geq h), \ where \ c(a_i) \geq c(a_j) \ if \ i < j$$

Since the coverage of one's publications varies from one source to another, one's h-index varies depends on whether the calculation is based on Google Scholar, the Web of Science, Scopus, or anything else.

For instance, as of August 8, 2017, Loet Leydesdorff, an active and productive researcher in scientometrics and several other fields, has a total of 36,474 citations for his hundreds of publications that we can find on Google Scholar. His h-index on Google Scholar is 86. By definition, among his numerous publications, 86 of them have at least 86 citations. In fact, many of his publications have much higher citations. In particular, two of his joint papers with Etzkowitz on a triple helix model of university-industry-government relations have been cited 6357 and 3367 times, way above the h-index of 86. The h-index is very simple in that it tags the productivity and the citations of a scientist with a single number. Broadly speaking, the higher the h-index, the more likely the scientist has made influential contributions to research.

On the other hand, the simplicity of the h-index also means that it does not represent some of the important aspects of a scientist's productivity or citations. Considering the complexity of citation distributions in reality, it is unlikely that any simplistic indicator can provide a comprehensive coverage of the underlying phenomenon that is so complex and dynamic. Scientists with the same h-index can still differ significantly before and after their the hth most cited publication. Scientists may have significantly different research profiles and yet still have the same h-index. For example, one researcher may have published exactly h papers and each of them has received h citations, which would make his/her h-index to be h. Another researcher may have published much more than h articles, say 10 times of h, but has a small number k of exceedingly influential and highly cited papers, $k \ll h$. As in with Loet Leydesdorff's case, his highest single paper citation is 6357, which is about 74 times of his h-index of 86. Although from the skewed distributions of citations we know that the former scenario is less likely to occur, the diversity within the class of scientists with the same h-index tends to be too large to be reliable for any evaluative purposes. After all, the h-index is biased towards researchers who have a sustained productivity as well as a long-lasting scholarly impact.

The g-Index

Many factors that influence citations may be used to normalize indicators such as the h-index. The academic age t of a researcher can be defined as the number of years since the first publication of a peer reviewed article, Hirsch proposed a normalized h-index m, which is the ratio of h to t. The stability of the m-index has been questioned, especially when the scientist is in his/her earlier career.

The h-index does not preserve any citation information about articles that are in the group of articles above the h citation mark, nor does it tell us anything about the size of the group below h. The h-index divides the publications of a scientist into two groups. One contains articles that have at least h citations, whereas the other contains articles that have fewer citations. Leo Egghe (2006) introduced the g-index as an enhanced modification of the h-index by taking into account the citations of the highly cited group. Similarly as in the h-index, the g-index divides the entire set of articles published articles into two groups using a single number g such that the top g highly cited articles as a whole have at least g^2 citations.

$$\sum_{i=1}^{g} c(a_i) \geq g^2$$

Alternatively, the g-index can be expressed in terms of the average citation of the g top cited articles' citations.

$$g \leq \frac{\sum_{i=1}^{g} c(a_i)}{g}$$

There are numerous ways one can normalize citation-based indicators such as the h-index and the g-index. For example, *Publish or Perish* normalizes the h-index by dividing the original citation counts by the number of co-authors first and then calculates the h-index on the author-normalized citation counts. Given the skewed citation distribution, instead of using the average of the g top cited articles' citations, one may consider using the median of the g citations or define an indicator G using a cumulative density function.

Other Measures

A key criterion of an indicator of scholarly impact of a scientific article should reflect how many researchers it has reached and how many people's thinking and behaviors have been changed. Thus, the number of citations an article has received or the number of citations a journal has received is commonly used measures. At the global level, Fig. 4.15 shows a dual-map overlay visualization of a set of publications on Terrorism research. There two maps in the visualization, hence it is called a dual-map visualization (Chen and Leydesdorff 2014). The map on the left is a set of citing journals organized according to their citing patterns, the map on the right is a set of cited journals positioned according to how similar they are cited. The curves represent citations from a citing journal on the left to a cited journal on the right.

Figure 4.16 shows some of the salient referential connections between clusters of citing and cited journals. For example, articles in this dataset frequently appeared in journals relevant to psychology, education, and health. These articles frequently cited references in similar types of journals. There are 17,276 such instances, which is equivalent to a z-score of 8.423. The strong pathway is visualized as a thick line. Some of the most cited journals are shown in Fig. 4.17.

Figure 4.18 depicts the distributions of citations by year of publication in Scientometrics (2010–2014). As expected, these distributions are strongly skewed towards the lower end of the citation scale. Most articles have zero or few citations, although highly cited articles do exist.

Figure 4.19 depicts the average number of references per paper in Terrorism (1982–2017). The thin solid line in green shows the average number of references per paper of the article type with citations. The dash-and-dot line in green shows the average number of references from articles without citations. Both lines are steadily increasing over time and the solid green line has about 15 references more on average. The thick solid line in blue represents the average number of references

Fig. 4.15 A dual-map overlay visualization of the terrorism2017 dataset (N = 14,656 articles and reviews)

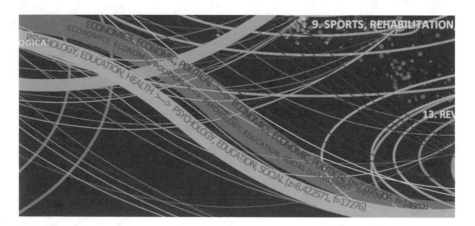

Fig. 4.16 The main field-level citation paths include Psychology|Education|Health to Psychology| Education|Social   (z = 8.423,   f = 17,276),   Economics|Economic|Political → Economics| Economic|Political (z = 7.075, f = 14,602)

from review articles with citations, whereas the dashed line in blue represents the average of references from review articles with no citations. Reviews with citations have cited more references than reviews with no citations. We cannot draw conclusions on any possible causal relations between references and citations. Although some journalists indeed attempted to make more shocking headlines by claiming such relations, we believe one has to examine the nature of citations to avoid picking up the wrong end of the stick. We refer to the number of references and many similar types of indicators as extrinsic factors as opposed to intrinsic ones when one aims to explain the scholarly impact (Chen 2012; Onodera and Yoshikane 2015).

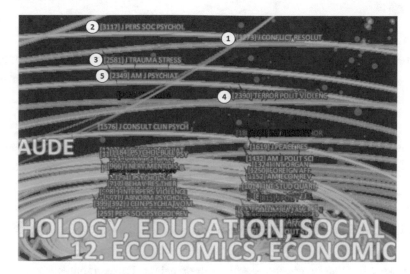

Fig. 4.17 Some of the most cited journals: 1. *Journal of Conflict Resolution*, 2. *Journal of Personality and Social Psychology*, 3. *Journal of Traumatic Stress*, 4. *Terrorism and Political Violence*, and 5. *The American Journal of Psychiatry*

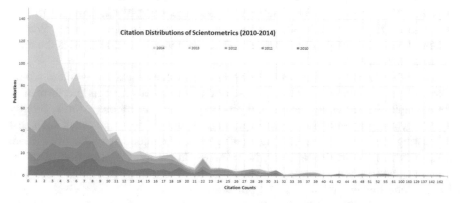

Fig. 4.18 Distributions of citations by year of publication in *Scientometrics* (2010–2014)

Normalization of Metrics

We all know that, to be fair, we should avoid comparing apples with oranges. Similarly, one should only pick on someone of his own size; otherwise, he would be considered either a bully or a coward. In weightlifting, athletes are grouped by their body mass. There are eight male divisions and eight female divisions. Men's weight classes include the 56 kg (123 lb) class, 62 kg (137 lb) class, and the highest 105 kg and over class. Athletes compete with others in the same class. In contrast, swimmers with longer arms have definite advantages over other

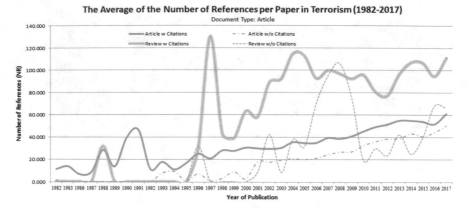

Fig. 4.19 Distributions of the average number of references per paper in Terrorism (1982–2017)

swimmers, but they compete regardless their height. Furthermore, there are four different styles in Olympic swimming: breaststroke, butterfly, backstroke, and free style. Different styles differ in their speed. One would expect it will take a swimmer a longer time to complete 100 m in breaststroke than in butterfly. On the other hand, we wouldn't be surprised if an Olympian swimmer's breaststroke is faster than a high schooler's freestyle. Given all these variabilities, we may still demand answers to questions that may sound like comparing apples with oranges after all. Who is the most powerful weightlifting athlete? Which swimmer's world record is the most remarkable?

Inevitably, scientists often find themselves in similar situations—others would like to compare their performance as a scientist with other scientists' performance, for example, for recruiting, tenure and promotion, and prestigious awards. Strictly speaking, every scientist is unique in numerous and fundamental ways such that comparing scientists based on quantitative measures alone may be even more ridiculous than comparing apples with oranges. In reality, the attraction of quantitative assessments is so strong that we will have to deal with a wide variety of issues along this line of inquiry and practice.

Research indicators, or academic indicators, are numeric figures that can give us a sense of something that maybe otherwise intangible. For example, a researcher's resume routinely includes the number of journal articles publishes, the number of presentations made at international conferences, the total amount of research grants secured, and the number of prestigious awards received. More recent years, researchers include additional indicators such as the number of citations to their publications in the Web of Science, the number of citations on Google Scholar, or relatively more mysterious h-index.

In addition to the evaluation of individual researchers' performance, their productivity and their scholarly impact, groups of researchers, institutions, and nations as well as journals and disciplines are subject to various evaluative assessments in a growing number of countries. It is important to understand the basics of commonly used indicators of research productivity and scholarly impact, especially their strengths and weaknesses.

Distributions of Citation Counts

The simplest indicator of a scientist's productivity is perhaps the number of articles he/she has published. Suppose one has published 400 articles and the other 200. Then the former clearly has a higher productivity. However, here is the first twist, what if we learn that the 400 publications together have received fewer citations than the 200 publications? In terms of the utility, who is more effective? Should we modify our assessment of the productivity based on the new information? Even if they have received the same number of citations N, the citation per paper rate (CPP) for the former is lower than the latter (N/400 < N/200). The efficiency of the latter is twice of that of the former.

It has been long realized that different disciplines of science may have drastically different citation rates. For example, mathematics is well known for its low citation rate, whereas biomedicine has the reputation of a high citation rate. Thus, being cited by 5 times may be not a big deal for a biomedical scientist, but it probably means a lot more to a mathematician. The differences between mathematics and biomedicine are probably much more profound than that between apples and oranges!

The age of a publication is also a known factor that may significantly influence the amount of citations. The diffusion of information takes time. The longer a publication has been exposed to the scientific community, the more likely it will be noticed and subsequently cited.

Normalization is a term that has been overloaded with multiple meanings. In our context, the term normalization refers to a transformation process that aims to eliminate or reduce the biases due to the heterogeneities between disciplines and between different durations of disclosure. The central idea of normalization is simple: how does the performance of our scientist compare with a *typical* scientist if everything else remains to be equal? As it turns out, in most of the cases it may not as straightforward as we wish to find our *typical* guy.

The distribution of citations is skewed. It means that the average number of citations does not evenly divide the distribution. Rather, one side of the mean may have a lot more instances than the other side. It would be nice and neat if citations are normally distributed. Then we can measure how far away an observed value from the average—the central tendency theory. We would be able to compare our observed value with the average. We would be able to look up the probability of observing a given value and we would be able to see how hard an achievement it might be.

A reference set is the term used by some researchers to refer to the baseline group to be taken into account. Once the performance of the reference set has been taken into consideration, their bias can be minimized or eliminated. In the early years, the average of citations in a reference set was used in initial attempts to normalize citations. However, it would work nicely only if citations follow a normal distribution.

The skewness of citations, or the skewness of science, is discussed in detail by Seglen (1992). First, the article age contributes to the skewness. The citedness of scientific articles changes with their age. Citations usually peak in the third year after the publication, then citations will decline steadily over time. The decline is considered to do with the obsolescence of the content. Seglen concluded that neither productivity nor citedness can adequately serve as general indicators of scientific quality and that the skewness shown in these indicators are probably in common in other indicators or potential indicators of scientific quality. After all, the evidence is more than sufficient that a small number of scientists contributed a lion share of the major advances of science.

Citation counts are a measure of utility rather than a direct measure of scientific quality. Citations measure the degree of attention from the scientific community. In this sense, citations measure the degree of perturbations to the complex system of scientific beliefs held by scientists as a whole. Direct measurements of scientific quality should characterize the core of scientific advances in terms of the novelty and the potential of transformative change.

Cross-field normalization of citation counts is primarily motivated by the inevitable fact that scientists from different fields of study are subject to quantitative evaluation from time to time. The general idea is to identify the scientific field in which a scientist should be evaluated so that the performance of the scientific field can be used to serve as a baseline reference. Slightly different terminologies have been used to refer to the baseline, including a reference standard or a reference set of publications.

Ideally, if there is a readily available classification system of scientific publication, then it is probably a good idea to consider utilizing the existing classification system. The most widely used such systems is the Subject Categories from the Web of Science. Each article indexed in the Web of Science is assigned with one or more subject category terms, for example, astronomy and astrophysics, artificial intelligence, and psychology. The research of a computer scientist specialized in artificial intelligence should be assessed in this particular context. Similarly, the research of a psychologist should be evaluated with peer researchers in the same subject category of psychology.

In an influential study published in PNAS, Radicchi et al. (2008) focused on the normalization of the citation performance of single publications. Given an article a in a particular field of research F, they considered the average number c_{mean} of citations received by all N articles b_1, ..., b_N in the field F published in the same year y, $F(y = year(a))$, as a normalized citation indicator c_f with reference to the particular field. Note that our notations may differ from those in Radicchi et al.'s original paper.

$$c_f(a) = \frac{c(a)}{c_{mean}} = \frac{c(a)}{\frac{\sum_{i=1}^{N} c_i(b_i)}{N}}, \text{ where } b_i \in F(y = year(a))$$

Radicchi et al. utilized the Subject Categories in the Web of Science as the definition of a scientific field. They found the chance of having a particular value of c_f is the same across distinct fields determined by the subject categories for articles published in the same year. More specifically, they found that the rescaled probability distribution $c_{mean}P(c, c_{mean})$ of the relative indicator c_f follows a lognormal distribution with a variance σ^2 of 1.3. If a random variable X has a lognormal distribution, then it means that ln(X) is normally distributed with μ as the mean and σ as the standard deviation. More specifically, $X = \exp(\mu + \sigma Z)$, where ln (X) = $\mu + \sigma Z$ is normally distributed.

What is remarkable about the finding is that a wide variety of subject categories such as allergy, astronomy & astrophysics, biology, mathematics, and tropical medicine appear to have the same property.

A lognormal distribution is defined by the following probability density function (PDF):

$$PDF(x) = \frac{1}{x} \bullet \frac{1}{\sigma\sqrt{2\pi}} e^{-\frac{(\ln x - \mu)^2}{2\sigma^2}}$$

where $ln(x)$ follows a normal distribution, μ is the mean, and σ is the standard deviation. In Radicchi et al.'s study, the equation $\sigma^2 = -2\mu$ reduces the number of fitting parameters to 1. A lognormal distribution with the same mean and the same standard deviation as the one in Radicchi et al.'s paper is shown in Fig. 4.20.

The results obtained by Radicchi et al. is very strong because it suggests that the rescaled lognormal distribution is independent of particular fields of study. On the other hand, when Radicchi et al. experimented with the universal characteristics of citation distributions across scientific fields, their study left out some common and

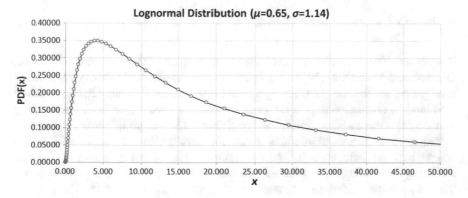

Fig. 4.20 A lognormal distribution

potentially significant categories, notably the multidisciplinary sciences category, which includes the most prestigious journals such as Science, Nature, and PNAS. Furthermore, their calculations exclude uncited articles.

It is now generally agreed that citation counts from different fields should not be directly compared with each other. To a lesser degree, it is also realized that one should be very careful when comparing citations of articles published in different years as well as publications of different types such as original research, review papers, editorials, and letters. In fact, scientometricians have studied a large number of factors that may influence how many citations research articles may get, when they are likely to peak, and how soon they may begin to decay. In a 2012 article on predictive effects of structural variations in a network of cited references, we distinguish factors as intrinsic and extrinsic. Intrinsic factors reflect the semantic and structural characteristics of the underlying scientific activity, whereas extrinsic factors do not have direct connections.

Examples of intrinsic indicators include structural variation metrics such as the ones we developed in our study of predictive effects of structural variations on citations. to measure the transformative potential of an article based on whether and to what extent it introduces novel and potentially groundbreaking links. The modularity change rate, for example, measures the degree to which a newly published article alters the structure of the network of scientific knowledge in terms of the change of modularity scores. Each newly published article brings us a set of references it cites. This set of references casts new lights on the existing network of scientific knowledge, which may be organized with cited references as nodes and co-citation relations as connecting links. The newly casted sub-network may introduce unprecedented links as well as reinforce existing ones with reference to the baseline network. The modularity of a network measures the degree to which the network is modularized. In other word, a network with a high modularity is organized in terms of a number of rather self-contained sub-networks. Interconnections between these sub-networks are minimal. In contrast, a network with a low modularity involves a considerable number of interwoven sub-networks.

Influential Factors on Citations

Researchers have identified some of the major sources of the skewness of science. Onodera and Yoshikane (2015), for example, published a study that systematically investigated several factors affecting citation rates. Ludo Waltman (2016) reviewed the literature on citation impact indicators, including a section on issues concerning normalizing citation-based indicators. In Table 4.13, we group some of the most commonly seen factors of citations in several broadly defined categories. Citation counts may be influenced by various factors about the authors of an article, including the productivity of the author, the academic age of the author, citations the author has received so far, and how the author connects with others in the academic network of collaborators.

Table 4.13 Factors that may influence citations of a scientific publication

Category	Factors on citation counts
Author	Productivity Reputation, citedness Gender Discipline Institution, Country Academic age: the number of years since the publication of the first peer reviewed article, the number of years since the first Ph.D. degree Academic network: eigenvector centrality, betweenness centrality
Article	Citations to date Altmetrics: Downloads, Views, Tweets Accessibility: Open access Visibility: Journal Impact Factor Co-authors: the number of co-authors, their diversity in author attributes Document type: original research, review, letter, etc. Extrinsic properties: the number of pages, the number of figures and equations Exposure: Duration since its publication date Language
References	The number of cited references The diversity of the references in terms of journals and disciplines The novelty of co-cited references
Discipline	The scientific field or fields to which the article belongs
Quality	Significance of research questions Rigor of methodology Clarity of presentation

Quantitative measures such as the citations and the number of cited references are relatively easy to handle. Factors that are of quality in nature are much more challenging to define (Hicks et al. 2015, Zhu et al. 2015). For example, identifying the scientific field that an article belongs too requires a substantial level of domain knowledge even with existing taxonomies of a domain. The significance of research questions requires a good understanding of a subject area, sometimes, more than one. Developing indicators of quality is an ongoing and challenging research in its own right (Ding et al. 2014, Wang et al. 2013).

Improvements of Impact Factors

The Journal Impact Factor is probably one of the most widely used and misused indicators of scholarly impact. In its original form, given a journal J, its impact factor $IF(J)$ is defined as the ratio of the citations to the citable items published in the previous two years c_{-1} and c_{-2} over the total number of citable items s_{-1} and s_{-2} within the same time frame.

$$IF(J) = \frac{c_{-1} + c_{-2}}{s_{-1} + s_{-2}}$$

The calculation of the impact factor over a two-year time span can be easily extended to a 5-year span or an arbitrary k-year impact factor.

$$IF_k(J) = \frac{\sum_{i=-k}^{-1} c_i}{\sum_{i=-k}^{-1} s_i}$$

Loet Leydesdorff is among the first to argue that the calculation should be done in a different order (e.g., Leydesdorff 2012; Leydesdorff et al. 2011). Instead of summing up the citations and citable items separately first and then taking the ratio, a more reasonable calculation should take the average citation per citable items in each year first and then calculate the average over the number of years.

$$NIF_k(J) = \frac{\sum_{i=-k}^{-1} \frac{c_i}{s_i}}{k}$$

The new impact factor (NIF) then becomes a k-year moving average of the annual citation rate. The original impact factor is the ratio of two averages, whereas the NIF is the average of citation ratios. Which one is more appropriate? What difference does it make? These questions are in fact part of a more profound debate in cross-field normalization.

Earlier citation normalization such as the Crown Indicator are calculated as the ratio of the mean of observed citation rates (OCR) over the mean of expected citation rates (ECR), which resemble to the way the original IF is calculated.

$$\frac{Mean(OCR)}{Mean(ECR)} = \frac{\frac{\sum_{i=1}^{n_{obs}} c_i}{n_{obs}}}{\frac{\sum_{j=1}^{n_{exp}} C_j}{n_{exp}}}$$

where $\{c_i\}$ are observed citations and $\{C_j\}$ are expected citations computed from a reference set such as all the publications from a field, i.e. biology or mathematics. In contrast, the more recently recommended citation normalization is the mean of the ratio of OCR to ECR:

$$Mean\left(\frac{OCR}{ECR}\right) = \frac{\sum_{j=1}^{n} \frac{c_j}{C_j}}{n}$$

The Mean(OCR)/Mean(ECR) is a division of two means. Using the Mean(OCR/ECR) has an advantage over the former—it comes with a standard deviation, which is additional information that is not available from the division of two means. Researchers have recognized the advantages of replacing the rate of averages with the average of rates.

The choice of using the mean of observed citation counts or the mean of the expected citation counts has also been subject to criticisms on the ground that the mean is no longer representative in a skewed distribution, which citation distributions typically fall into this category. Instead, the median would be a better choice. An ideal indicator should reflect the shape of the distribution and it should provide a metric that is independent from fields of study, the age of the article, and other major factors.

One of the most appealing indicators proposed in recent years is perhaps the approach that ranks articles on a percentile scale. It is proposed by Leydesdorff et al. (2011). The rank of an article is defined as the percentage of papers in the reference set that have citations fewer than the citation of the paper. The percentile is then rounded as an integer as the rank. Most cited 1% papers on the top of Mount Kilimanjaro should belong to the 99 percentile class. Given an article a, the probability that it belongs to the 99 percentile class is way below one in a million, considering that the size of the Web of Science as the reference set is about 50 million, depending on particular subscriptions.

The rank of articles in the kth percentile class can be expressed as the cumulated relative frequencies $p(r)$ weighted by their corresponding rank r:

$$R(k) = 1 \cdot p(1) + 2 \cdot p(2) \mid \ldots \mid k \cdot p(k) - \sum_{r=1}^{k} r \cdot p(r)$$

where f_r is the number of articles in the rth bin (there are 100 bins; one for each percentile class), $p_r = f_r/n_r$ and $n_r = \sum_i f_i$. The maximum weight is 100, which appears in R(100). The minimum weight is 1. Leydesdorff et al. gave an example of R(6) = 1*0.5 + 2*0.25 + 3*0.15 + 4*0.05 + 5*0.04 + 6*0.01 = 1.91.

Note that the range of R is not [0, 1]. One will need additional information to tell whether 1.91 is large or small. A further improvement can scale the range to the unit interval [0, 1] so that it is instantly clear about the position of 0.89 on a scale of [0, 1].

$$R(k) = \frac{\sum_{r=1}^{k} r \cdot p(r)}{\sum_{r=1}^{100} r \cdot p(r)}$$

More generally, in addition to work with percentiles, one can extend it to an arbitrary number of classes, for example, with 1000 bins or 100,000 bins, especially when dealing with a large number of articles at the disciplinary level. The more finer sliced bins we use, the more accurate the indicator tracks the underlying distribution. This line of reasoning leads to an ideal indicator I in an integral form, which suggests that when necessary, one can use finer grained bins to improve the accuracy of the indicator with reference to the underlying distribution. Here the $p(x)$ is the probability density function.

$$I(c) = \int_{-\infty}^{c} p(x)dx$$

The value of the cumulative density function at an arbitrary level of citation count c is between 0 and 1. It reaches the maximum of 1 when the probabilities of all sorts of scenarios are accounted for. In this way, the metric of quality is both intuitive and field-independent. Questions concerning quantifying scientists' performance can be answered in terms of the cumulative probability of observing a particular level of performance. If the performance of a mathematician has the cumulative probability of 0.90, then we know that this is a better performance of a molecular scientist with a cumulative probability of 0.80 in terms of scholarly publications.

Furthermore, given an article with citations of c, the cumulative density function will return a value between 0 and 1. The value can be considered as a rarity measure. The rarer a citation frequency, the harder it is to achieve and thus the more excellent it is.

In summary, cross-field normalization of citation-based indicators of scholarly impact has produced many indicators. However, researchers continue to refine the normalization procedures to reduce various biases that may be originated from the delineation of disciplinary boundaries or from the way to estimate expected levels of citation with reference to year of publication as well as relevant fields. Researchers have identified a large number of potential factors (See Table 4.13). We need to further develop our understanding of the magnitudes of the effects of these factors and how they interact at multiple levels of granularity. Most normalizations focus on a very small number of factors. It remains to be found to what extent existing normalizations preserve the order of articles in terms of their relative positions in their own crowd. Normalization should transform the values of apples and oranges into numbers within [0, 1].

Science Mapping

In this section, we will illustrate some of the important concepts with a collection of 17,731 papers on science mapping. A systematic review of science mapping published in 2017 is based on this dataset (Chen 2017). 17,721 of the 17,731 records are successfully loaded into a database. The following examples are based on the 17,721 records (Fig. 4.21). The dataset contains 14,794 articles (83.48%), 1861 proceeding papers, 1034 review, and a few items of other types such as book review, editorial, and book chapter. A copy of the dataset is downloadable from the ResearchGate project of the book.

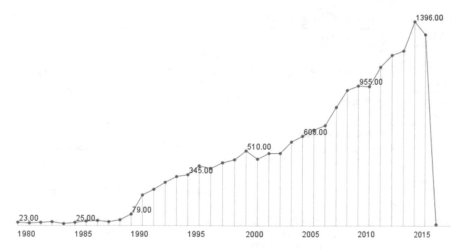

Fig. 4.21 The number of records in the dataset of Science Mapping (1980–2017)

Exploring the Science Mapping Dataset with CiteSpace's Database

We first loaded the dataset to a MySQL database on the localhost through an interface provided by CiteSpace. We demonstrated the example with MySQL queries such that interested readers can practice with their own datasets.

Table 4.14 illustrates the information stored in the Articles table of the wos database regarding the 2006 publication on CiteSpace. Each record from the Web of Science has a unique ID such as WOS:000234932600008. Similarly, a record from Scopus can be converted to the same format. The Scopus ID contains the DOI of the article, which appears to make the rest of the long string redundant. The information from the two sources has some discrepancies, which are highlighted in the table. For instance, the author name is Chen, Cm in the Web of Science, but Chen, C in Scopus. The journal title is abbreviated slightly different. More interestingly, citation counts differ substantially: 331 in the Web of Science and 503 in Scopus. A quick inspection of citing articles' sources reveals that many of the Scopus records are from conferences such as ISSI 2007 (8 papers), 2009 (8), 2013 (4), and 2015(3). These conferences, to our best knowledge, are not included in the Web of Science. This discrepancy in citation counts underlines practical issues one should consider for mixing citation records from distinct sources.

Table 4.15 lists the index terms assigned to the article. The author of the article did not provide any keywords. The index terms are algorithmically assigned as so-called KeywordPlus in the Web of Science. The keywords assigned by Scopus such as knowledge domain visualization and scientific literature are more accurate than the keywords under the Web of Science. The nearest term from the Web of

Table 4.14 Information stored in the Articles table of the wos database

Field	Example from the Web of Science	Example from Scopus (Format Converted)
id	433075	258910
uid	WOS:000234932600008	Scopus:2-s2.0-33644531603&doi = 10.1002%2fasi.20317
project	sciencemapping17731	scopus651
author	Chen, Cm	Chen, C
title	CiteSpace II: Detecting and visualizing emerging trends and transient patterns in scientific literature	CiteSpace II: Detecting and visualizing emerging trends and transient patterns in scientific literature
abstract	This article describes the latest development ……	This article describes the latest development ……
source	JOURNAL OF THE AMERICAN SOCIETY FOR INFORMATION SCIENCE AND TECHNOLOGY	Journal of the American Society for Information Science and Technology
j9	J AM SOC INF SCI TEC	J AM SOC INF SCI TECHNOL
volume	57	57
issue	3	3
nr	61	61
bp	359	359
ep	377	377
page	359–377	359–377
dt	Article	Article
doi	10.1002/asi.20317	10.1002/asi.20317
year	2006	2006
citations	331	503

Science is domain visualization. Other keywords on the list are more related to the case studies included in the article than CiteSpace as a tool as the focus of the paper.

The type sc in the last two rows of the table stands for Subject Category (SC). Two subject categories are assigned to the paper, namely Computer Science and Information Science & Library Science. It is not surprising that many articles published in the journal involve these two subject categories. In the Science Mapping dataset, Computer Science is the second largest subject category, whereas Information Science & Library Science is the third largest one.

Major Subject Categories in Science Mapping

In this section, we will explore several aspects of the Science Mapping with reference to the need for cross-field normalization and cross-time normalization. Corresponding MySQL queries are included for interested readers to replicate the results if they wish.

Table 4.15 The same article is indexed differently in different sources

Web of Science			Scopus		
id	Keyword	Type	id	Keyword	Type
2901483	Triassic mass extinction	id	1223053	Knowledge domain visualization	id
2901484	Domain visualization	id	1223054	Scientific literature	id
2901485	Terrorist attack	id	1223055	Algorithm	id
2901486	Science	id	1223056	Computer programming language	id
2901487	Paradigm	id	1223057	Information retrieval	id
2901488	Knowledge	id	1223058	Information science	id
2901489	Network	id	1223059	Research	id
2901490	City	id	1223060	Natural sciences computing	id
2901491	September-11	id			
2901492	Technology	id			
2901493	Computer Science	sc			
2901494	Information Science & Library Science	sc			

The number of records distributed per year in the dataset is shown in Fig. 4.21. The volume steadily increases. In 2015 alone, there are 1396 publications in the dataset. In 2000, the number of 510. The plot is generated in CiteSpace with the following MySQL query.

```
SELECT year, count(year)
FROM articles
WHERE project='sciencemapping17731'
GROUP BY year
ORDER BY year
```

Science Mapping is a field of interdisciplinary research. The dataset involves 149 distinct Web of Science Subject Categories. The Subject Categories of each record are stored in the keywords table. The following MySQL query finds the number of distinct subject categories. Each subject category is considered as a field of study. Researchers commonly identify the fields of study in terms of the Subject Category classification system.

```
SELECT count(distinct(keyword))
FROM keywords
WHERE project='sciencemapping17731'
AND type='sc';
```

The top 10 largest subject categories in the science mapping dataset are shown in Table 4.16. The largest subject category is Engineering, which has 4387 publications (24.8% of the entire dataset). The second largest one, Computer Science, has

Table 4.16 The number of articles distributed in subject categories

Publications	% of 17,721	Keyword
4387	24.7559	Engineering
3467	19.5644	Computer Science
2075	11.7093	Information Science & Library Science
1080	6.0945	Physics
1076	6.0719	Business & Economics
708	3.9953	Environmental Sciences & Ecology
623	3.5156	Telecommunications
605	3.4140	Optics
599	3.3802	Science & Technology—Other Topics
538	3.0359	Materials Science

3467 publications. The third one, Information Science & Library, has 2075 publications.

```
SELECT count(*), count(*)/17721, keyword
FROM keywords
WHERE project='sciencemapping17731'
AND type='sc'
GROUP BY keyword
ORDER BY count(*) DESC limit 10;
```

One may not anticipate to see Engineering appearing as the largest subject category in this dataset; after all, Science Mapping should be more closely related to computer science and information science. The following query lists the top 20 most frequent keywords assigned to Engineering papers in this dataset.

```
SELECT count(*), k2.keyword
FROM
    keywords AS k1,
    keywords AS k2
WHERE
    k1.project='sciencemapping17731' AND
    k2.project='sciencemapping17731'
AND k1.uid=k2.uid
AND k1.type='sc' AND k2.type!='sc'
AND k1.keyword='Engineering'
GROUP BY k2.keyword
ORDER BY count(*) DESC
LIMIT 20;
```

As shown in Table 4.17, Engineering papers are related to time-domain analysis, frequency-domain analysis, scattering, electromagnetic scattering, and information visualization. Although information visualization is semantically connected to

Table 4.17 Top 20 keywords associated with papers from the Engineering subject category

Engineering		Computer Science		Information Science	
Count (*)	Keyword	Count (*)	Keyword	Count (*)	Keyword
488	Time-domain analysis	635	**Citation analysis**	748	**Citation analysis**
268	Frequency-domain analysis	594	Information visualization	525	**Science**
258	Time domain analysis	481	**Science**	217	**Bibliometrics**
255	System	234	Visualization	196	**Journal**
253	Frequency domain analysis	223	**Network**	193	**Impact**
221	Model	198	**Bibliometrics**	170	**Network**
189	Design	191	System	166	**Indicator**
162	Simulation	189	Model	150	**Citation**
109	Algorithm	167	**Impact**	141	**Publication**
98	Scattering	157	**Information**	131	**Information-science**
91	Information visualization	149	**Journal**	118	**Co-word analysis**
89	Identification	147	**Indicator**	114	Scientometrics
86	Performance	146	Design	112	Library
73	Stability	141	**Citation**	109	**Information**
71	Vibration	117	**Pattern**	108	Impact factor
70	Equation	116	**Co-word analysis**	104	**Pattern**
67	Domain analysis	115	**Publication**	100	Index
65	Dynamics	108	Visual analytics	95	h-index
62	Electromagnetic scattering	99	Knowledge	93	Web
61	Wave	97	**Information-science**	90	Cocitation analysis

science mapping, the inclusion of papers on time-domain analysis and frequency-domain analysis appears to be a side effect of the set of queries used to retrieve the 17,731-record dataset from the Web of Science. In particular, domain analysis is one of the sub-topics in Science Mapping. Apparently, domain analysis is a term that is also used in Engineering for a completely different subject. When using CiteSpace, our advice to how to handle such unanticipated and potentially irrelevant topics is to proceed to the network analysis stage without attempting to eliminate the potentially irrelevant records. There are at least two good reasons for deferring any actions to eliminate any records prematurely:

The suspicious irrelevancy at this stage is based on our current knowledge. If we conclude the irrelevancy without further investigation, we may lose the opportunity to learn anything new from the process. After all, there may exist profound connections that we are simply not aware of.

The best time to eliminate irrelevant data is probably after we have a chance to inspect the resultant network model. It is much easier to identify an isolated sub-network in a visualization of the network than try to determine the relevancy from the dataset of such complexity.

The top 20 keywords for Computer Science and Information Science are quite consistent, including common ones such as citation analysis, science, bibliometrics, impact, network, and indicator. Common keywords are highlighted in the table. Unique keywords in Computer Science include information visualization, visualization, design, and visual analytics, whereas unique keywords in Information Science include scientometrics, h-index, and cocitation analysis.

Many publications are indexed with multiple subject categories. For example, there are 455 publications in common between Engineering and Computer Science, 1556 shared publications between Computer Science and Information Science. Interestingly, while Engineering, Physics, Telecommunications, Materials Science, and Environmental Science and Technology overlap one another, Information Science does not overlap with any of them within this dataset. To compute the number of overlapping records between two subject categories, one can use the following query by substituting K1.keyword and K2.keyword accordingly.

```
SELECT count(*)
FROM
    keywords AS K1,
    keywords AS K2
WHERE
    K1.project='sciencemapping17731' AND
    K2.project='sciencemapping17731'
AND K1.type='sc' AND K2.type='sc'
AND K1.uid=K2.uid AND
    K1.keyword='Information Science & Library Science'
AND
    K2.keyword='Computer Science';
```

The total number of papers in Information Science & Library Science is 2075, apart from 1556 papers that are jointly indexed as Computer Science papers, there are only 519 papers that do not share the Computer Science category. This is an indication of the role of computer science in Science Mapping.

Citation Distributions

Based on our earlier discussions, one would expect that citation rates are field-dependent as well as time-variant. One would also expect that the number of references cited by an article varies across distinct subject categories. Using the query below, we can find that the average of citations of the dataset is 16.79, the minimum citations is 0, and the maximum is 1547.

```
SELECT avg(citations), min(citations), max(citations)
FROM articles
WHERE project='sciencemapping17731';
```

Figure 4.22 shows a log-log plot of the frequencies of citations per paper in Science Mapping. Citation counts are log-transformed, so are the frequencies of citations. Since *log*(citations) is not defined for zero citations, a common practice is to add 1 to the citation count of each paper. As expected, papers with zero citations are most common, whereas highly cited papers are increasingly unusual.

The total number of references cited by the 17,731 articles is 672,899, of which 508,564 are distinct. On average, each publication in the dataset has 37.95 references. Publications in the dataset received a total of 297,529 citations across publications indexed in the Web of Science. On average, each paper has a citation count of 16.78.

In addition to the average over the entire dataset, to what extent does a particular subject category differ from the overall dataset? Using the following query, we can find the average number of citations and cited references specifically for a particular subject category.

```
SELECT
      avg(citations),
      avg(nr),
      keywords.year
FROM
      articles,
      keywords
WHERE
      articles.project='sciencemapping17731' AND
      keywords.project='sciencemapping17731' AND
      articles.uid=keywords.uid AND
      keywords.keyword='Computer Science'
GROUP BY keywords.keyword, keywords.year;
```

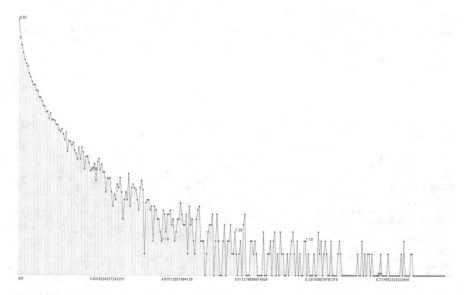

Fig. 4.22 A log-log plot of the frequencies of citations per paper in Science Mapping (1980–2017)

The average number of citations per paper and the average of references per paper of the largest four subject categories show that Engineering and Physics papers have an average of citations per paper about 12, whereas Computer Science, Information Science & Library Science papers have a citation count of 16. Furthermore, for papers in Information Science & Library Science, they have 42 references on average (Table 4.18).

Figure 4.23 illustrates the differences between the four largest subject categories in the science mapping dataset in terms of the average number of references per paper and the average number of citations per paper. The curves in green plot the average number of citations per paper, whereas those in red represent the average number of references per paper. Overall, the red curves show an upward trend. It means that the average number of references per paper is increasing over the years regardless of subject categories. There are a few outliers of papers in Information Science & Library Science. In 1999, Wilson CS for example published a paper that cited 491 references. In 2004, Phillips LI cited 400 references in a single paper. More recently, Guimaraes cited 346 references in a paper and Waltman cited 342. These papers are review papers. Engineering and Physics papers in this dataset have a lower average number of cited references per paper, whereas Computer Science and Information Science have about 10–15 more references on average. The average is steadily increasing for both groups of subject categories. The growth rates are the same because the four lines are essentially parallel to one another.

The lines representing the average numbers of citations are more complex than their reference counterparts, although they diminish towards the present time because recently published papers are yet to receive their citations. Citations of Engineering papers are relatively smooth over the years. In contrast, citations of Information Science & Library Science fluctuated over time, but their citation average is higher than that of Engineering. The earliest outliers include a 1981 paper by Howard White with 406 citations and a 1989 paper by Macroberts. Other prominent papers include Callon, Small, and Chen from Information Science. Papers by Holten, Shneiderman, and Bostock respectively are from the subject category of Computer Science, more precisely, from information visualization and visual analytics.

Table 4.18 The average number of citations per paper and that of references per paper are both field-dependent

Subject category	Papers	Average (Citations)	Average (References)
Overall	17,721	16.7896	37.9718
Engineering	4387	12.3855	24.8849
Computer Science	3467	16.1554	39.5953
Information Science & Library Science	2075	16.4308	42.1667
Physics	1080	12.0820	27.7991

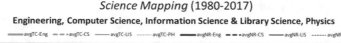

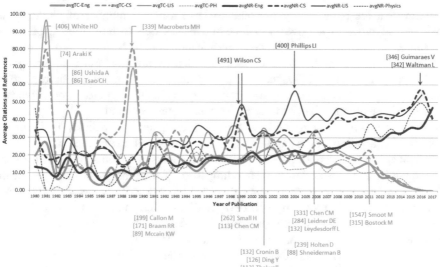

Fig. 4.23 Trends of the number of references and the number of citations of the four largest subject categories

The following query searches for most highly cited papers in a particular year from a specific subject category. Using it along with the plots such as the one shown in Fig. 4.23, one can identify landmark works in Science Mapping as well as general trends in terms of the average number of citations and the average number of references.

```
SELECT
        citations, author, doi, k.keyword, a.year
FROM
        articles AS a, keywords AS k
WHERE
        a.project='sciencemapping17731' AND
        k.project='sciencemapping17731' AND
        a.uid=k.uid AND
        (k.keyword='Engineering' OR k.keyword='Computer Science') AND
        (a.year=1981 OR a.year=1983 OR a.year=1984 OR a.year=1989)
ORDER BY citations DESC LIMIT 30;
```

Citation Normalization Over Time

Articles that are published earlier tend to have a higher number of citations on average. In order to remove or reduce the biases due to how long a paper is available to potential citers, the age of publication should be taken into account.

Figure 4.24 depicts the cumulative citation density function. For instance, given an article published in 2000, the probability that the article has no more than 5 citations (s5) is much lower than the probability that a 2015 article. In other words, a 2000 paper is more likely to get more than 5 citations than a 2015 paper. The formula in the figure suggests how one may estimate a probability in terms of relative frequencies. For example, c_0 is the number of items that have a zero citation, c_1 is the number of papers with citation counts of 1, and so on. Thus $i* c_i$ is the sub-total of the citations corresponding to c_i. If there are $c_5 = 11$ papers with $i = 5$ citations each, these 11 papers collectively received 55 citations. If the entire set of publications is allocated to 100 evenly divided bins, then this method is very close to the percentile indicator proposed by Loet Leydesdorff and his coauthors. Furthermore, our indicator has two additional advantages:

The percentile-based indicator is an approximation to the cumulative citation density function in its integral form. Realizing its connection to the integral form, one can easily improve the approximation by using an arbitrarily large number of bins. In effect, we are taking the limit of the discrete sum of the citations over bins. With a sufficiently large number of data points and a sufficiently large number of bins, the estimate can be arbitrarily close to the integral value.

Our indicator is scaled to [0, 1], which makes it independent of its range and thus easy to understand and compare with other fields. Instead of wondering where a scientist with an indicator of 1.91 would be positioned on an irregular scale, The unit range of [0, 1] simplifies the interpretation and comparison.

Figure 4.25 depicts the probabilities of articles published in a particular year having c citations between 2000 and 2015 in Science Mapping. The citation probability distributions of articles published in the first 11 years (2000–2010) resemble to normal distributions with the highest probability is around p50, which is the middle of the [0, 100] scale and near-to-zero probabilities towards both ends. The probability curves of articles published in the recent five years (2011–2015) are increasingly higher towards the lower end of the citation scale. It appears that, in general, the peak of a citation distribution steadily shifts from left to right and the overall distribution is stabilized approximately after five years of publication. We suspect that the rate of the settlement is likely to be field-dependent.

In addition to the fluctuations of citation probability, we smooth the citation probabilities with 5-year average citation probabilities between 2000 and 2015. As shown in Fig. 4.26, the trends become more apparent—citation probability distributions are gradually shifted from low-citation probabilities to average—and higher-citation probabilities. The citation probability distribution of articles published in the recent five years has a substantial weight on the left, i.e. the probability of having few citations is relatively high. The two citation distributions of articles

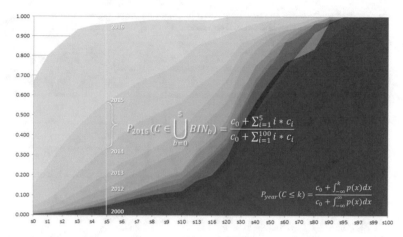

Fig. 4.24 Cumulative relative citations by year of publication

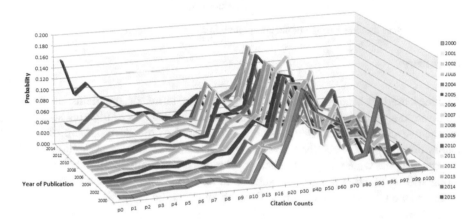

Fig. 4.25 Probabilities of articles published in each year having citations c in Science Mapping

published more than five years ago overlap considerably with one another, suggesting a relatively stable distribution. Normalizing citations over time is reasonably reliable for publications more than 5 years old. In contrast, citation probabilities fluctuated considerably more with articles published less than 5 years old. The key to citation normalization over time is to account these factors.

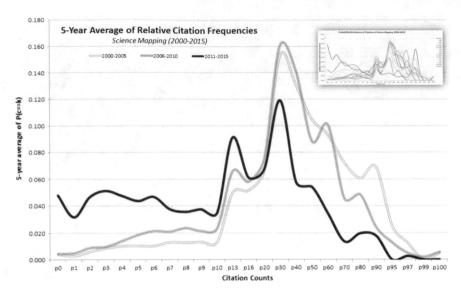

Fig. 4.26 5-year moving average of citation probabilities

Summary

Citation-based indicators should be normalized in terms of the fields of study involved, the year of publication. There are distinct advantages of utilizing standard cumulative citation probability functions as opposed to the development of indicators that may not share the universality in terms of their interpretability. More importantly, the wide variety of indicators should be taken into account collectively along with qualitative analyses of science to serve the purposes of research evaluation as well as learning the state of the art of scientific research.

References

Banerjee S, Pedersen T (2002) An adapted Lesk algorithm for word sense disambiguation using WordNet. In: Gelbukh A (ed) CICLing 2002, LNCS 2276. Springer, Heidelberg, pp 136–145
Bonacich P (1972) Factoring and weighting approaches to status scores and clique identification. J Math Sociol 2(1):113–120
Bookstein A, Klein ST, Raita T (1998) Clumping properties of content-bearing words. JASIS 49 (2):102–114
Chen C (2006) CiteSpace II: Detecting and visualizing emerging trends and transient patterns in scientific literature. J AM SOC INF SCI TEC 57(3):359–377. doi:10.1002/asi.20317
Chen C (2012) Predictive effects of structural variation on citation counts. J Am Soc Inform Sci Technol 63(3):431–449

Chen C (2014) The fitness of information: quantitative assessments of critical evidence. Wiley, Hoboken

Chen C (2017) Science mapping: a systematic review of the literature. J Data Inf Sci 2(2):1–40

Chen C, Leydesdorff L (2014) Patterns of connections and movements in dual-map overlays: A new method of publication portfolio analysis. J Assoc Inf Sci Technol 65(2): 334–351

Deerwester S, Dumais T, Landauer T, Furnas G, Harshman R (1990) Indexing by latent semantic analysis. J Am Soc Inf Sci 41(6): 391–407

Ding Y, Rousseau R, Wolfram D (eds) (2014) Measuring scholarly impact: methods and practice. Springer, Heidelberg. doi:10.1007/978-3-319-10377-8

Dunning T (1993) Accurate methods for the statistics of surprise and coincidence. Comput Linguist 19(1):61–74

Egghe L (2006) Theory and practise of the g-index. Scientometrics 69(1):131–152. doi:10.1007/s11192-006-0144-7

Freeman LC (1977) A set of measuring centrality based on betweenness. Sociometry 40:35–41

Fries CC (1952) The structure of English. Harcourt Brace, New York

Hicks D, Wouters P, Waltman L, Rijcke Sd, Rafols I (2015) Bibliometrics: The Leiden Manifesto for research metrics. Nature 520(7548):429–431. doi:10.1038/520429a

Hirsch JE (2005) An index to quantify an individual's scientific research output. Proc Natl Acad Sci USA 102(46):16569–16572. doi:10.1073/pnas.0507655102

Hirst G, St-Onge D (1998) Lexical chains as representations of context for the detection and correction of malapropisms. In: Fellbaum C (ed) WordNet: an electronic lexical database. The MIT Press, Cambridge, MA

Jiang JJ, Conrath DW (1997) Semantic similarity based on corpus statistics and lexical taxonomy. In: Proceedings of international conference research on computational linguistics (ROCLING X), Taiwan

Kleinberg J (2002) Bursty and hierarchical structure in streams. In: Proceedings of the 8th ACM SIGKDD international conference on knowledge discovery and data mining, pp 91–101

Leacock C, Chodorow M (1998) Combining local context and WordNet similarity for word sense identification. In: Fellbaum C (ed) WordNet: an electronic lexical database, Chapter 11. The MIT Press, Cambridge, MA

Leontief WW (1941) The structure of American economy, 1919–1929. Harvard University Press

Lesk M (1986) Automatic sense disambiguation using machine readable dictionaries: how to tell a pine cone from an ice cream cone. In: Proceedings of the 5th annual international conference on systems documentation (SIGDOC'86). Toronto, Ontario, Canada, pp 24–26

Leydesdorff L (2012) Alternatives to the journal impact factor: I3 and the top-10% (or otp-25%) of the most-highly cited papers. Scientometrics 92:355–365

Leydesdorff L, Bornmann L, Mutz R, Opthof T (2011) Turning the tables on citation analysis one more time: principles for comparing sets of documents. J Am Soc Inform Sci Technol 62 (7):1370–1381. doi:10.1002/asi.21534

Lin D (1998) An information-theoretic definition of similarity. In: ICML '98 Proceedings of the fifteenth international conference on machine learning, pp 296–304, 24–27 July 1998

Miller GA, Charles WG (1991) Contextual correlates of semantic similarity. Lang Cogn Process 6 (1):1–28

Onodera N, Yoshikane F (2015) Factors affecting citation rates of research articles. J Assoc Inf Sci Technol 66(4):739–764

Piffer D (2012) Can creativity be measured? An attempt to clarify the notion of creativity and general directions for future research. Thinking Skills Creativity 7(3):258–264. doi:10.1016/j.tsc.2012.04.009

Rada R, Bicknell E (1989) Ranking documents with a thesaurus. JASIS 40(5): 304–310

Radicchi F, Fortunato S, Castellano C (2008) Universality of citation distributions: toward an objective measure of scientific impact. Proc Natl Acad Sci 105(45):17268–17272

Resnik P (1995) Using information content to evaluate semantic similarity in a taxonomy

Seeley JR (1949) The net of reciprocal influence: a problem in treating sociometric data. Can J Psychol 3:234–240

Seglen PO (1992) The skewness of science. J Am Soc Inf Sci 43(9):628–638

Shannon CE (1948) A mathematical theory of communication. Bell Syst Tech J 27:3379–3423

Shwed U, Bearman PS (2010) The temporal structure of scientific consensusformation. American Sociological Review 75(6):817–840

Tversky A (1977) Features of similarity. Psychol Rev 84(4):327–352

Wang D, Song C, Barabási A-L (2013) Quantifying long-term scientific impact. Science 342 (6154):127–132

Wu Z, Palmer M (1994) Verbs semantics and lexical selection. In: Proceedings of the 32nd annual meeting on association for computational linguistics. Las Cruces, New Mexciao, pp 133–138, 27–30 June 1994

Zafarani R, Abbasi MA, Liu H (2014) Chapter 3, Network measures. In: Social media mining: an introduction. Cambridge University Press. http://dmml.asu.edu/smm/chapters/SMM-ch3.pdf

Zhu X, Turney P, Lemire D, Vellino A (2015) Measuring academic influence: not all citations are equal. J Assoc Inf Sci Technol 66(2):408–427

Chapter 5
Representing Biomedical Knowledge

Abstract We introduce the structures and features of several widely known and inspirational resources for representing concepts and semantic relations in biomedical knowledge, namely MeSH, ULMS, SemRep, and Semantic MEDLINE. Many examples in subsequent chapters make use of these resources.

Introduction

A remarkable project named the Semantic Knowledge Representation (SKR)[1] is maintained by the National Institutes of Health (NIH). The homepage of the project provides access to tools such as SemRep (Rosemblat et al. 2013a, b) and SemMed, and resources such as SemMedDB (Kilicoglu et al. 2012).

According to the information provided on the SKR project website, SemRep, considered a core resource, extracts semantic predications from text. The SemRep tool was originally developed for biomedical research based on the UMLS knowledge sources. Efforts have been made to extend the domain from biomedicine to areas such as influenza epidemic preparedness, health promotion, and health effects of climate change.

The SKR project maintains a database of 84.6 million SemRep predications extracted from all MEDLINE citations. This is the database behind the Semantic MEDLINE web application. A particularly active area of research with SKR is related to literature-based discovery using semantic predications.

---

[1]https://skr3.nlm.nih.gov/index.html.

© Springer International Publishing AG 2017
C. Chen and M. Song, *Representing Scientific Knowledge*,
https://doi.org/10.1007/978-3-319-62543-0_5

MEDLINE

MEDLINE is the U.S. National Library of Medicine (NLM) premier bibliographic database.[2] It covers over 24 million references to journal articles in life sciences and concentrates on biomedicine. The majority of the journals covered by MEDLINE are selected based on the recommendation of an advisory committee of external experts called the Literature Selection Technical Review Committee (LSTRC). Currently, MEDLINE has more than 5600 journals from worldwide in 40 languages.

MEDLINE covers articles published from 1966, although it does include some pre-1966 records. MEDLINE is the primary component of PubMed, provided by the NLM National Center for Biotechnology Information (NCBI). MEDLINE records are indexed with Medical Subject Headings (MeSH).

MEDLINE refers to its records as citations, which may cause confusions in the context of citation analysis or citation-based indicators. A MEDLINE citation is in fact the bibliographic record of a publication without information concerning the references it cites or the number of times it has been cited by other publications. In contrast, a bibliographic record from the Web of Science or Scopus may contain information on specific references that are cited by the corresponding article and the number of times the article itself has been cited by others within the scope. Thus, in this book, we will avoid using the term citation in the MEDLINE sense. Rather, when we use the term citation, it is meant to be in the Web of Science or Scopus sense.

MeSH

Medical Subject Headings (MeSH) is a controlled vocabulary thesaurus maintained by the NLM. It consists of sets of terms naming descriptors in a hierarchical structure that permits searching at various levels of specificity.

The main purpose of MeSH is to index MEDLINE articles using the controlled vocabulary ("Descriptors" in NLM's term) and the thesaurus ("Entry terms" in NLM's term); thus MeSH can be used for cataloging the articles. During the process of indexing articles (after reading full versions), MeSH concepts are assigned to each MEDLINE article. When MeSH terms are assigned to MEDLINE documents, around 3–5 MeSH terms are set to "MajorTopic" which represents the document very well. We use MeSH terms assigned to the MEDLINE documents since we believe that MeSH terms (especially MeSH descriptors, assigned as "MajorTopic") represent documents more precisely.

MeSH is downloadable from https://www.nlm.nih.gov/mesh/filelist.html. MeSH descriptors are arranged in both an alphabetic and a hierarchical structure. At the

---

[2]https://www.nlm.nih.gov/pubs/factsheets/medline.html.

Fig. 5.1 The hierarchical structure of MeSH descriptors

most general level of the hierarchical structure consists of 16 broad categories and are very broad headings such as "Anatomy" or "Chemicals and Drugs" (See Fig. 5.1). More specific headings are found at more narrow levels of the thirteen-level hierarchy, such as "Ankle" and "Conduct Disorder." There are 27,883 descriptors in 2016 MeSH with over 87,000 entry terms that assist in finding the most appropriate MeSH Heading.

For example, "Vitamin C" is an entry term to "Ascorbic Acid." In addition to these headings, there are more than 232,000 Supplementary Concept Records (SCRs) within a separate file. Generally SCR records contain specific examples of chemicals, diseases, and drug protocols. They are updated more frequently than descriptors. Each SCR is assigned to a related descriptor via the Heading Map (HM) field. The HM is used to rapidly identify the most specific descriptor class and include it in the citation.

In addition to browsing the MeSH descriptors, MeSH terms can be searched. For instance, if you wat to search MeSH with the query "Raynaud disease," then you go to or searchable at https://meshb.nlm.nih.gov/search and type in "Raynaud disease." (see Fig. 5.2).

The MeSH search returns the results for "Raynaud disease" in the following (Fig. 5.3).

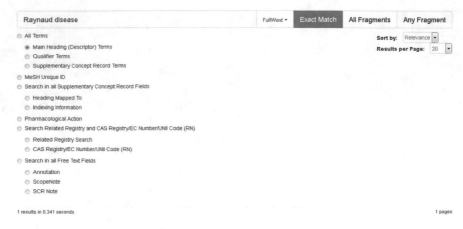

Fig. 5.2 The search page of MeSH

Fig. 5.3 The result page of the query "Raynaud disease."

On this result page, you can navigate the hierarchy of MeSH descriptors by clicking on the link "MeSH Tree Structures."

ULMS

Unified Medical Language System (UMLS), provides a mechanism for integrating all the major biomedical vocabularies including MeSH. UMLS is a set of files and software that brings together many health and biomedical vocabularies and standards to enable interoperability between computer systems. UMLS consists of three knowledge sources; Metathesaurus, Semantic Network, and SPECIALIST lexicon. Metathesaurus as a core is organized by concepts (meaning), synonymous terms are clustered together to form a concept, and concepts are linked to other concepts by means of various types of relationships to provide the various synonyms of

Addison's disease	SNOMED CT	PT	363732003
Addison's Disease	MedlinePlus	PT	T1233
Addison Disease	MeSH	PT	D000224
Primary Adrenal Insufficiency	MeSH	EN	D000224
Primary hypoadreanlism syndrome, Addison	MedDRA	LT	10036696
...	...		

C0001403   Addison's disease

Fig. 5.4 An example of Metathesaurus concept

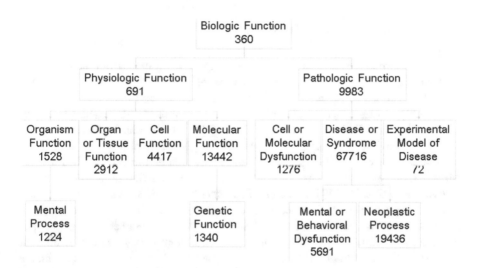

Fig. 5.5 Biologic function hierarchy

concepts and to identify useful relationships between different concepts. All concepts are assigned to at least one semantic type as a category (See Fig. 5.4).

Semantic network consists of 135 semantic types, which are broad subject categories in 2 hierarchies and assigned to all Metathesaurus concepts. It also contains 54 semantic relationships that are defined as useful, important links between types. They are typically hierarchical "is-a" and associative relations. The main goal of semantic network is to categorize the Metathesaurus and enhance meaning of concepts. An example of semantic network is illustrated in Fig. 5.5.

For example, the term *Raynaud Disease* has a semantic type [*Disease or Syndrome*], and *Fish Oils* has a semantic type [*Biologically Active Substance*]. Currently, there are 135 semantic types. Each semantic type has at least one relationship with other semantic types. At this time of writing, there are 54 relations. Both the semantic types and semantic relationships are hierarchically organized (Fig. 5.6).

Since most MeSH terms from MEDLINE documents, are included into UMLS Metathesaurus Concepts, we know the semantic types of MeSH terms. Thus, given

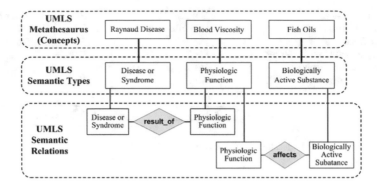

Fig. 5.6   An illustrative example of the UMLS

two MeSH terms, we can derive the relationship between them from their semantic relation. Figure 5.6 shows the relationships of concepts, semantic types, and semantic relations of *Raynaud Disease*, *Blood Viscosity* and *Fish Oils*.

UMLS provides a number of applications as part of the UMLS package. Several key applications include MetamorphoSys, UTS Metathesaurus Browser, UTS Semantic Network Browser, and UMLS APIs. MetamorphoSys is the UMLS installation wizard and UMLS customization tool included in each UMLS release. MetamorphoSys enables local installation of UMLS Metathesaurus, the Semantic Network and SPECIALIST Lexicon. UTS Metathesaurus Browser is a web interface for browsing UMLS concepts and relationships. UTS Semantic Network Browser is an interface for the UMLS Semantic Network—a set of broad subject categories, or semantic types, that provide a consistent categorization of all concepts represented in the UMLS, and a set of useful and important relationships, or semantic relations, that exist between semantic types. UMLS APIs provide programmatic access to the UMLS Metathesaurus.

SemRep

SemRep is a rule-based automatic NLP program developed by NLM that extracts semantic predications (subject-relation-object triples) from biomedical free text (Rindflesch and Fiszman 2003). SemRep uses underspecified syntactic analysis and structured domain knowledge from UMLS. SemRep relies on syntactic analysis based on the SPECIALIST Lexicon and the MedPOS tagger (Smith et al. 2004). MetaMap helps to map noun phrases in the sentences to UMLS Metatheaurus concepts. SemRep interpreted the semantic relationships (syntactic indicators in the sentence, such as verbs, nominalizations, prepositions, etc.) between two concepts in the sentences based on dependency grammar rules and ontology (i.e., an extended version of the UMLS Semantic Network). SemRep represents semantic knowledge from each sentence in citations as the format of semantic predications.

Fig. 5.7 System architecture
of SemRep

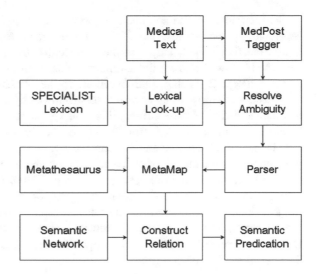

Subject and object arguments of each predication are concepts from the UMLS Metathesaurus and the relation (in uppercase) is a relation from the UMLS Semantic Network.

The overall system architecture of SemRep is illustrated in Fig. 5.7.

SemRep can be run interactively or in batch mode using the SKR Scheduler (https://ii.nlm.nih.gov/Interactive/UTS_Required/semrep.shtml). If SemRep is used in either an interactive or batch mode, you need to have an UMLS account, which can be applied for at https://uts.nlm.nih.gov//license.html. SemRep program is also available as a stand-alone program on Linux platform.

For example, if you have the following sample input from PubMed and feed it into the SemRep interactive mode: "the aim of this study was the characterization of the specific effects of alprazolam versus imipramine in the treatment of panic disorder with agoraphobia and the delineation of dose-response and possible plasma level-response relationships." The output generated by SemRep looks like following:

```
00000000.tx.1 The aim of this study was the characterization of the specific
effects of alprazolam versus imipramine in the treatment of panic disorder with
agoraphobia and the delineation of dose-response and possible plasma level-
response relationships.
00000000.tx.1|relation|C0002333|Alprazolam|orch,phsu|orch|||TREATS|C0030
319|Panic Disorder|mobd|mobd||
00000000.tx.1|relation|C0002333|Alprazolam|orch,phsu|orch||||compared_with|
C0020934|Imipramine|orch,phsu|orch||
00000000.tx.1|relation|C0020934|Imipramine|orch,phsu|orch|||TREATS|C0030
319|Panic Disorder|mobd|mobd||
```

SemRep output falls into three categories: text, entity, and relation. All fields are separated by the vertical bar ("|"). Certain fields can be empty, although in non-production output, they may be represented by non-empty placeholders, as described below. There are five common fields: 1. SE: designates that the output is from SemRep; 2. PMID; 3. Subsection: If the utterance begins with one of a specified set of strings of uppercase letters followed by a colon (see Appendix A for a complete listing of these strings) this field will contain that string; otherwise it is blank; 4. ti: if the utterance is from the title of the citation; ab if the utterance is from the abstract of the citation; and 5. Sentence ID: an integer indicating the utterances position within the title/abstract. There are a number of remaining output fields for relation in SemRep: CUI of the subject concept (C0002333), Preferred name of the subject concept (Alprazolam), Semantic Type(s) of the subject concept (orch,phsu), Subject Semantic Type used for the relation (orch), Predicate (TREATS), CUI of the object concept (C0030319), Preferred name of the object concept (Panic Disorder), Semantic Type(s) of the object concept (mobd), and Object Semantic Type used for the relation (mobd).

Extracting Semantic Predications

SemRep extracts semantic predications from unstructured text of biomedical publications. A semantic predication is a triple of subject, relation, and object, for example, as in HIV-CAUSES-AIDS. Each bibliographic record in MEDLINE contains various metadata of the corresponding publication, including its title, abstract, authorship, and MeSH terms assigned to the article. Each sentence may come from either a title or an abstract. For each sentence, SemRep may find none or several semantic predications for reasons that we will discuss shortly in more detail.

The subject and the object arguments of a semantic predication are concepts defined in the UMLS metathesaurus. The relation, or the predicate, is defined in the UMLS Semantic Network.

There are three ways to make use of SemRep: use its interactive mode, use its match mode, or install SemRep on Linux on your own computer and use its standalone version as you wish.

The Interactive Mode

Figure 5.8 shows the user interface of SemRep's interactive mode. Users need to login first.

First, copy and paste the text to the text window of the Interactive SemRept and submit the request to SemRep to process (Fig. 5.9).

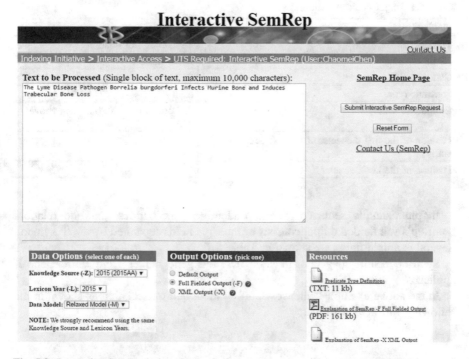

Fig. 5.8 The user interface of the interactive SemRep

Interactive SemRep

Contact Us

Indexing Initiative > Interactive Access > UTS Required: Interactive SemRep (User:ChaomeiChen)

Text to be Processed (Single block of text, maximum 10,000 characters):

The Lyme Disease Pathogen Borrelia burgdorferi Infects Murine Bone and Induces Trabecular Bone Loss

SemRep Home Page

Submit Interactive SemRep Request

Reset Form

Contact Us (SemRep)

Data Options (select one of each)

Knowledge Source (-Z): 2015 (2015AA) ▼

Lexicon Year (-L): 2015 ▼

Data Model: Relaxed Model (-M) ▼

NOTE: We strongly recommend using the same Knowledge Source and Lexicon Years.

Output Options (pick one)

○ Default Output
● Full Fielded Output (-F) ❷
○ XML Output (-X) ❷

Resources

Predicate Type Definitions
(TXT: 11 kb)

Explanation of SemRep -F Full Fielded Output
(PDF: 161 kb)

Explanation of SemRep -X XML Output

Fig. 5.9 Interactive SemRep's interface

SemRep returns the results in the format you choose from the three options: default, full fielded output, and XML. Figure 5.10 illustrates the output in the full fielded output.

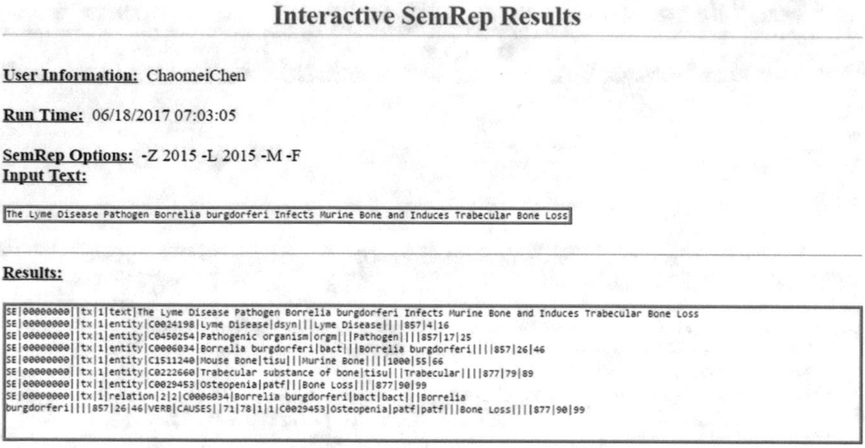

Interactive SemRep Results

User Information: ChaomeiChen

Run Time: 06/18/2017 07:03:05

SemRep Options: -Z 2015 -L 2015 -M -F
Input Text:

```
The Lyme Disease Pathogen Borrelia burgdorferi Infects Murine Bone and Induces Trabecular Bone Loss
```

Results:

```
SE|00000000||tx|1|text|The Lyme Disease Pathogen Borrelia burgdorferi Infects Murine Bone and Induces Trabecular Bone Loss
SE|00000000||tx|1|entity|C0024198|Lyme Disease|dsyn|||Lyme Disease||||857|4|16
SE|00000000||tx|1|entity|C0450254|Pathogenic organism|orgm|||Pathogen||||857|17|25
SE|00000000||tx|1|entity|C0006034|Borrelia burgdorferi|bact|||Borrelia burgdorferi||||857|26|46
SE|00000000||tx|1|entity|C1511240|Mouse Bone|tisu|||Murine Bone||||1000|55|66
SE|00000000||tx|1|entity|C0222660|Trabecular substance of bone|tisu|||Trabecular||||877|79|89
SE|00000000||tx|1|entity|C0029453|Osteopenia|patf|||Bone Loss||||877|90|99
SE|00000000||tx|1|relation|2|2|C0006034|Borrelia
burgdorferi|||857|26|46|VERB|CAUSES||71|78|1|1|C0029453|Osteopenia|patf|patf|||Bone Loss||||877|90|99
```

Fig. 5.10 The results from SemRep in the full fielded output format

Table 5.1 Major fields and illustrative values

Field	Value
Type of result	Text, entity, or relation
Name of entity	Lyme disease
CUI of entity	C0024198
Semantic type	dsyn
Confidence score	857 out of 0–1000, but often above 250
Position of the first character of the text denoting entity	4
Position of the last character	16

In the example sentence, SemRep identified six entities and one relation. SemRep's full fielded output consists of multiple fields separated by the '|' symbol. Several example fields are shown in Table 5.1.

SemRep results may contain gene IDs and gene names from EntrezGene if applicable.

In the above example sentence, SemRep identified one relation as shown in the following full fielded format:

```
SE|00000000||tx|1|relation|2|2|C0006034|Borrelia
burgdorferi|bact|bact|||Borrelia
burgdorferi||||857|26|46|VERB|CAUSES||71|78|1|1|C0029453|Osteopeni
a|patf|patf|||Bone Loss||||877|90|99
```

Simply speaking, the relation defines a semantic predication that the bacteria Borrelia burgdorferi causes bone loss, with a confidence score of 877 out of 1000.

In general, semantic predications are more structured and more accessible than unstructured text. Representing scientific knowledge in terms semantic predications has many appealing advantages. On the other hand, as we can see, a sentence may not use the name of an entity that can be resolved in the UMLS metathesaurus. It may not use any UMLS concepts at all. It can be simply about a subject matter beyond the scope of UMLS. SemRep does not address sentences that do not match closely with UMLS concepts and semantic types. Furthermore, as we will explain later, the extent to which semantic predications can reliably represent the meanings in the original scientific assertions may vary considerably.

The Batch Mode

The batch mode of SemRep is essentially similar to the interactive mode, except that you upload a file as the input of your request and SemRep will notify you via email on the completion of the job. You should have your files ready as the first step. The input format for SemRep is straightforward. Each line in your file should represent one record, starting with the ID of the record and the text. The two fields should be separated by the "|" symbol.

CiteSpace provides a function for you to convert bibliographic records in the Web of Science format to the format for SemRep requests. Follow the Data> Import/Export menu, select the WoS tab and you will see a button for reformatting files for SemRep. A resultant file would look like the following. The record ID is the value of the UT field in the Web of Science format. The text field contains the title and the abstract of the record (Fig. 5.11).

Once the input files are ready, you can upload these files one by one for SemRep. You need to provide your full email address to receive notifications on the completion of the job.

According to the instructions,[3] one of the options for SemRep is whether to use MetaMap's strict or relaxed model (Fig. 5.12). The strict model is a subset of the relaxed model. It means that the relaxed model contains more UMLS concepts than the strict model. For example, both concepts Arms and legs (C0015385) and Disease or syndrome of heart (C0018799) are in the relaxed model, but none in the strict model. We recommend that the relaxed model should be used in order to increase the recall from SemRep. In other words, using the relaxed model may allow SemRep to identify as many entities as possible.

---

[3]https://metamap.nlm.nih.gov/Docs/FAQ/DataModels.pdf.

```
1  A1995RA19100018|CHARACTERIZATION OF A HELICOBACTER-PYLORI NEUTROPHIL-ACTIVATING PROTEIN. Helicobacter pylori-associated
   gastritis is mainly an inflammatory cell response. In earlier work we showed that activation of human neutrophils by a
   cell-free water extract of H. pylori is characterized by increased expression of neutrophil CD11b/CD18 and increased
   adhesiveness to endothelial cells. The work reported here indicates that the neutrophil-activating factor is a
   150,000-molecular-weight protein (150K protein). Neutrophil proadhesive activity copurified with this protein, which is a
   polymer of identical 15K subunits. Specific antibody, prepared against the purified 15K subunit, neutralized the proadhesive
   activity of the pure protein and of water extracts obtained from different strains of H. pylori. The gene (napA) for this
   protein (termed HP-NAP, for H. pylori neutrophil-activating protein) was detected, by PCR amplification, in all of the H.
   pylori isolates tested; however, there was considerable strain variation in the level of expression of HP-NAP activity in
   vitro. HP-NAP could play an important role in the gastric inflammatory response to H. pylori infection.
2  A1995QJ52400012|MOLECULAR-CLONING AND CHARACTERIZATION OF THE NONTYPABLE HAEMOPHILUS-INFLUENZAE-2019 RFAE GENE REQUIRED FOR
   LIPOPOLYSACCHARIDE BIOSYNTHESIS. The lipooligosaccharide (LOS) of nontypeable Haemophilus influenzae (NTHi) is an important
   factor in pathogenesis and virulence. In an attempt to elucidate the genes involved in LOS biosynthesis, we have cloned the
   rfaE gene from NTHi 2019 by complementing a Salmonella typhimurium rfaE mutant strain with an NTHi 2019 plasmid library. The
   rfaE mutant synthesizes lipopolysaccharide (LPS) lacking heptose, and the rfaE gene is postulated to be involved in
   ADP-heptose synthesis. Retransformation with the plasmid containing a 4 kb of NTHi DNA isolated from a reconstituted mutant
   into rfaE mutants gave wild-type LPS phenotypes. Sodium dodecyl sulfate-polyacrylamide gel electrophoresis analysis confirmed
   the conversion of the rfaE mutant LPS to a wild-type LPS phenotype. Sequence analysis of a 2.4-kb Bg/II fragment revealed two
   open reading frames. One open reading frame encodes the RfaE protein with a molecular weight of 37.6 kDa, which was confirmed
   by in vitro transcription and translation, and the other encodes a polypeptide highly homologous to the Escherichia coli HtrB
   protein. These two genes are transcribed from the same promoter region into opposite directions. Primer extension analysis of
   the rfaE gene revealed a single transcription start site at 37 bp upstream of the predicted translation start. site. The
   upstream promoter region contained a sequence (TA AAAT) homologous to the -10 region of the bacterial sigma(70)-dependent
   promoters at an appropriate distance (7 bp), but no sequence resembling the consensus sequence of the -35 region was found.
   These studies demonstrate the ability to use complementation of defined LPS defects in members of the family
   Enterobacteriaceae to identify LOS synthesis genes in NTHi.
3  A1995RB29800006|PHARMACOKINETIC, SAFETY, AND ANTIVIRAL PROFILES OF ORAL GANCICLOVIR IN PERSONS INFECTED WITH
   HUMAN-IMMUNODEFICIENCY-VIRUS - A PHASE I/II STUDY. A phase I/II study evaluated the pharmacokinetics, tolerability, and
   antiviral activity of oral ganciclovir in persons infected with human immunodeficiency virus (HIV). Oral bioavailability
   ranged from 2.6% to 7.3%. The mean maximum serum concentration achieved at 1000 mg every 8 h was 1.11 mu g/mL, and mean
   trough level was 0.54 mu g/mL. The time to maximum serum drug concentration was 1.0-2.9 h, with a serum half-life of 3.0-7.3
   h, suggesting prolonged oral absorption. Serious adverse events were uncommon. Decreased cytomegalovirus (CMV) shedding was
   observed from all sites. The median days (by dosage) to retinitis progression assessed by retinal examination after
   initiation of oral ganciclovir were 62 (1000 mg every 8 h), 148 (500 mg every 3 h), 75 (750 mg every 3 h), 148 (1000 mg every
   3 h), and 139 (2000 mg every 8 h). Thus, oral ganciclovir has pharmacokinetic, toxicity, and antiviral profiles that may
   prove beneficial for both maintenance therapy of CMV retinitis and prevention of CMV disease in HIV-infected persons.
```

Fig. 5.11 The input file format for SemRep's batch mode

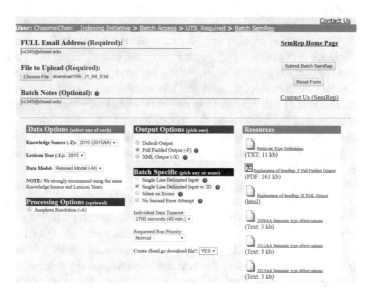

Fig. 5.12 The user interface of the batch mode SemRep

The batch mode of SemRep is managed by a scheduler (Fig. 5.13). Requests from multiple remote users are scheduled to complete. As shown in the screenshot of the scheduler batch job status, there are 145 workstations available. The average processing speed is also shown to the users.

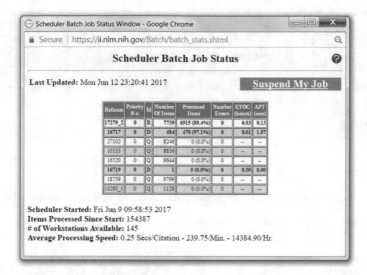

Fig. 5.13 Scheduler batch job status

Once a batch job is completed, the user will be notified with a URL where several files related to the job can be downloaded. SemRep output several files. Table 5.2 lists a few commonly used ones.

The job summary reports the number of items processed, the number of errors reported, and how long it took to complete. And interestingly, it also reports how long it would have taken if the scheduler was not used. We did a test on a dataset of bibliographic records on infectious disease from 1991 till June 2017. SemRep batch scheduler processed 194,059 records requested through multiple jobs. These jobs altogether took 12.099 h to complete. According to the scheduler, it would have taken 30 days if the batch scheduler was not used. Processing ∼200,000 records in 12 h is indeed a very convenient options for users.

Semantic MEDLINE

Another very valuable resource made available by the SKR project of the NIH is Semantic MEDLINE. Semantic MEDLINE is a prototype web application (Figs. 5.14 and 5.15). It integrates PubMed search, natural language processing, automatic summarization, and network visualization.

Table 5.2 Output files from SemRep

File name	Description
text	The original input file
text.out	Batch results file
text.out.ERR	Reported errors
text.out.SUMMARY	A summary of the job

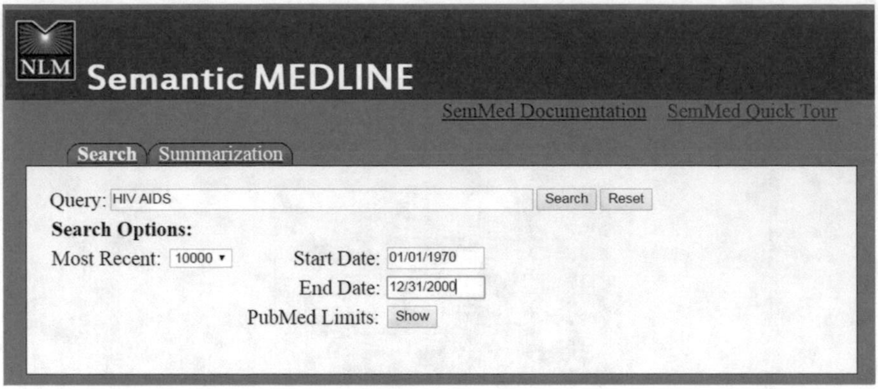

Fig. 5.14 Semantic MEDLINE's search page

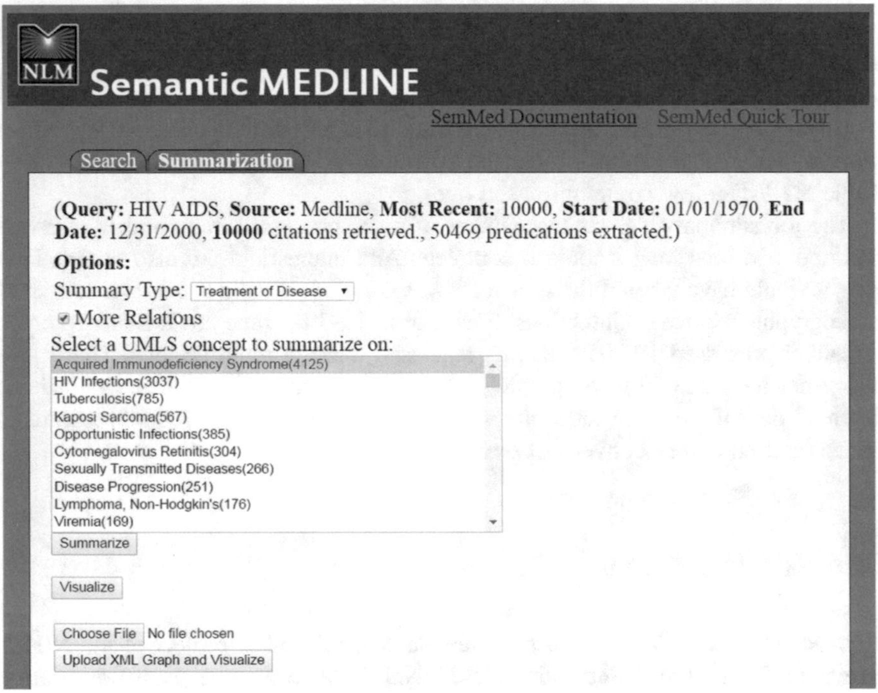

Fig. 5.15 Semantic MEDLINE's result page

Note that Semantic MEDLINE, or SemMed, uses the term citation differently from how we use it elsewhere in the book. The SKR project uses the term citation to refer to the metadata of a scientific article. It does not include any information concerning the citations made or received by the article. In contrast, we use the term citation in the same way as the Web of Science. A citation refers to a specific part of a bibliographic record rather than the entire record. Moreover, MEDLINE records do not contain information regarding neither the references cited by an article nor how many times the article has been cited.

Semantic MEDLINE currently supports four types of summarization, namely Treatment of Disease, Substance Interactions, Diagnosis, and Pharmacogenomics. The default summary type Treatment of Disease focuses on drugs and procedures for the treatment of a disease. These drugs and procedures are commonly referred to as therapeutic interventions. The second summary type Substance Interactions focuses on substances.

The visualization function of the Semantic MEDLINE requires the Adobe Flash plug-in. It can be easily installed if your browser does not have one installed. Semantic predications extracted from PubMed search results form a network, or graph as Semantic MEDLINE calls it. The visualization in Semantic MEDLINE has an upper limit of 1000 predications.

Nodes in the network are entities, which can be the subject or the object of a semantic predication. Links in the network are the semantic relations such as CAUSES, PREVENTS, or TREATS. Semantic MEDLINE's visualization page displays the network (Fig. 5.16). The user can interact with the visualized graph, including dragging a node, panning the canvas, and zooming in and out. The graph

Fig. 5.16 A radial layout visualization of a network based on 1000 predications on HIV and AIDS

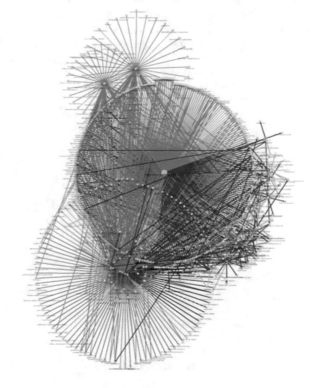

control pane is placed at the bottom of the display. The user can select the layout from four options: spring, node-link tree, circle, and radial. A layout is a configuration of the positions of the nodes. The same network may appear differently in different layouts.

We searched "HIV AIDS" on PubMed within Semantic MEDLINE within the range between 1970 and 2000. This range includes the most active period of research on HIV and AIDS. The search was limited to the top 10,000 most recent results returned from PubMed, which is the maximum allowed by Semantic MEDLINE. The 10,000 MEDLINE records led to a total of 50,469 predications. The most frequently appeared UMLS concept in this set of predications is Acquired Immunodeficiency Syndrome, which appeared 4125 times. HIV infections appeared 3037 times. If we selected the Treatment of Disease as the summary type, the summarizer found 4327 predications. The visualization will be limited to 1000 predications.

The user can select relations to retain in the network from a list of relations. Each relation is assigned a color. For example, the visualized network in Fig. 5.17 shows semantic predications such as HIV-1 CAUSES Dementia and HIV-1 CAUSES Kaposi Sarcoma. If we click on the Kaposi Sarcoma node, the informational panel on the right will show information regarding this concept. For example, the concept Kaposi Sarcoma's CUI is C0036220 of a semantic type of neop and appeared in 56 predications. The user can follow links pointing to additional resources such as OMIM. OMIM—Online Mendelian Inheritance in Man, is a comprehensive, authoritative compendium of human genes and generic phenotypes. It is freely available and updated daily. According to OMIM, Kaposi sarcoma (KS) is an invasive angioproliferative inflammatory condition that occurs commonly in men

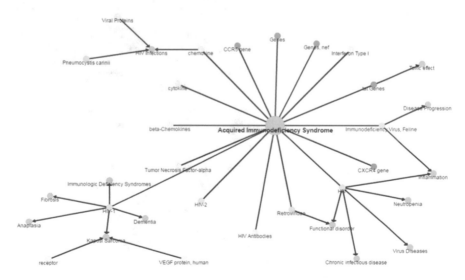

Fig. 5.17 The network contains the CAUSES relations only (lines in red)

infected with human immunodeficiency virus and it is connected to genes such as Interleukin-6 (IL6).

The user can save the entire network to a file in XML by right-clicking on the display. The XML file can be processed for visualization and analysis using graph visualization tools such as Gephi or importing to graph databases such as Neo4j. Alternatively, if the user has a similar type of network generated elsewhere, it is possible to upload the network in XML format to the visualization page.

There are a few things that would be desirable with the Semantic MEDLINE. First, the user can explore the original text of a semantic predication and go to the record's paper on PubMed. However, this function is currently separated from the visualization page, where the user may access further information on entities and relations in UMLS metathesaurus or other resources, but not the original text where a predication comes from. Second, the visualization is convenient for filtering out unwanted relations. However, the user is essentially dealing with the network at the level of individual nodes rather than any aggregations of nodes or subgraphs. Facilitating the analysis of the visualized network at higher levels of aggregation is desirable. Topological properties and temporal properties would also be useful to consider. In part, given that the user can export the network in XML, one may reformat the network file and analyze it using software tools specialized for network modeling and visualization. Regardless, the Semantic MEDLINE provides a valuable service for us to explore scientific knowledge in biomedicine.

References

Kilicoglu H, Shin D, Fiszman M, Rosemblat G, Rindflesch TC (2012) SemMedDB: A PubMed-scale repository of biomedical semantic predications. Bioinformatics 28(23):3158–3160. doi:10.1093/bioinformatics/bts591

Rindflesch TC, Fiszman M (2003) The interaction of domain knowledge and linguistic structure in natural language processing: interpreting hypernymic propositions in biomedical text. J Biomed Inform 36(6):462–477

Rosemblat G, Resnick MP, Auston I, Shin D, Sneiderman C, Fizsman M, Rindflesch TC (2013a) Extending SemRep to the public health domain. J Am Soc Inform Sci Technol 64(10):1963–1974. doi:10.1002/asi.22899

Rosemblat G, Shin D, Kilicoglu H, Sneiderman C, Rindflesch TC (2013b) A methodology for extending domain coverage in SemRep. J Biomed Inform 46(6):1099–1107

Smith L, Rindflesch TC, Wilbur WJ (2004) MedPost: a part-of-speech tagger for biomedical text. Bioinformatics 20(14):2320–2321

Chapter 6
Text Mining with Unstructured Text

Abstract This chapter introduces computational techniques that enable us to extract concepts, relations, and other patterns from text documents, and from scientific publications in particular. After targets of interest have been extracted and annotated, text mining techniques can be applied to identify higher-order patterns and trends that may not be obvious from individual documents. The basic concepts and the general procedure for applying these tools to the study of scientific publications are explained with illustrative applications.

Natural Language Processing

The ultimate goal of Natural Language Processing (NLP) is to capture meaning from an input of words (sentences, paragraphs, pages, etc.) in the form of a structured output (which varies greatly depending on the application) so that further analytics can be applied to the output of NLP.

There are a variety of approaches for NLP, which can be classified into three approaches: symbolic, statistical, and hybrid. The symbolic approach to NLP is based on human-developed rules and lexicons, which is based on a set of accepted rules of speech within a given language that are materialized and recorded by linguistic experts for computer systems to follow. The statistical approach is based on observable and recurring examples of linguistic phenomena. Models based on statistics recognize recurring themes by mathematical analysis of large text corpora. The system can develop its own linguistic rules that it will use to analyze future input and/or the generation of language output by identifying trends in large samples of text. The hybrid approach is a combination of the symbolic and statistical approaches. This approach starts with generally accepted rules of language and tailors them to specific applications from input derived from statistical inference.

© Springer International Publishing AG 2017
C. Chen and M. Song, *Representing Scientific Knowledge*,
https://doi.org/10.1007/978-3-319-62543-0_6

Modeling and Analytic Tools

There are many open source as well as commercial NLP tools. In Table 6.1, we only listed some of well-known, open source tools that are pertinent to the objectives of this book.

Table 6.1 A list of well-known, open source NLP tools

Tool	Description	Platform
Stanford's CoreNLP	A pipeline framework of tools for processing English, Chinese, and Spanish. Includes tools for tokenization (splitting of text into words), part of speech tagging, grammar parsing (identifying things like noun and verb phrases), named entity recognition, sentiment analysis, and more. There are several spin-off projects based on Stanford's CoreNLP http://nlp.stanford.edu/software/corenlp.shtml	Java
GATE and Apache UIMA	GATE combined with UIMA provides a placeholder for building complex NLP workflows which need to integrate several different processing steps. In these cases, a framework like GATE or UIMA is a good option for standardizing and abstracting much of the repetitive work that goes into building a complex NLP application https://gate.ac.uk/	Java
Natural language toolkit	Similar to the Stanford CoreNLP, it includes capabilities for tokenizing, parsing, and identifying named entities as well as many more features http://www.nltk.org/	Python
Apache Lucene and Solr	While originally targeted at solving Information Retrieval problems, Lucene and Solr contain a number of powerful tools for working with text ranging from advanced string manipulation utilities to powerful and flexible tokenization libraries to blazing fast libraries for working with finite state automatons http://lucene.apache.org/	Java
Apache OpenNLP	Using a different underlying approach than Stanford's CoreNLP, the OpenNLP project is an Apache-licensed suite of tools to do tasks like tokenization, part of speech tagging, parsing, and named entity recognition. While not necessarily state of the art anymore in its approach, it remains a solid choice that is easy to get up and running http://opennlp.apache.org/	Java
ScalaNLP	ScalaNLP is the umbrella project for several libraries, including Breeze and Epic. Breeze is a set of libraries for machine learning and numerical computing. Epic is a high-performance statistical parser and structured prediction library http://www.scalanlp.org/	Scala

(continued)

Table 6.1 (continued)

Tool	Description	Platform
Snowball	Snowball is a string processing language designed for creating stemming algorithms for use in Information Retrieval in many different languages including English, French, Spanish, Portuguese, Italian, Romanian, German, Dutch, Swedish, Norwegian, and Danish http://snowball.tartarus.org	Java C
Deeplearning4j	Deeplearning4j is designed to be used in a big scale setting in business environments, rather than as a research tool. It is a Java-based, industry-focused, commercially supported, distributed deep-learning framework https://deeplearning4j.org/	Java
Torch	Torch is written in Lua, and used at NYU, Facebook AI lab and Google DeepMind. It claims to provide a MATLAB-like environment for machine learning algorithms. Lua is easily to be integrated with C so within a few hours' work, any C or C ++ library can become a Lua library." With Lua written in pure ANSI C, it can be easily compiled for arbitrary targets http://torch.ch/	Lua
TensorFlow	TensorFlow is an open source library for numerical computation using data flow graphs (which is all that a Neural Network really is). Originally developed by the researchers on the Google Brain Team within Google's Machine Intelligence research organization, the library has since been open sourced and made available to the general public https://www.tensorflow.org/	Python

Information Extraction

Information extraction (IE) is a research topic to automatically extract target information from unstructured text. IE is involved in two major tasks: entity extraction and relation extraction. Extracting entities such as people, organizations, locations, times, dates, prices, etc. from unstructured text. Entities are objects that often the major nouns in texts. Extracting relations is associated with identifying the relation between two entities. Example relation types are located in, employed by, part of, and is associated with.

Extracting Entities from Text

Extracting entities is mainly studied in the field of Named Entity Extraction (NEE) or Named Entity Recognition (NER). The NER problem is a tagging task, similar to part-of speech (POS) tagging. Thus, if entity extraction is carried out by a supervised learning approach, the task is typically uses sequence classifiers like

Hidden Markov Models (HMMs) or Conditional Random Fields (CRFs). In that case, features used to train classifier usually include words, POS tags, word shapes, orthographic features, gazetteers, etc.

With the huge amount of accessible biomedical literature nowadays, extracting entities from the literature has been receiving more and attention. Entity extraction from biomedical literature can be used to automatically extract useful biomedical information, particularly those key concepts dealing with genes, proteins, diseases and associations among them. The information extracted from biomedical literature has notable potential to automate database construction in biomedicine, with minimal human effort. Entity extraction can also be useful in other areas. Query suggestion, for instance, is another important application area, where concepts can be output as correction suggestions for misspelled queries. Entities are also widely used for text categorization tasks. Most of text categorization techniques are based on word and/or phrase analysis of the text. It has been shown that concept-based text categorization can help improve the precision of clustering documents by topic. Also entity extraction is important and useful in areas like automatic text summarization, information retrieval, question answering, and so forth.

There are various approaches to handle the entity extraction problem. Most of them are based on statistical features such as word counting, inverse document frequency (IDF) as well as semantic features. Attempting to automatically extract useful biomedical information from web accessible biomedical literature, particularly the key concepts dealing with genes, proteins, drugs and diseases and associations among these concepts, Fu et al. (2002) developed a system called VCGS (Vocabulary Cluster Generating Systems) that automatically extracts and determines associations among tokens from biomedical literature. They used three local databases to validate tokens extracted that are gene names or protein fragments. Both statistical and semantic features were used for token extraction. They proposed a clustering algorithm to identify specific groups of tokens, collectively represented as centroids, which are different from each other in terms of their separation as individual clusters. Similarly, Shehata et al. (2007) exploited the semantic structure at both sentence and document level. Their model combined the selected statistical features and the conceptual ontological graph representation that they built. Majoros et al. (2003) proposed a method of improving the quality of automatically extracted noun phrases by employing prior knowledge during the hidden Markov model (HMM) training procedure for the part-of-speech tagger. They modified the basic Markov model tagger with states corresponding to part-of-speech tags and an alphabet of symbols corresponding to individual words.

External ontologies and thesauri are also widely used for concept extraction tasks in biomedical domain. Rindflesch et al. (2000) developed a system that extracts information about drugs and genes relevant to cancer from the biomedical literature. Two external ontologies were used to build their system: the MEDLINE database of biomedical citations and abstractions and the Unified Medical Language System (UMLS), which provides syntactic and semantic information about the terms identified in the biomedical abstracts. Zhou et al. (2006) introduced an approximate dictionary lookup technique to capture significant words rather than

```
The [European Commission ORG] said on Thursday it
disagreed with [German MISC] advice.
Only [France LOC] and [Britain LOC] backed
[Fischler PER] 's proposal .

"What we have to be extremely careful of is how
other countries are going to take [Germany LOC]
's lead", [Welsh National Farmers ' Union ORG]
( [NFU ORG] ) chairman [John Lloyd Jones PER]
said on [BBC ORG] radio .
```

Fig. 6.1 An example of entity extraction results for location, organization, and person

all words in a concept name. They also used UMLS as the dictionary to train the significance score of each word to biological concepts containing that word. A set of simple rules were applied to identify the boundary of a concept candidate and their experimental results show that their approach can dramatically improve the extraction recall while maintaining the precision.

Given unstructured text, the goal of a NER tool is to tag the sequence of words denoting a target entity type. For instance, the below example show the results of NER tagging. For the task of extracting four types of entities such as ORG, LOC, PER, MISC, the below examples shows that a NER classifier tags the words or phrases predicted to be an entity (Fig. 6.1).

Extracting Entities from Biomedical Literature

To demonstrate the process of entity extraction, we use our concept extraction system, which is publicly available at http://informatics.yonsei.ac.kr/tsmm/uncertainty_book/ConceptExtraction.zip.

First, we start with a parsing procedure using biomedical Named Entity Recognition (NER) software to extract key entities from the input text. In particular, we utilize Lingpipe's NER API and a statistical model trained on the Genia corpus. In addition, we use a biomedical domain ontology to map the extracted entities into concepts. We choose the Unified Medical Language System (UMLS) installed on our server as a MySQL database of biomedical concepts and relationships between them. UMLS offers a semantic network that allows retrieving higher level semantic types of a 'is-a' link nature. Since such semantic types are usually general enough for humans to interpret, we will use them in the final stages of the algorithm to extract meaningful concept descriptions from text documents.

The mapping is based on matching entities to corresponding concept strings used as concept labels in the database. We only use exact string matching because most named entity strings are defined exactly the same way as concept labels in the database. Although this leads to a lower recall, the precision remains very high. The number of concepts matched is usually sufficient for building a good graph representation of a document.

After mapping the extracted entities, the next step is to build concept graphs. Nodes in a graph represent concepts in a document. Edges represent their relationships. For each concept node, we search for additional related nodes so as to enhance the concept extraction process later on. We query the UMLS database, resultant concepts are added as new nodes to the graph unless they already exist, in which case we add the relation only.

During this process, some concepts might occur repeatedly as commonly related concepts. We keep track of the occurrence count and use it later on in a weighting scheme. At this point, graphs include many concept nodes that may or may not be related to one another. Within the same graph, the domain similarity of concepts varies since a single graph may include more than one general idea from the text. We then group similar concepts using the k-Medoids method to compact a group of concepts into the most representative one for the final extraction phase.

k-Medoids is a variant of the popular k-Means clustering algorithm. The main difference is that in k-Medoids, the center chosen for each cluster is one of the existing data elements in the set as opposed to finding a mean value for each cluster. We divide each graph into k clusters, where k is chosen so that it matches to either the number of author-provided keyword labels or the number of the top labels generated by the KEA package (http://www.nzdl.org/Kea).

Initially, concepts are added to the clusters randomly. In the following steps, the algorithm tries to find a better medoid candidate for each cluster based on concept weights. The weights are calculated for each concept based on (1) the average distance to all other nodes in the cluster and (2) the concept occurrence count mentioned earlier. The distance between two concepts is a compound value based on their text similarity and the relationship between them. Next, the distance between each node and the medoid of each cluster is calculated. If a node is closer to another cluster other than the one it currently belongs to, it will be placed in the new cluster. This process is repeated until all medoids in the graph are fixed. The process is summarized as follows:

1. Apply NER to extract biomedical named entities
2. Map entities to UMLS concepts using string matching and add concept weights
3. Add related nodes (parents and synonyms)
4. Use k-Medoids to group the top k concepts and extract the medoids

   - Node Distance is calculated based on text similarity and on relationships
   - Concept occurrence frequency is also used in the medoid calculation score

Supposed the sample text from a text file called input.txt contains the following sentences:

The occurrence of subsequent neoplasms has direct impact on the quantity and quality of life in cancer survivors. We have expanded our analysis of these events in the Childhood Cancer Survivor Study (CCSS) to better understand the occurrence of these events as the survivor population ages.

Use the following command to extract named entities from the text. In addition, the graph object is generated in the concept extraction project directory for visualization.

```
java -Xms64m -Xmx12550m -cp .:./bin:./lib/* ce.Main4.
```

The extracted named entities are listed as follows:

processing input.txt
named entitiy: subsequent neoplasms(subsequent neoplasms)
named entitiy: cancer survivors(cancer survivors)
named entitiy: Childhood Cancer Survivor Study(Childhood Cancer Survivor Study)
adding related...
Concept: childhood cancer survivor study
named entitiy: CCSS(CCSS)
adding related...
Concept: ccss
named entitiy: survivor population(survivor population)
named entitiy: neoplasms(neoplasms).

Extracting Relations from Text

An important step to understand human natural language automatically is relation extraction. If we can turn unstructured text into structured by annotating semantic information in a programmatic way, knowledge buried in the sheer volume and heterogeneity of data can be available to create new values for humanity. The reliable, accurate relation extraction is not a trivial task.

Examples of relations are person-affiliation and organization-location. Existing named entities recognizers (NER) (e.g., Bikel et al. 1999; Finkel et al. 2005) can automatically label data with high accuracy. However, the computer needs to know how to recognize a piece of text having a semantic property of interest in order to make a correct annotation. Thus, extracting semantic relations between entities in natural language text is an important step towards natural language understanding applications.

A relation is defined in the form of a tuple $t = (e1, e2, ..., en)$ where the ei are entities in a pre-defined relation r within document D. Most relation extraction systems focus on extracting binary relations. Examples of binary relations include located-in (CMU, Pittsburgh), father-of (Manuel Blum, Avrim Blum). It is also possible to go to higher-order relations as well. For example, in the sentence "At codons 12, the occurrence of point mutations from G to T were observed" exists a 4-ary biomedical relation. The biomedical relationship between a type of variation,

Table 6.2 The sample relation types in the news domain

Relations		Examples	Types
Affiliations	Personal organizational artifactual	Married to, mother of, spokesman for, president of, owns, invented, produces	PER → PER PER → ORG (PER| ORG) → ART
Geospatial	Proximity directional	Near, on outskirts, southeast of	LOC → LOC LOC → LOC
Part-of	Organizational political	A unit of, parent of annexed, acquired	ORG → ORG GPE → GPE

its location, and the corresponding state change from an initial-state to an altered-state can be extracted as point mutation(codon, 12, G, T).

Depending on the domain that relation extraction is applied to, the list of relation types will be determined. For example, in relation extraction for the news articles, the following would be an example of relation types (Table 6.2).

Another example of relation types is from the Automatic Content Extraction (ACE) program held in 2003. The goal of the program is to develop technology to automatically infer from human language data the entities being mentioned, the relations among these entities that are directly expressed, and the events in which these entities are involved. Data sources include audio and image data in addition to pure text, and Arabic and Chinese in addition to English. One of the tasks offered by ACE 2003 was relation extraction called the relation detection and characterization task (RDC). This task requires detection and characterization of relations between (pairs of) entities. There are four general types of relations, some of which are further sub-divided, yielding a total of 24 types/subtypes of relations:

ROLE: relates a person to an organization or a geopolitical entity
subtypes: member, owner, affiliate, client, citizen
PART: generalized containment
subtypes: subsidiary, physical part-of, set membership
AT: permanent and transient locations
subtypes: located, based-in, residence
SOCIAL: social relations among persons
subtypes: parent, sibling, spouse, grandparent, associate.

To discover the hidden knowledge from the unstructured text, NLP techniques were adopted to reveal the relation extraction patterns, which splits the sentences into word or presents syntactic structures (Zhou and He 2008; Bui et al. 2010). Bui et al. (2010) extract the drug-mutation relation from PubMed abstract by applying a rule-based approach. To extract the relation, they justify the two rules. The first rule is <keyword, relation keyword> pattern which is mostly common in sentences. The second rule is <relation, keyword1, keyword2> which calculates the distance and number of occurrences in the phrase. Also, Koike et al. (2005) present an extraction method from the biomedical text by using a shallow parser. NLP helps to assign Gene Ontology ID to PubMed abstracts and then they use shallow parsing approaches to break down and analyze the sentences. After parsing the sentences, they extract ACTOR-OBJECT relation from the sentence structure. Huang et al. (2006) propose a new approach, a hybrid method using shallow parsing and pattern matching, to extract relation between two proteins from biomedical literature. They use rule-based shallow parsing that defines heads of each chunk and processes appositive and coordinative structures. The result indicates that pattern matching is remarkably improved with shallow parsing.

Several researches adopted feature based approaches to extract the relation on the biomedical text. To extract the relation on Protein-Protein Interaction (PPI), Song et al. (2011) propose the relation extraction technique called PPISpotter which is a combination of active learning and semi-supervised SVM techniques. They extract features from MEDLINE records by using NLP techniques. Chowdhury et al. (2011) extract the relation on drug—drug interaction (DDI). They employ the feature-based method which uses different the feature selection technique compared to Song et al.'s study (2011). Their features are word features, morpho-syntactic features, trigger words, and negation. Using SVM classifier with selected features, they evaluate their performance of DDI extraction.

Many researches employ different feature based approaches to Protein-Protein interaction extraction including Song et al.'s study (2011). Lin et al. (2011) extract the PPI relation by using a multiple kernels learning based approach which ensembles the feature-based kernel, tree kernel and graph kernel. Furthermore, they propose a lexical feature-based technique which considers not only bag of word features but also n-gram features. Yang et al. (2010) propose a BioPPISVMExtractor to extract PPI from the biomedical literature which is based on SVM classifier. They select various features including word features, keyword features, protein names distance feature and link path features. Also, Chen et al. (2011) propose PPIEor to extract PPI pairs from the biological literature. They use SVM classifier to extract features based on clause parsing output. Features they use include word feature, distance feature and location feature.

There are a set of common steps that are involved in relation extraction in biomedical literatures. Figure 6.2 illustrates how the task of relation extraction can be carried out in a common scenario.

For binary classification of relation where only true and false labels exist, the feature set for relation extraction are generated by the following three techniques: Named Entity Recognition, Shallow Parsing, and Negation. Of course, other types of features may be used, but for the simplicity reason, we used those three representative features for the tutorials.

Named Entity Recognition

The Named Entity Recognition (NER) technique automatically extracts pre-defined Named Entities (NEs) like gene, protein, and cell in text. It tags each word whether it is located in the starting or ending position, or outside the target entity. Most

Fig. 6.2 An architecture of a typical supervised relation extraction system

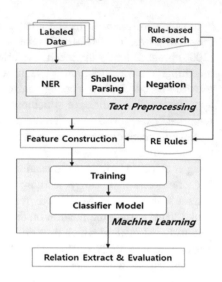

corpora for relation extraction provide NE annotations that have information about target entities in a given text. We extract NEs by using a LingPipe tool introduced earlier in this chapter.

Shallow Parsing

Shallow parsing, also known as text chunking, splits sentence into phrases, such as Noun Phrase (NP), Verb Phrase (VP), Prepositional Phrase (PP), and Adverb Phrase (ADVP). This shallow parsing result gives us an important clue to extract relation in that relation of between entities is usually expressed in [entity1…verb… entity2] structure in a sentence. We apply shallow parsing to all sentences by the Apache OpenNLP toolkit (https://opennlp.apache.org/).

Negation

The negation technique examines whether a sentence is negated or not by finding negation terms ('neither', 'not', etc.) and a negation scope. In relation extraction, negation terms change the relation judgment in an opposite direction. We use NegEx (Chen et al. 2011) toolkit for negation detection.

We combine a rule-based approach with a machine learning (ML) based approach in order to efficient relation extraction. In particular, the hybrid framework consists of the rule-based feature selection and the ML-based classification algorithm.

Feature Construction from Defined Rules

In a rule-based approach, rules are constructed by combination of complex factors such as sentence structure, relation keyword, distance of between entities, grammatical relation and so on. Since those factors appear differently along with relationship type, it can be treated as variables for statistical algorithm. Those factors are the clue that predicted the difference to differ between relationships involving sentence and the others. Our approach is to choose features that represent the key clue for extracting relation in a rule-based approach.

For relation extraction, we used seven features using relation keyword, negation, distance of between two entities, location, order of entities and relation keyword etc. The seven features are as follows:

1. Predicate: a main verb that is located inside or nearest two entities. It must be found in the BioVerb list
2. Predicate POS: part-of-speech of predicate
3. Number of left words: the number of words in the left side of the first appearance of a named entity in a sentence
4. Number of right words: the number of words in the right side of the last appearance of a named entity in a sentence
5. Number of words in between entities: the number of words in the first left named entity and the extreme right named entity in a sentence.
6. Negation: sentence is negated or not
7. LinkPath: link path between two named entities exists or not.

ML-Based Classification

The pattern matching method finds a sentence in accordance with the predefined patterns. It could give more accurate results if patterns are precisely articulated. However, at the same time, it is very difficult to detect matches due to wide variations. Hence, we apply a machine learning algorithm for relation extraction with the rule-based feature set. We treated relation extraction as binary classification task. A sentence is classified depending on whether relation between entities exists or not. We use the WEKA toolkit for classification algorithms.

Given the following input,

11218788 Larsen J, Arnberg A, Brosen K: [Tramadol and oxazepam. Ugeskr Laeger. 2001 Jan 22; 163(4):458-60. Effect on pulmonary function in elderly patients with chronic obstructive lung disease]. Many patients with chronic obstructive pulmonary disease (COPD) suffer from osteoporotic pain as a result of glucocorticoid treatment and nervous symptoms partly related to their lung disease. There seems to be some reluctance to treat these patients with an opioid or benzodiazepine. Upon request, the Drug Information Centre in Odense made an extensive literature search on the subject. No documentation was found that tramadol additionally depresses the respiration in patients with COPD, nor has oxazepam in clinically relevant doses been found to exacerbate their lung disease. The clinical effect is subject to large interindividual variability, and the use of these drugs should, to a greater extent, rest on experience with the individual patient. There seems to be no reason to maintain a priori this rigoristic reluctance to use tramadol and/or oxazepam in patients with COPD.

The following results are produced:

ID 11218788 ANSWER N LEFT ENTITY tramadol RIGHT ENTITY COPD

The above report means that no relation between tramadol and COPD is predicted. For readers who are interested in reproducing the procedure, download the tool at http://informatics.yonsei.ac.kr/tsmm/uncertainty_book/RE.zip. Once you download and un-compress it, change to RE and run the following command:

```
java -Xms64m -Xmx15550m -cp .:./bin:./lib/* evaluation. PolyDrugGeneEvaluation
```

Well-Known Relation Extraction Tools

There are several tools that do relation extractions including PKDE4J, OpenIE, Stanford CoreNLP OpenIE, GATE, etc. PKDE4J will be explained in the next chapter. Here a well-accepted relation extraction tool, OpenIE, will be described.

Open IE

Open IE, standing for the Open Information Extraction (Open IE) system, was developed by the Etzioni's group at University of Washington. Open IE takes natural language sentences as an input and extracts relations in text. For example, consider the following sentence:

> The U.S. president George W. Bush gave his speech on Friday to hundred thousands of people.

There are many binary relations in this sentence that can be expressed as a triple (A, B, C) where A and B are arguments, and C is the relation between those arguments. Since Open IE is not aligned with an ontology like WordNet, the extracted relation is a phrase of text. The following list shows binary relations extracted from the sentence above:

1. (George W. Bush, is the president of, the U.S.)
2. (George W. Bush, gave, his speech)
3. (George W. Bush, gave his speech, on Friday)
4. (George W. Bush gave his speech, to hundred thousands of people).

The first result above is a "noun-mediated extraction", because the extraction has a relation phrase is described by the noun "president". The above results show that an n-ary extraction represents them in an informative way. Here is a possible list of the n-ary relations in the sentence:

(George W. Bush, is the president of, the U.S.)
(George W. Bush, gave, [his speech, on Friday, to hundred thousands of people])

To use the Open IE system, first install the system by typing the following command in the UNIX like prompt: sbt compile. Open IE uses Java 7 SDK and the sbt build system. The sbt command makes downloading dependencies and compiling very simple. The sbt command results in the jar file called "openie-assembly.jar" that contains all required dependent libraries. Once the jar file for Open IE is ready, execute the following command:

```
java -jar openie-assembly.jar.
```

The Open IE system takes one sentence per line unless the argument "–split" is specified. If the argument "–split" is presented, the input text will be split into sentences. Input can be fed into Open IE either as a file (an option first argument) or in an interactive mode where you type sentences interactively. Results will be written to the console unless a second option argument is specified for an output file.

Open IE takes a number of command line arguments. All of available arguments are displayed if you run java -jar openie-assembly.jar–usage. There are several interesting arguments. The first argument is "–binary" that generates the triple output. The second argument is "–split" that partitions the input text into sentences. The third argument is "–ignore-errors" that allows for Open IE to continue to execute even if an exception is encountered. Regarding the output format, There are two formats: simple and column. The argument "–format simple" enables to make ease of reading whereas a columnated format is used for machine processing.

Extracting Semantic Predications with SemRep

SemRep, standing for Semantic Knowledge Representation, is an automatic program that extracts semantic predications (subject-predicate-object triples) from biomedical free text (Rindflesch and Fiszman 2003). SemRep was developed at developed at National Library of Medicine. SemRep uses MetaMap to map noun phrases to UMLS concepts. Through its rule-based summarization system, it maps the syntactic elements to semantic network predicates. About 36.7 millions of sentences extracted from titles and abstracts of PubMed generate the predication analysis. SemRep detects about 12.7 millions of unique predicate instances and 58 unique predicate types (Aronson 2001).

Semantic predications are extracted based on the UMLS knowledge sources where subject and object are UMLS Metathesaurus concepts and the predicate to a relation type in the UMLS Semantic Network. SemRep extracts a wide range of predicates including (1) clinical medicine such as TREATS, DIAGNOSES, and PROCESS_OF, (2) substance interactions such as INTERACTS_WITH, INHIBITS, and STIMULATES, (3) genetic etiology of disease such as ASSOCIATED_WITH and CAUSES, and pharmacogenomics such as AFFECTS, AUGMENTS, and DISRUPTS. SemRep can be run interactively or in batch mode using the SKR Scheduler. SemRep program is also available as a stand-alone program on Linux platform.

For example, given the input text:

dexamethasone is a potent inducer of multidrug resistance associated protein expression in rat hepatocytes

SemRep generates three semantic predications as follows:

- Dexamethasone STIMULATES Multidrug Resistence—Associated Proteins
- Multidrug Resistance—Associated Proteins PART_OF Rats
- Hepatocytes PART_OF Rat

SemRep is part of the SKR project that maintains a database of 84.6 million SemRep predications extracted from all MEDLINE citations (Hristovski et al. 2006). SKR stands for Semantic Knowledge Representation which is available at https://skr.nlm.nih.gov/. The SemRep database supports the Semantic MEDLINE Web application, which integrates PubMed search, SemRep predications, automatic summarization, and data visualization. The goal of the application is to assist users to manage the results of PubMed searches. Output is visualized as an informative graph with links to the original MEDLINE citations. Convenient access is also provided to additional relevant knowledge resources, such as Entrez Gene, the Genetics Home Reference, and UMLS Metathesaurus.

As a tool designed for automatic identification of semantic predication from biomedical literature, SemRep operates by applying a set of linguistic rules to sentences found in MEDLINE abstracts. Semantic relations identified by SemRep (Ahlers et al. 2007; Hristovski et al. 2006) have been used in literature-based discovery (LBD) (Wilkowski et al. 2011), among many other approaches to mining information from biomedical literature. Biomedical articles on which SemRep is designed to operate contain explicit and implicit mentions of relationships between various medical concepts. For example a TREATS relation between a medication and a disorder may be found in a single sentence in a MEDLINE citation containing the following text: "Metamorphosia associated with topiramate for migraine prevention."

One example of using SemRep for semantic predications is creating RDF triples. It would be interesting to observe whether semantic predications can be leveraged to create network representations (will leverage the OWL-NETS abstraction to create networks containing only the mechanisms of interest). Supposed that we take

Table 6.3 Concept table

CONCEPT_ID	CUI	TYPE	PREFERRED_NAME	GHR	OMIM
1844	C0003873	META	Rheumatoid Arthritis	NULL	180300:604302
1276072	215	ENTREZ	ABCD1	NULL	NULL

Table 6.4 CONCEPT_SEMTYPE table

CONCEPT_SEMTYPE_ID	CONCEPT_ID	SEMTYPE	NOVEL
2628	1844	dsyn	Y
1481123	1276072	gngm	Y

Table 6.5 PREDICATION table

PREDICATION_ID	PREDICATE	TYPE
87120	PROCESS_OF	semrep

drug repurposing such as Rapamycin, Tamoxifen as a use scenario. There was a previous effort that SemRep was converted into RDF.[1]

The following words and phrases can be used to search for Tamoxifen for Repurposing in PubMed: Tamoxifen (C0039286), Bipolar Disorder (C0005586), Manic (C0338831), Protein Kinase C (C0033634), Protein Kinase C Inhibitor (C1514555). RDF "schema" for the SemRep predications consists of the following three:

1. UMLS CUI, relationship, UMLS CUI—annotation triple
2. UMLS CUI, rdfs:label, <preferred term>—label triple
3. UMLS CUI, umls:semtype, <semantic type>—semantic type triple

An example for the SemRep annotation is as follows:
Protein Kinase C Inhibitor TREATS Bipolar Disorder

umls:C1514555 umls: TREATS umls:C0005586
umls:C1514555 rdfs: label "Protein Kinase C Inhibitor"
umls:C0005586 rdfs:label "Bipolar Disorder"
umls:C1514555 umls: semtype "mobd"
umls:C0005586 umls: semtype "phsu"
(mobd = Mental or Behavioral Dysfunction)
(phsu = "Pharmacologic Substance")

The transformation of the SemRep tables into triples results in the following tables (Tables 6.3, 6.4, 6.5 snd 6.6).

---

[1]https://github.com/OHDSI/KnowledgeBase/tree/master/LAERTES/SemMED.

Table 6.6 PREDICATION_ARGUMENT table

PREDICATION_ARGUMENT_ID	PREDICATION_ID	CONCEPT_SEMTYPE_ID	TYPE
176604	87120	2628	S
176605	87120	21437	O

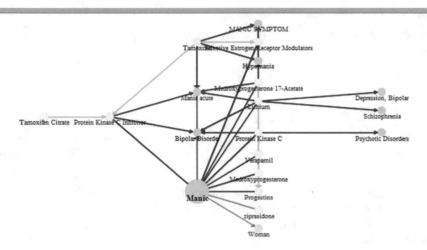

Fig. 6.3 A resulting RDF graph for drug reposition

The example query for the concept table would look like "column = CUI

With prefix umls: if TYPE = META
With prefix e.g if TYPE = ENTREZ
(ignore other types)

For label: object column = PREFERRED_NAME (literal)"
The example query for the concept_semtype table would look like "Subject column = CONCEPT_ID
    With prefix umls: or e.g.: depending on what was assigned in the CONCEPT table
For semantic type: column = SEMTYPE (literal)"
For a given PREDICATION_ID
Subject = PREDICATION_ARGUMENT.CONCEPT_SEMTYPE_ID   (where TYPE = S)
Predicate = PREDICATION.PREDICATE
Object = PREDICATION_ARGUMENT.CONCEPT_SEMTYPE_ID   (where TYPE = O)

    Given these transformed semantic predications, we are now able to visualize a RDF graph like Fig. 6.3.

Topic Modeling

Topic modeling methods are known to be useful for analyzing and summarizing large scale textual data in an unsupervised manner. Topic modeling have been applied in various different data sources including biomedical data, images, videos, and social media (Blei 2012). The goal of topic modeling is to group sets of words which are co-occurred within texts as topics by giving high probability for the words under same topics. The prominent feature of topic modeling lies in its ability not to require any training datasets which normally demand tremendous manual efforts of annotating or labeling and make quality of output heavily depend on training datasets.

Among the topic modeling algorithms, Latent Dirichlet Allocation (LDA) is the simplest and most well-accepted algorithm. LDA is a generative model (e.g., Naïve Bayes) which is a full probabilistic model of all the variables. In generative modeling, data is derived from a generative process which defines a joint probability distribution of observed and hidden variables. It contrasts to discriminative model (e.g., linear regression) which only models the conditional probability of unobserved variables on the observed variables. In LDA, the observed variables are words in the documents, and the hidden variables are topics. It follows the assumption that authors first decide a number of topics for an article, then pick up words related to these topics to write the article. In LDA, all documents in the corpus have the same set of topics, but each document has different portions of those topics (Blei 2012).

The basic assumption of LDA is that one document contains multiple topics and each of those requires specific words to describe them. For example, a paper, entitled "Artificial Intelligence in Biomedical Literatures", discussed the application of artificial intelligence algorithms to discover hidden associations between biological entities. Words such as neural network, autoencoder, neuron are from the topic of artificial intelligence; disease, gene, and protein are used to describe the biomedical topic; bioinformatics, genomics, cheminformatics are used for the topic of computer applications in biology.

Topic modeling algorithms aim to capture topics from a corpus automatically by using words observed in documents to infer the hidden topic structure (e.g., document topic distribution, and word topic distribution). The number of topics is usually decided by perplexity and can be heuristically set between 20 and 300 (Blei 2012). Topics are represented by distributions of words in the entire collection. Each document is generated by selecting a distribution over the topics. For each word, a topic assignment is chosen. In addition, the word from the corresponding topic is chosen.

The perplexity is often used to measure how a probability distribution fits a set of data. The perplexity is the inverse of the geometric mean per-word likelihood and is used to evaluate the models. A lower perplexity means a better model (Blei et al. 2003). The inference mechanics in topic models are independent of languages and contents. They capture the statistical structure of using language to represent thematic content. LDA approximates its posterior distribution by using inference (e.g., Gibbs sampling) or optimization (e.g., variational methods). The detailed explanations of how to run LDA with Mallet are provided in the later section.

Latent Semantic Indexing

Latent Semantic Indexing (LSI) tries to overcome the problems of lexical matching by using statistically derived conceptual indices instead of individual words for retrieval (Deerwester et al. 1990). LSI assumes that there is some underlying or latent structure in word usage that is partially obscured by variability in word choice. A truncated singular value decomposition (SVD) is used to estimate the structure in word usage across documents. Retrieval is then performed using the database of singular values and vectors obtained from the truncated SVD. Performance data shows that these statistically derived vectors are more robust indicators of meaning than individual terms.

The SVD projection is computed by decomposing the document-by-term matrix $A_{t \times d}$ into the product of three matrices, $T_{t \times n}$, $S_{n \times n}$, $D_{d \times n}$:

$$A_{t \times d} = T_{t \times n} S_{n \times n} (D_{d \times n})^T$$

where t is the number of terms, d is the number of documents, n = min(t, d), T and D have orthonormal columns, i.e.,

$$TT^T = D^T D = I,$$

$$\text{rank}(A) = r,$$

$$S = \text{diag}(\sigma_1, \sigma_2, \ldots, \sigma_n), \ \sigma_i > 0 \, for \ 1 \leq i \leq r, \ \sigma_j = 0 \, for \ j \geq r+1.$$

We can view SVD as a method for rotating the axes of the n-dimensional space such that the first axis runs along the direction of largest variation among the documents, the second dimension runs along the direction with the second largest variation and so forth. The matrices T and D represent terms and documents in this new space. The diagonal matrix S contains the singular values of A in descending order. The ith singular value indicates the amount of variation along the ith axis.

By restricting the matrixes T, S and D to their first $k < n$ rows, one obtains three truncated matrices $T_{t \times k}$, $S_{k \times k}$, $(D_{d \times k})^T$. Their product $\hat{A}$ is the best square approximation of A by a matrix of rank k in the sense defined in the equation $||\Delta = A - \hat{A}_2||$

$$\hat{A}_{t \times k} = T_{t \times k} S_{k \times k} (D_{d \times k})^T$$

Choosing the number of dimensions (k) for $\hat{A}$ is an interesting problem. While a reduction in k can remove much of the noise, keeping too few dimensions or factors may lose important information. As discussed in (Deerwester et al. 1990) using a test database of medical abstracts, LSI's performance improved considerably after 10 or 20 dimensions, peaked between 70 and 100 dimensions, and then began to diminish slowly. This pattern of performance—initial large increases and slowly decreases—is observed with other datasets as well. Eventually the performance becomes the same as standard vector methods because, with k = n factors, $\hat{A}$ is the same as the original term by document matrix A. The fact that LSI works well with a relatively small (compared to the number of unique terms) number of dimensions or factors k shows that these dimensions are, in fact, capturing a major portion of the meaningful structure (Berry et al. 1995).

There are several open source packages available for LSI, including

1. Text Mining Library for Latent Semantic Analysis at http://tml-java.sourceforge.net/
2. Weka at http://weka.sourceforge.net
3. airhead-research (a.k.a. s-space) at https://code.google.com/archive/p/airhead-research/

Out of these three packages, airhead-research is used for describing how LSI can be used due to its simplicity and robustness compared to the other two packages. airhead-research was developed in Java and provides an API. Thus, the provided LSI functionalities in airhead-research can be used in two ways: (1) API and (2) command line.

To use airhead-research, download and uncompress its Java package. airhead-research supports a variety of options. For instance, the argument "-n" or "–dimensions <int>" sets how many dimensions to use for the LSA vectors. Another basic option is "-p" or "–preprocess <class name>", which specifies an instance of a transform class to use in preprocessing the word-document matrix compiled by LSA prior to computing the SVD. More advanced options include "-S" or "–svdAlgorithm", which specifies manually a particular SVD algorithm should be used internally. Valid options are SVDLIBC, MATLAB, OCTAVE, JAMA and COLT. Since LSA will select the fastest algorithm available, use this option only when it is necessary.

Depending on the number of options to be used, several combinations of options can be used. For example, in order to remove stop words from the corpus while processing, the following command can be used:

```
java -Xmx8g -jar lsa.jar -d corpus.txt -F exclude=stopwords.txt my-lsa-output-no-stopwords.sspace.
```

To generates a 500-dimension LSA space, use the following command:

```
java -Xmx8g -jar lsa.jar -d corpus.txt -n 500 my-lsa-output-500dim.sspace.
```

Table 6.7 The list of related terms to "farm"

Term	Relevance score
Hay	0.64
Farmer	0.87
Farming	0.78
Farmland	0.72
Landowner	0.67
Cattle	0.66
Homestead	0.65
Agricultural	0.65

To generates an LSA space with known compound words, use the following command:

```
java -Xmx8g -jar lsa.jar -d corpus.txt -C my-list-of-ngrams.txt my-lsa-output-with-ngrams.sspace
```

Once the LSI model is built, the user can query the model with a term of interest. For instance, for the query "farm," the LSI model returns a list of related terms (Table 6.7).

Latent Dirichlet Allocation (LDA)

As described in the topic modeling section, LDA is a type of generative, probabilistic model for the latent topic layer (Blei et al. 2003). For a document d, a multinomial distribution θ_d over topics is sampled from a Dirichlet distribution with parameter α. For each word w_{di}, a topic z_{di} is chosen from the topic distribution. A word w_{di} is generated from a topic-specific multinomial distribution $\phi_{z_{di}}$. The probability of generating a word w from a document d is:

$$P(w|d, \theta, \phi) = \sum_{z \in T} P(w|z, \phi_z) P(z|d, \theta_d)$$

Therefore, the likelihood of a document collection D is defined as:

$$P(Z, W|\Theta, \Phi) = \prod_{d \in D} \prod_{z \in T} \theta_{dz}^{n_{dz}} \times \prod_{z \in T} \prod_{v \in V} \phi_{zv}^{n_{zv}}$$

where n_{dz} is the number of times that a topic z has been associated with a document d, and n_{zv} is the number of times that a word w_v has been generated by a topic z. The model can be explained as: to write a paper, an author first decides what topics and then uses words that have a high probability of being associated with these topics to write the article.

For the tutorial of how to do topic modeling, we introduce Mallet that was developed by McCullum and his team at University of Massachusetts Amherst. Mallet is a Java-based tool that provides various techniques including statistical natural language processing, document classification, clustering, topic modeling, information extraction, and other machine learning applications to text.

To use Mallet for topic modeling, download it at http://mallet.cs.umass.edu/download.php. The latest version is 2.0.8. Once MALLET has been downloaded and installed, the next step is to import text files into MALLET's internal format. The following instructions assume that the documents to be used as input to the topic model are in separate files, in a directory that contains no other files. For detailed information on how to import data in MALLET, we refer the reader to instructions available at http://mallet.cs.umass.edu/import.php.

Once the MALLET package is successfully installed, it is ready to use. Simply change to the MALLET directory and run the following command:

```
bin/mallet import-dir --input data/topic-input --output topic-input.mallet  --keep-sequence --remove-stopwords
```

The input data is assumed to be under the MALLET packgage's data sub directory called "topic-input." To learn more about options available in MALLET, use the argument "–help". To build a topic model, use the train-topics command, assuming that documents are formatted properly for MALLET. For example, the following command will create 100 topics and save the trained topic model, again, assuming that the MALLET instance object has already been created with the input data:

```
bin/mallet train-topics --input topic-input.mallet --num-topics 100 --output-state topic-state.gz
```

If you want to know more about available options in MALLET, use the option-help to get a complete list of options for the train-topics command. There are several options that are frequently used when you run Mallet for topic modeling (See Table 6.8).

You may download the sample input file from http://informatics.yonsei.ac.kr/tsmm/uncertainty_book/ISI_Abstract_original.txt and generate the same results as shown in Table 6.9. The following command includes several options such as the number of topics (10), the number of iterations (1000), applying the stopword list to create the MALLET instance object. The "keep-sequence" option in the command denotes that the input text is converted into a sequence of features, and it is normal that topic modeling in MALLET assumes that the input is converted to a feature sequence.

```
bin/mallet import-dir --input ISI_Abstract_original.txt --output topic-input.mallet
--num-topics 10 --num-iterations 1000 --keep-sequence --remove-stopwords
```

Table 6.8 The list of core options available in MALLET

Option	Description
–input [FILE]	Use this option to specify the MALLET collection file you created in the previous step
–num-topics [NUMBER]	The number of topics to use. The best number depends on what you are looking for in the model. The default (10) will provide a broad overview of the contents of the corpus. The number of topics should depend to some degree on the size of the collection, but 200 to 400 will produce reasonably fine-grained results
–num-iterations [NUMBER]	The number of sampling iterations should be a tradeoff between the time taken to complete sampling and the quality of the topic model
–optimize-interval [NUMBER]	This option turns on hyperparameter optimization, which allows the model to better fit the data by allowing some topics to be more prominent than others. Optimization every 10 iterations is reasonable
–optimize-burn-in [NUMBER]	The number of iterations before hyperparameter optimization begins. Default is twice the optimize interval
–output-model [FILENAME]	This option specifies a file to write a serialized MALLET topic trainer object. This type of output is appropriate for pausing and restarting training, but does not produce data that can easily be analyzed
–output-state [FILENAME]	Similar to output-model, this option outputs a compressed text file containing the words in the corpus with their topic assignments. This file format can easily be parsed and used by non-Java-based software. Note that the state file will be GZipped, so it is helpful to provide a filename that ends in.gz
–output-doc-topics [FILENAME]	This option specifies a file to write the topic composition of documents. See the–help options for parameters related to this file
–output-topic-keys [FILENAME]	This file contains a "key" consisting of the top k words for each topic (where k is defined by the–num-top-words option). This output can be useful for checking that the model is working as well as displaying results of the model. In addition, this file reports the Dirichlet parameter of each topic. If hyperparamter optimization is turned on, this number will be roughly proportional to the overall portion of the collection assigned to a given topic
–inferencer-filename [FILENAME]	Create a topic inference tool based on the current, trained model. Use the MALLET command bin/mallet infer-topics–help to get information on using topic inference

Table 6.9 The number of topics and top terms generated by LDA

Topic no.	Top terms
0	Cell cells tissue study engineering differentiation bone potential nanofiber regeneration factors culture scaffolds critical scaffold mechanical control stem increase
1	Structure based technology polymer design fabricated carbon band high performance size circuit interconnect materials capacity electrical interconnects significantly improved
2	Graphene surface growth electronics epitaxial electronic material electron use layer magnetic layers high scattering chemical multilayer demonstrated surfaces landau
3	New formation effects vascular important known cells elsevier shown number factor development reserved rights sod network reduced vegf role
4	Energy power zno applications piezoelectric potential voltage output approach mechanical low cmos flexible area current density thin reduce crystal
5	Used results function different data show large small including webs elements delivery indicate application aligned electrospun activity direct resistance
6	Model using paper process well proposed mems structures parameters present simple fabrication response range presented linear stochastic mode nonlinear
7	Method time two order zoning system efficiency found optimization algorithm significant provides higher optical films experimental compared study air
8	Properties effect solar using devices temperature carrier charge morphology doping silicon transport film interface than transfer provide bulk device
9	Expression complex gene novel rna changes essential hsp genes stress specific functional cmr rnai species cofactors redox predicted patterns

Semantic Networks and Ontology

A semantic network is a propositional knowledge structure consisting of a set of nodes that are selectively connected to each other by links labeled by the relationship between each pair of connected nodes (Stillings et al. 1987). Semantic networks as a representation of knowledge have been in use in artificial intelligence (AI) research in a number of different areas. Some of the first uses of the nodes-and-links formulation were in the work of Collins and Quillian (1969), where the networks acted as models of associative memory. Their work centers on how natural language is understood and how the meanings of words can be captured in a machine.

Building a semantic network was previously done manually, which requires experts to put a significant amount of time and effort. Therefore, automatic construction of a semantic network was a recent, focal point of the semantic web community (Harrington 2009; Harrington and Wojtinnek 2011). One of the recent efforts for automatic construction of a semantic network is a hybrid set of classification systems based on weakly, distant or semi-supervised learning systems. These systems require a smaller set of training material that focuses on either two independent categories or utilizes two different classification methods. After the intermediate classifiers are run on non-annotated documents, the results are analyzed and the documents that best represent of the categories are added to the training data to improve the classifier. This process is repeated until some predefined condition is met (Aggarwal and Zhai 2012). While words contain a lot of information about the document under inspection, they also create a high-level of dimensionality and ambiguity. Different words can be used to describe the same meaning (synonyms), for example *earth* and *dirt*. Using both words as separate terms in a VSM creates a high-level of dimensionality. We can use natural language processing (NLP) techniques to recognize and consolidate synonyms to reduce dimensionality but a second problem arises. Some words, like *earth,* have multiple meanings (polysemy) (Aggarwal and Zhai 2012) and these meanings can be domain dependent. It is in these cases that information extraction techniques such as concept hierarchies can be used to determine appropriate meaning (Feldman and Dagan 1995).

Concept hierarchies are created through analyzing the relationships of tokens found in a document. Relationships can be defined manually, based on token distributions, or specified through background knowledge. Zheng et al. (2009) defined a concept as a set of words, usually noun phrases, which have semantic relationships. Feldman and Sanger (2007) emphasize that a concept hierarchy can be used to describe a document which contains one or more concept nodes going from a more generalized meaning to more specific meaning. Representing a document as a set of concepts, or concept signature, provides a richer representation which, when used with clustering techniques, makes the resulting index scheme more useful (Zheng et al. 2009). The explosive growth in digital content emphasizes the need to develop automated management (organizational) and access (discovery) tools to support the processing of digital content for information access systems. Organization of this generally unstructured content requires one to identify the scope, concepts, and purpose of the resource and then analyze the relationships of the concepts to provide an overall understanding of the document (Tseng et al. 2007).

Early text classification schemes were built on labor intensive training sets that were used to model the predefined categories to be identified and required sufficient text in the document being classified to ensure good accuracy (Zelikovitz and Hirsh 2000). Due to these challenges research started to explore the use of background knowledge, that is, domain-specific heuristics that can be used as constraints to reduce the ambiguity of natural languages and help in the feature selection process. Taxonomies, controlled vocabularies, and ontologies are various types of formalized specification that provide a conceptualization of a domain of interest (Gruber 1995). It is generally agreed that a controlled vocabulary is the most basic form of background knowledge. It can be used for keyword or concept identification. Taxonomies take controlled vocabularies and identify relationships between concepts, such as an "is-a" relation that is used to identify synonyms of terms. Ontologies are the most complex of the three specifications and add on to taxonomies additional domain specific rules. An ontology contains a shared, controlled vocabulary which models a specific domain with the definition of concepts and their properties and relations.

WordNet

WordNet is considered by most an implementation of the general English language ontology (Miller et al. 1990). It identifies words and word phrases, includes morphological and semantic relationships, and identifies a hierarchy of relationships (hypernym and hyponym). It has been used in query expansion (Hsu et al. 2008), text classification (Elberrichi et al. 2008), and text clustering (Hotho et al. 2003). Using WordNet's background knowledge, text documents are analyzed for concepts based on relationships between terms. Common linguistic relationships are antonyms (opposite meaning), synonyms (similar meaning), hypernyms ("IS-A" generalization of a term), hyponyms (more specific meaning of a term), holonyms ("PART-OF" relationship), and meronyms ("HAS-A" relationship). These relationships are shown in Fig. 6.4.

Hypernym relationships form a directional "IS-A" connection between two terms that moves from a specific meaning to a more generalized one ("Earth IS-A planet"). Many studies have been performed to automatically extract these relationships from unstructured text, such as in (Snow et al. 2004). Unlike hypernyms, terms which are synonyms can replace each other and still hold a similar meaning. For example, "sunshine" and "sunlight" terms may be used interchangeably in a sentence without significant loss of meaning. Meronyms are a bit more complex. Girju et al. (2006) defined six types of meronyms which WordNet consolidates three categories; *member-of* (faculty HAS-A professor), *stuff-of* (tree HAS-A wood), and *part-of* (solar system HAS-A sun). Additionally, Girju et al. identifies the *part-of* category as the most prominently used while Miller et al. (1990) indicate meronym transitivity may be optional as one moves away from the original relationship. For example, "Earth HAS-A moon" but the "plant HAS-A moon" relationship is optional (not all planets have moons).

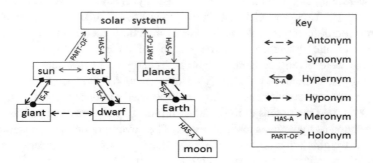

Fig. 6.4 Concept map using natural language relationships

WordNet has been used in numerous document-clustering experiments. Some of the earliest uses of WordNet in text categorization supported techniques to address effectively the classification of low frequency categories (Rodriguez et al. 1997). Green (1999) used WordNet's hypernym and hyponym links to build lexical chains to analyze the similarity between information in different paragraphs. Hotho et al. (2003) showed utilizing background knowledge (i.e., relationships) between terms improved document-clustering. Hung and Wermter (2004) present three text vector representations, two of which used hypernym as concepts to improve classification accuracy. Zheng et al. (2009) used WordNet relationships with noun phrases to analyze clustering improvements. Wang and Taylor (2007) used WordNet to capture hypernym relations in short text documents creating clusters of concepts called concept forests to represent a document. Elberrichi et al. (2008) used WordNet to create a concept vector format they compared to traditional bag-of-word vector representation. Except for Zheng et al. (2009), all these methods use single term analysis (using synonyms) and calculate term frequency from hypernyms. In fact, many of the papers listed suggest using more than one relationship as a future area research.

Accurately identifying concepts for categorization purposes is fraught with time-consuming manual analysis by content experts and librarians. A digital library catalog/index must represent the digital content and reflect the expectations of its users. Automating this process requires new techniques in concept extraction to analyze any size document and capture main concepts based on the appropriate domain. In this paper, we describe an extension to existing natural language and machine-learning techniques to improve the accuracy of extracting concepts from small text based resources and grouping them appropriately.

The selection of terms is a critical first step in concept generation. Terms with multiple meanings (polysemy) create ambiguity, while a term that is similar (synonyms) to others or have a degree of generalization (hypernym) can strengthen the importance of a concept. For these reasons, term frequency calculations often use hypernym and synonym information once ambiguity is resolved. We also use this approach in our algorithm but the novelty of our approach is the inclusion of meronyms. The choice of meronyms comes from the idea of finding mechanisms to improve frequency measures for significant terms in short text documents without over constraining larger documents. Some meronyms studies have been conducted as outlined by Yang and Callan (2009). Basu et al. (2001) developed a set of measures for different lexical relationships, including meronyms to identify the average semantic difference (i.e., the weight of an edge between two terms). Meronyms were given the same weight as hypernyms in this study. Girju et al. (2006) suggest techniques for identifying meronyms for the specific use of incorporating them into taxonomies so they may be used in concept extraction. Zheng et al. (2009) used meronyms as the relationship to support clustering and found it to be not as good as hypernyms and holonyms. The novelty of our study examines the effects of weighing meronyms differently than synonyms or hypernyms when incorporating them into a frequency count for text characterization.

In addition to a general English ontology, domain specific ones exist. In the realm of education, there are many used to define guidelines for knowledge goals. Strand Map Benchmarks is a representation of the AAAS' Project 2061, a "statement of what all students should know and be able to do in science, mathematics, and technology by the end of grades 2, 5, 8, and 12." ... "It provides educators with sequences of specific learning goals they can use to design a core curriculum".

The basic statistics of WordNet 3.0 are provided as follows (Table 6.10–6.11):

By and large, WordNet can be used in two ways. First approach is use WordNet online. WordNet is accessible online at http://wordnetweb.princeton.edu/perl/webwn. Once you type in a query and choose options for displaying results, WordNet returns the matched results (Fig. 6.5).

The second option is to download and install WordNet to a local machine. Depending on the operating system, you need to download different version. The most recent Windows version of WordNet is 2.1, released in March 2005. Yes, it has been a long time. For the Unix or Linux OS, version 3.0 is available for download, which was released in December, 2006. However, database files are updated to the version 3.1 and can substitute for the 3.0 files on the Unix or Linux OS.

Regarding database files, the following standoff files provide further semantic information to supplement the WordNet 3.0:

- Semantically annotated gloss corpus
- Evocation database
- Morphosemantic Links (Semantic relations between morphologically related nouns and verbs)
- Teleological Links (an encoding of typical activity for which artifact was intended)

Table 6.10 Number of POS, words, Synsets, and sense pairs

POS	Unique strings	Synsets	Total word-sense pairs
Noun	117798	82115	146312
Verb	11529	13767	25047
Adjective	21479	18156	30002
Adverb	4481	3621	5580
Totals	155287	117659	206941

Table 6.11 Polysemy information

POS	Monosemous words and senses	Polysemous words	Polysemous senses
Noun	101863	15935	44449
Verb	6277	5252	18770
Adjective	16503	4976	14399
Adverb	3748	733	1832
Totals	128391	26896	79450

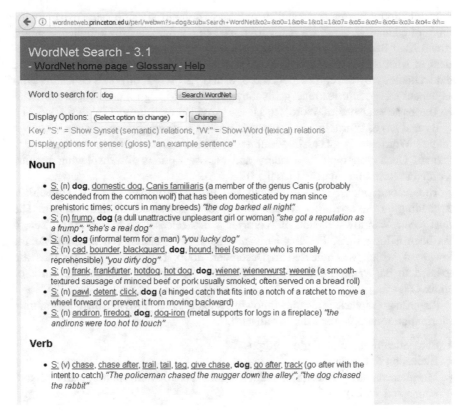

Fig. 6.5 The homepage of WordNet search

- "Core" WordNet (5000 more frequently used word senses)
- Logical Forms (logical forms for glosses)

WordNet can be utilized with NLTK, the Python based text mining tool.
WordNet is a NLTK corpus reader, and it is imported with the following import statement:

```
>>> from nltk.corpus import wordnet
```

To examine a word with the NLTK WordNet module, we can use the NLTK function called synsets(). This function has an optional pos argument which lets you constrain the part of speech of the word:

```
>>> wn.synsets('dog') # doctest: +ELLIPSIS +NORMALIZE_WHITESPACE
[Synset('dog.n.01'), Synset('frump.n.01'), Synset('dog.n.03'), Synset('cad.n.01'),
Synset('frank.n.02'), Synset('pawl.n.01'), Synset('andiron.n.01'), Synset('chase.v.01')]
>>> wn.synsets('dog', pos=wn.VERB)
[Synset('chase.v.01')]
```

Table 6.12 Statistics of BabelNet 3.7

1971744856	Total number of RDF triples
745859932	Total number of Babel senses
380239084	Total number of lexico-semantic relations
40709194	Total number of glosses (textual definitions)
13801844	Total number of Babel synsets
10767833	Total number of images
7735448	Total number of named entities
6393568	Total number of other forms
6066396	Total number of concepts
2948668	Total number of Babel synsets with at least one picture
2675385	Total number of Babel synsets with at least one domain
743296	Total number of compounds
271	The number of languages

BabelNet

BabelNet is a very large multilingual encyclopedic dictionary and semantic network (Navigli and Ponzetto 2012). It integrates the largest multilingual Web encyclopedia with the most popular computational lexicon of English such as WordNet, other lexical resources such as Wiktionary, OmegaWiki, Wikidata, and the Open Multilingual WordNet. The integration is performed by an automatic linking algorithm and by filling in lexical gaps with the aid of machine translation algorithms. The result is an encyclopedic dictionary that provides Babel synsets including concepts and named entities lexicalized in many languages and connected with large amounts of semantic relations.

BabelNet's current version is 3.7, which includes many feature such as FrameNet (lexical units), more than 2500 Babel synsets identified as key concepts, mappings with several versions of WordNet integrated, more than 2.6 million Babel synsets labeled with domains, more than 625 million new senses, 6.4 million surface forms for Babel synsets, and 3.5 million YAGO external links (Table 6.12). BabelNet also provides both Java and HTTP RESTful APIs.[2]

BabelNet can be utilized in two ways. The first method is to use the web interface of BabelNet. The second method is to use Java API or REST API. For the REST API, one can query BabelNet through an HTTP interface that returns JSON. The user can append the *key* parameter to the HTTP requests as shown in the examples below. To obtain an API key please read the page. All requests must be executed using the GET method and they should include Accept-Encoding: gzip as the header in order to obtain compressed content. The example of a REST API is as follows:

---

[2]http://babelnet.org/download.

https://babelnet.io/v4/getVersion?key={key}.

where the {key} denotes the API key obtained after signing up to BabelNet. Another example is to retrieve the IDs of the Babel synsets (concepts) denoted by a given word:

https://babelnet.io/v4/getSynsetIds?word={word}&langs={lang}&key={key}.

In this example, there are seven options that can be added to the REST API (Table 6.13).

The results of the ID retrieval REST API are shown in Table 6.14.

Another example of to retrieve the senses of a given word from BabelNet using the REST API. Table 6.15 shows a list of options.

Table 6.13 Options available for ID retrieval in BabelNet API

Name	Description
word	The word you want to search for
langs	The language of the word. Accepts multiple values
filterLangs	The languages in which the data are to be retrieved. Default value is the search language and accepts not more than 3 languages except the search language
pos	Returns only the synsets containing this part of speech (NOUN, VERB, etc.). Accepts only a single value
source	Returns only the synsets containing these sources (WIKT, WIKIDATA, etc.). Accepts multiple values
normalizer	Enables normalized search
key	API key obtained after signing up to BabelNet

Table 6.14 The results of the ID retrieval REST API

```
[
{"id":"bn:15409009n","pos":"NOUN","source":"BABELNET"},
{"id":"bn:00046063n","pos":"NOUN","source":"BABELNET"},
{"id":"bn:03345344n","pos":"NOUN","source":"BABELNET"},
{"id":"bn:00055685n","pos":"NOUN","source":"BABELNET"},
{"id":"bn:01204395n","pos":"NOUN","source":"BABELNET"},
{"id":"bn:02131227n","pos":"NOUN","source":"BABELNET"},
{"id":"bn:02799103n","pos":"NOUN","source":"BABELNET"},
{"id":"bn:15586454n","pos":"NOUN","source":"BABELNET"},
{"id":"bn:00088150v","pos":"VERB","source":"BABELNET"},
{"id":"bn:00355636n","pos":"NOUN","source":"BABELNET"},
{"id":"bn:00090750v","pos":"VERB","source":"BABELNET"},
{"id":"bn:00071669n","pos":"NOUN","source":"BABELNET"},
{"id":"bn:02363694n","pos":"NOUN","source":"BABELNET"},
{"id":"bn:01610649n","pos":"NOUN","source":"BABELNET"},
{"id":"bn:03783607n","pos":"NOUN","source":"BABELNET"},
{"id":"bn:01683382n","pos":"NOUN","source":"BABELNET"},
{"id":"bn:15010220n","pos":"NOUN","source":"BABELNET"}]
```

Table 6.15 Options available for word sense retrieval of BabelNet API

Name	Description
word	The word you want to search for
lang	The language of the word. Required
filterLangs	The languages in which the data are to be retrieved. Default value is the search language and accepts not more than 3 languages except the search language. Example
pos	Returns only the synsets containing this part of speech (NOUN, VERB, etc.). Accepts only a single value
source	Returns only the synsets containing these sources (WIKT, WIKIDATA, etc.). Accepts multiple values
normalizer	Enables normalized search
key	API key obtained after signing up to BabelNet

https://babelnet.io/v4/getSenses?word={word}&lang={lang}&key={key}.

The results of the word sense retrieval REST API are shown as follows.

```
[
    {
        "lemma":"Simians_in_Chinese_poetry",
        "simpleLemma":"Simians_in_Chinese_poetry",
        "source":"WIKIRED",
        "sensekey":"",
        "sensenumber":0,
        "frequency":1,
        "position":1,
        "language":"EN",
        "pos":"NOUN",
        "synsetID":{"id":"bn:15409009n","pos":"NOUN","source":"BABELNET"},
        "translationInfo":"",
        "pronunciations":{"audios":[],"transcriptions":[]},
        "bKeyConcept":false
    },
    {
        "lemma":"Simians_(Chinese_poetry)",
        "simpleLemma":"Simians",
        "source":"WIKI",
        "sensekey":"",
        "sensenumber":0,
        "frequency":10,
        "position":1,
        "language":"EN",
        "pos":"NOUN",
        "synsetID":{"id":"bn:15409009n","pos":"NOUN","source":"BABELNET"},
        "translationInfo":"",
        "pronunciations":{"audios":[],"transcriptions":[]},
        "freebaseId":"0_frc72",
        "YAGOURL":"Simians_(Chinese_poetry)",
        "bKeyConcept":false
    },
    ...
]
```

Deep Learning

Deep Learning is a subfield of machine learning concerned with algorithms inspired by the structure and function of the brain called artificial neural networks. An artificial intelligence function that imitates the workings of the human brain in processing data and creating patterns for use in decision making. Deep learning is a subset of machine learning in Artificial Intelligence (AI) that has networks which are capable of learning unsupervised from data that is unstructured or unlabeled.

One of the most common AI techniques used for processing Big Data is Machine Learning. Machine learning is a self-adaptive algorithm that gets better and better analysis and patterns with experience or with new added data. If a digital payments company wanted to detect the occurrence of or potential for fraud in its system, it could employ machine learning tools for this purpose. The computational algorithm built into a computer model will process all transactions happening on the digital platform, find patterns in the data set, and point out any anomaly detected by the pattern.

Deep learning, a subset of machine learning, utilizes a hierarchical level of artificial neural networks to carry out the process of machine learning. The artificial neural networks are built like the human brain, with neuron nodes connected together like a web. While traditional programs build analysis with data in a linear way, the hierarchical function of deep learning systems enables machines to process data with a non-linear approach. A traditional approach to detecting fraud or money laundering might rely on the amount of transaction that ensues, while a deep learning non-linear technique to weeding out a fraudulent transaction would include time, geographic location, IP address, type of retailer, and any other feature that is likely to make up a fraudulent activity. The first layer of the neural network processes a raw data input like the amount of the transaction and passes it on to the next layer as output. The second layer processes the previous layer's information by including additional information like the user's IP address and passes on its result. The next layer takes the second layer's information and includes raw data like geographic location and makes the machine's pattern even better. This continues across all levels of the neuron network until the best and output is determined.

Recently, deep learning approaches have obtained very high performance across many different NLP tasks. These models can often be trained with a single end-to-end model and do not require traditional, task-specific feature engineering. There several good reasons for using deep learning for NLP problems. First is that it is quite suitable for learning representation. Hand crafting features is time-consuming. The features are often both over-specified and incomplete. The work has to be done again for each task/domain, etc. We must move beyond handcrafted features and simple ML. Humans develop representations for learning and reasoning. Our computers should do the same. Deep learning provides a way of doing this. Second, Current NLP systems are incredibly fragile because of their atomic symbol representations. Distributed representation enabled by deep learning based NLP can relax this problem. Learned word representations help enormously in NLP. They provide a powerful similarity model for words. Distributional similarity based word clusters greatly help most applications.

Distributed representations can do even better by representing more dimensions of similarity. Distributed representations deal with the curse of dimensionality. Generalizing locally (e.g., nearest neighbors) requires representative examples for all relevant variations. Classic solutions: Manual feature design, assuming a smooth target function (e.g., linear models), Kernel methods (linear in terms of kernel based on data points). Neural networks parameterize and learn a "similarity" kernel. Third, deep learning is suitable for unsupervised feature and weight learning. Today, most practical, good NLP& ML methods require labeled training data (i.e., supervised learning). But almost all data is unlabeled. Most information must be acquired unsupervised. Fortunately, a good model of observed data can really help you learn classification decisions. Despite prior investigation and understanding of many of the algorithmic techniques before 2006, training deep architectures was unsuccessful. But since then, faster machines and more data help DL more than other algorithms. New methods for unsupervised pre-training have been developed (Restricted Boltzmann Machines = RBMs, autoencoders, contrastive estimation, etc.). More efficient parameter estimation methods. Better understanding of model regularization.

Word Embeddings

Word embeddings are one of the most well accepted deep learning algorithms that has been applied to NLP, which the original concept was introduced by Bengio et al. (2003). Word embedding algorithms are one of the best options to gain intuition about why deep learning is effective. Let's discuss the basic notion of word embeddings.

A word embedding W:words $\rightarrow$ Rn is a parameterized function mapping words in some language to high-dimensional vectors (perhaps 200 to 500 dimensions). For example, we might find:

$$W(''cat'') = (0.2, -0.4, 0.7, \ldots)$$
$$W(''mat'') = (0.0, 0.6, -0.1, \ldots)$$

Typically, the function is a lookup table, parameterized by a matrix, θ, with a row for each word: $W\theta(wn) = \theta n$. W is initialized to have random vectors for each word. It learns to have meaningful vectors, which can be used for advanced NLP tasks such as sentiment analysis or information retrieval. For example, one task we might train a network for is predicting whether a 5 g (sequence of five words) is valid. We can generate a number of 5 g from Wikipedia (e.g., "cat sat on the mat") and then make half of them invalid by switching a word with a random word (e.g., "cat sat **song** the mat"), since that will almost certainly make our 5 g nonsensical.

The model we train will run each word in the 5 g through W to get a vector representing it and feed those into another 'module' called R which tries to predict if the 5 g is valid or invalid, and it will result in the following:

$$R(W(''cat''), W(''sat''), W(''on''), W(''the''), W(''mat'')) = 1$$
$$R(W(''cat''), W(''sat''), W(''song''), W(''the''), W(''mat'')) = 0$$

In order to predict these values accurately, the network needs to learn good parameters for both W and R. Although it could be helpful in detecting grammatical errors in text, but what is interesting is to learn W. One way to understand the word embedding space is to visualize them with t-SNE, a sophisticated technique for visualizing high-dimensional data. Figure 6.6 shows the results of t-SNE with the word2vec model built on the news articles related to companies producing platform software and hardware.

This visualization of words helps us make sense of word associations. Similar words are close together. Another way to get at this is to look at which words are closest in the embedding to a given word. Again, the words tend to be quite similar. Figure 6.7 below shows how the similar words to the word "IBM" changed over time.

It may be adequate for a network to make words with similar meanings have similar vectors. If a word is replaced with a synonym (e.g., "a few people sing well" → "a *couple* people sing well"), the meaning of the sentence still remains the same. Thus, we may say that even if the input sentence has changed a lot, if W maps synonyms (like "few" and "couple") close together, from R's perspective not much changes are made. This implies many important points. There is the enormous number of possible 5 g whereas we have a comparatively small number of data

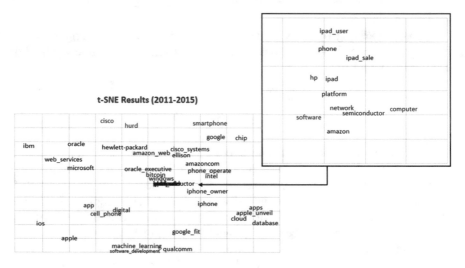

Fig. 6.6 t-SNE visualizations of word embeddings

2000-2005		2006-2010		2011-2015		2016-2017	
dell	0.959924936	oracle	0.79832989	buy_open-source	0.688270867	ibm_watson	0.820717812
unix	0.954558194	mainframe	0.77750206	idataplex	0.676672041	watson	0.808973908
solaris	0.952200234	sun	0.762942851	nirvanix_software	0.662971854	cognitive	0.803402245
server	0.950758517	orcl	0.756299734	reinvigorate_dell	0.660516798	ibm_cloud	0.757027686
datum_center	0.950342536	sap	0.742992461	ibm_provide	0.653238475	watson_internet	0.742083311
source	0.950308442	novl	0.738868535	bleed_market	0.652676821	cognitive_computing	0.741557717
move	0.948748887	patent_bottle	0.735488415	teradataexadata	0.651375115	watson_unit	0.734454691
operate_system	0.947559893	seven-year_patent	0.734089375	provide_vertical	0.64939189	ibm_collaborations	0.727057636
run	0.946791053	sun_microsystems	0.72926867	require_pace	0.648596585	ibm_research	0.726451933
proprietary	0.944843233	open-source_linux	0.726028204	j&o_professional	0.647573769	ibm_continue	0.723082423
support	0.94289434	seven-year	0.725172222	dell_bring	0.646723807	watson_cognitive	0.720367968
hewlett-packard	0.942260742	larry_ellison	0.724186003	boost_ibm	0.646304667	ibmcomoutthink	0.717018962
linux	0.940122664	ibm_corp	0.723489523	systems_hope	0.64626497	watson_developer	0.716962516
develop	0.939396739	software_giant	0.721525371	teradataexadata_hardware	0.645682216	watson_health	0.716135025
hardware	0.935241699	java_patent	0.718840897	hp_struggle	0.645025253	cognitive_solution	0.716092885
pc	0.933961868	oracle_acquire	0.715612531	ibm_build	0.644733489	watson_education	0.711184084
corporate	0.933870077	outlooksoft	0.715118825	software_ip	0.644324541	watson_iot	0.705735028
websphere	0.932072341	business_object	0.713380098	ibm_software	0.642227352	ibm_announce	0.704842031
system_software	0.931963742	maker_sco	0.713360071	deep-pocketed_player	0.642201006	cloud_video	0.704631567
improve	0.929755747	applix	0.708620787	market_computer	0.640904605	video_unit	0.703812957
sell_server	0.929618537	xensource	0.706225395	smb_apps	0.640653729	ibm_insight	0.702063084
oracle	0.928216755	giant_roil	0.703248084	highly-tested	0.640520751	data_services	0.697872102
proprietary_solaris	0.927893996	unix_maker	0.701602578		0.640065312	wwwibm	0.696583211
rational	0.926518321	telelogic	0.701261342	intel-compatible	0.639556468	watson_service	0.695548415
antitrust_lawsuit	0.926318586	january_purchase	0.701039195	prime_competitor	0.639511287	watson_beat	0.694848537
rate	0.926261961	sun-developed	0.700815618	intel-compatible_server	0.638401508	developer_partner	0.693343222
intense	0.926035345	sun-developed_java	0.700212479	power_pc-based	0.63667053	watson_iot	0.691154718
potential	0.923175871	prominent_name	0.696930885	ibm_power-based	0.636367381	imaging_collaborative	0.690888762
solaris_server	0.922847867	java_software	0.694627345	ibm_x	0.635502517	beat_jeopardy	0.690651357
java	0.92274034	oracle_buy	0.694197714	entire_x	0.634402633	@forbes_inhi	0.689501226

Fig. 6.7 Similar words associated with the word "IBM" over time

points to try to learn from. Similar words being close together allows us to generalize from one sentence to a class of similar sentences. This does not just switch a word for a synonym, but rather switch a word for a word in a similar class (e.g., "the wall is blue" → "the wall is *red*"). Further, we can change multiple words (e.g., "the wall is blue" → "the *ceiling* is *red*"). This is a benefit that W provides.

Word embeddings also allow us to generalize to new combinations of words. You've seen all the words that you understand before, but you haven't seen all the sentences that you understand before. So too with neural networks. Word embedding models can automatically organize concepts and learn implicitly the relationships between them, as during the training we did not provide any supervised information about what a capital city means.

Word embeddings exhibit an even more remarkable property: analogies between words seem to be encoded in the difference vectors between words. For example, there seems to be a constant male-female difference vector:

$$W("woman") - W("man") \simeq W("aunt") - W("uncle")$$
$$W("woman") - W("man") \simeq W("queen") - W("king")$$

Gender pronouns mean that switching a word can make a sentence grammatically incorrect. For instance, supposed that there are sentences like "*she* is the aunt" and "*he* is the uncle." (Similarly, "*he* is the King" but "*she* is the Queen." If one sees "*she* is the *uncle*," the most likely explanation is a grammatical error. If words are being randomly switched half the time, it seems pretty likely that happened here.

Mikolov et al. (2013) points out that the word embeddings learn to encode gender in a consistent way. Depending on the datasets that word embedding models

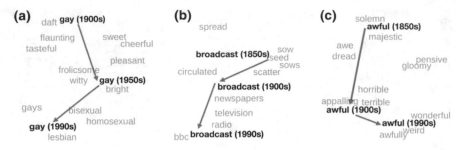

Fig. 6.8 The shift of meanings of words in HistWords. *Source* Hamilton et al. (2016)

like word2vec is built upon, there's probably a gender dimension. Same thing for singular vs plural.

HistWords is an interesting collection of tools and datasets for analyzing language change using word embeddings for historical text.[3] The semantic evolution of more than 30,000 words across four languages was modeled by historical word vectors. HistWords is maintained by a group of researchers at Stanford University, William L. Hamilton, Jure Leskovec, and Dan Jurafsky (2016). They found that the meanings of more frequently used words tend to be more stable over time than less frequently used words and that the meanings of polysemous, those words with multiple meanings, change at faster rates than others (see Fig. 6.8).

It's important to appreciate that all of these properties of *W* are *side effects*. This seems to be a great strength of neural networks: they learn better ways to represent data, automatically. Representing data well, in turn, seems to be essential to success at many machine learning problems. Word embeddings are just a particularly useful example of learning a representation.

There are several word embedding models that are publicly available. The most popular one is the Google news word2vec model.[4] The name of the model is called GoogleNews-vectors-negative300.bin.gz. Google published pre-trained vectors trained on part of Google News dataset (about 100 billion words). The model contains 300-dimensional vectors for 3 million words and phrases. The phrases were obtained using a simple data-driven approach described by Mikolov et al. (2013).

To use the word2vec program provided by Google, download it with svn checkout from http://word2vec.googlecode.com/svn/trunk/. Then compile word2-vec with 'make' from a Linux terminal window or linux emulator like cygwin, and then run the demo scripts: ./demo-word.sh and. /demo-phrases.sh.

---

[3]https://nlp.stanford.edu/projects/histwords/.
[4]https://code.google.com/archive/p/word2vec/.

Summary

Extracting entities and their relations from unstructured text is essential for text mining, topic modeling, constructing ontological structures, and deep learning. Resources such as WordNet and BabelNet play an instrumental role in a wide variety of applications. Deep learning, especially advances such as word2vec, has revitalized the interest in text analysis and document understanding. As demonstrated by the wide adoption of word2vec and distributional paradigms, the series of technical advances from LSI, LDA, to word2vec will continue to grow. With increasingly powerful and intuitive tools, one can tackle more challenging problems at a larger scale. In terms of Shneider's four-stage evolution model, quantitative studies of science as a field are likely to benefit profoundly from the stream of text modeling techniques.

References

Aggarwal CC, Zhai C (2012). Mining text data. Springer

Ahlers CB, Fiszman M, Demner-Fushman D, Lang F, Rindflesh TC (2007). Extracting semantic predication from MEDLINE citations for pharmacogenomics. In: Pacific symposium on biocomputing, pp 209–220

Aronson AR (2001) Effective mapping of biomedical text to the UMLS Metathesaurus: the MetaMap program. In: AMIA annual symposium proceedings, pp 17–21

Basu S, Mooney RJ, Pasupuleti K, Ghosh, J (2001) Evaluating the novelty of text-mined rules using lexical knowledge. In: Proceedings of the seventh ACM SIGKDD international conference on Knowledge discovery and data mining, ACM San Francisco, California, pp 233–238

Bengio Y, Ducharme R, Vincent P, Jauv C (2003) A neural probabilistic language model. J Mach Learn Res 3:1137–1155

Berry MW, Dumais ST, O'Brian GW (1995) Using linear algebra for intelligent information retrieval. SIAM Rev 37(4):573–595

Bikel DM, Schwartz RL, Weischedel RM (1999) An algorithm that learns what's in a name. Mach Learn 34:211–231

Blei DM (2012) Probablisitic topic models. Commun ACM 55(4):77–84. doi:10.1145/2133806. 2133826

Blei DM, Ng AY, Jordan MI (2003) Latent Dirichlet allocation. JMLR 3:993–1022

Bui QC, Nualláin BÓ, Boucher CA, Sloot PM (2010) Extracting causal relations on HIV drug resistance from literature. BMC Bioinform 11(1):101

Chen Y, Liu F, Manderick B (2011) Extract protein-protein interactions from the literature using support vector machines with feature selection, biomedical engineering, trends, research and technologies. In: Olsztynska S (ed). ISBN: 978-953-307-514-3

Chowdhury MFM, Abacha AB, Lavelli A, Zweigenbaum P (2011) Two different machine learning techniques for drug-drug interaction extraction. In: Challenge task on drug-drug interaction extraction, pp 19–26

Collins AM, Quillian MR (1969) Retrieval time from semantic memory. J Verbal Learn Verbal Behav 8:240–247

Deerwester S, Dumais ST, Landauer TK, Furnas GW, Harshman RA (1990) Indexing by latent semantic analysis. J Am Soc Info Sci 41(6):391–407

Elberrichi Z, Rahmoun A, Bentaalah MA (2008) Using WordNet for text categorization. Int Arab J Info Technol 5(1):16–24

Feldman R, Dagan I (1995) Knowledge discovery in textual databases (KDT). KDD

Feldman R, Sanger J (2007) The text mining handbook: advanced approaches in analyzing unstructured data. Cambridge University Press, Cambridge, UK

Finkel JR, Grenager T, and Manning C (2005) Incorporating non-local information into information extraction systems by gibbs sampling. In: ACL '05, proceedings of the 43rd annual meeting on association for computational linguistics, Association for Computational Linguistics Morristown, NJ, USA, pp 363–370

Fu Y, Bauer T, Mostafa J, Palakal M, Mukhopadhyay S (2002) Concept extraction and association from Cancer literature. Eleventh international conference info knowledge management (CIKM2002)/Fourth ACM interenational workshop web info data management (ACM WIDM 2002). McLean, VA, USA, pp 100–103

Girju R, Badulescu A, Moldovan D (2006) Automatic discovery of part-whole relations. Comput Linguist 32(1):83–135

Green SJ (1999) Building Hypertext Links By Computing Semantic Similarity. IEEE Trans Knowl Data Eng 11(5):713–730

Gruber TR (1995) Toward principles for the design of ontologies used for knowledge sharing? Int J Hum Comput Stud 43(5):907–928

Hamilton WL, Leskovec J, Jurfsky D (2016) Diachronic word embeddings reveal statistical laws of semantic change. In: Proceedings of the 54th annual meeting of the association for computational linguistics, Berlin, Germany, August 7–12, 2016, Association for Computational Linguistics, pp 1489–1501. http://aclweb.org/anthology/P16-1141

Harrington B (2009) ASKNet: automatically creating semantic knowledge networks from natural language Text, Ph.D. thesis, University of Oxford

Harrington B, Wojtinnek PR (2011) Creating a standardized markup language for semantic networks. In: Proceedings of the 5th ieee international conference on semantic computing

Hotho A, Staab S, Gerd S (2003) Wordnet improves text document clustering. In: Proceedings of the SIGIR 2003 semantic web workshop of the 26th annual international ACM SIGIR conference, Toronto, CA

Hristovski D, Friedman C, Rindflesch TC, Peterlin B (2006) Exploiting semantic relations for literature-based discovery. AMIA Annu Symp Proc 2006:349–353

Hsu MH, Tsai MF, Chen HH (2008) Combining WordNet and ConceptNet for automatic query expansion: a learning approach. In: Proceedings of the 4th asia information retrieval conference on information retrieval technology, Springer Harbin, China, pp 213–224

Huang M, Zhu X, Li M (2006) A hybrid method for relation extraction from biomedical literature. Int J Med Informatics 75(6):443–455

Hung C, Wermter S (2004) Neural network based document clustering using WordNet ontologies. Int J Hybrid Intell Syst 1(3–4):127–142

Koike A, Niwa Y, Takagi T (2005) Automatic extraction of gene/protein biological functions from biomedical text. Bioinformatics 21(7):1227–1236

Lin H, Yang Z, Li Y (2011). Protein-protein interactions extraction from biomedical literatures. biomedical engineering, trends, research and technologies. In: Olsztynska S (ed). ISBN: 978-953-307-514-3

Majoros WH, Subramanian GM, Yandell MD (2003) Identification of key concepts in biomedical literature using a modified Markov heuristic. Bioinformatics 19(3):402–407

Mikolov T, Sutskever I, Chen K, Corrado GS, Dean J (2013) Distributed representations of words and phrases and their compositionality. In: Advances in neural information processing systems, pp 3111–3119

Miller GA, Beckwith R, Fellbaum C, Gross D, Miller K (1990) WordNet: an on-line lexical database. Int J Lexicogr 3(4):235–244

Navigli R, Ponzetto SP (2012) BabelNet: The automatic construction, evaluation and application of a wide-coverage multilingual semantic network. Artif Intell 193:217–250

Rindflesch TC, Fiszman M (2003) The interaction of domain knowledge and linguistic structure in natural language processing: interpreting hypernymic propositions in biomedical text. J Biomed Inform 36(6):462–477

Rindflesch TC, Tanabe L, Weinstein JN, Hunter L (2000) EDGAR: extraction of drugs, genes and relations from the biomedical literature. Pac Symp Biocomput pp 517–528

Rodriguez MDB, Gomez-Hidalgo JG, Diaz-Agudo B (1997) Using WordNet to complement training information in text categorization. In: Milkov R, Nicolov N, Nikolov N (ed) Second international conference on recent advances in natural language processing (RANLP), John Benjamins Publishing, Stanford CA USA

Shehata S, Karray F, Kamel M (2007) A concept-based model for enhancing text categorization, KDD

Snow R, Jurafsky D, Ng AY (2004) Learning syntactic patterns for automatic hypernym discovery. In: Advances in neural information processing systems (NIPS 2004), Vancouver, British Columbia

Song M, Yu HJ, Han WS (2011) Combining active learning and semi-supervised learning techniques to extract protein interaction sentences. BMC Bioinform 12(Suppl):12

Stillings NA, Feinstein MH, Garfield JL, Rissland EL, Rosenbaum DA, Weisler SE, Baker-Ward L (1987) Cognitive science: an introduction. MIT Press, Cambridge, MA

Tseng YH, Lin CJ, Lin YI (2007) Text mining techniques for patent analysis. Inf Process Manage 43(5):1216–1247

Wang JZ, Taylor W (2007) Concept forest: a new ontology-assisted text document similarity measurement method. In: Proceedings of the IEEE/WIC/ACM international conference on web intelligence, IEEE Computer Society, pp 395–401

Wilkowski B, Fiszman M, Miller CM, Hristovski D, Arabandi S, Rosemblat G, Rindflesh TC (2011). Graph-based methods for discovery browsing with semantic predications. In: AMIA Annual Symposium Proceedings, pp 1514–1523

Yang H Callan J (2009) A metric-based framework for automatic taxonomy induction. In: Proceedings of the joint conference of the 47th annual meeting of the acl and the 4th international joint conference on natural language processing of the AFNLP, vol 1. Suntec, Singapore, Association for Computational Linguistics, pp 271–279

Yang Z, Lin H, Li Y (2010) BioPPISVMExtractor: a protein–protein interaction extractor for biomedical literature using SVM and rich feature sets. J Biomed Inform 43(1):88–96

Zelikovitz S, Hirsh H (2000) Improving short text classification using unlabeled background knowledge to assess document similarity. In: Proceedings of the seventeenth international conference on machine learning

Zheng HT, Kang BY, Kim HG (2009) Exploiting noun phrases and semantic relationships for text document clustering. Inf Sci 179(13):2249–2262

Zhou D, He Y (2008) Extracting interactions between proteins from the literature. J Biomed Inform 41(2):393–407

Zhou X, Zhang X, Hu X (2006) Maxmatcher: biological concept extraction using approximate dictionary lookup. In: PRICAI 2006 Aug 9-11, pp 1145–1149

Chapter 7
Literature-Based Discovery

Abstract Literature-Based Discovery (LBD) refers to a range of approaches that take a body of scientific literature as the input, apply a series of computational, manual, or a hybrid processes, and finally generate hypotheses that are potentially novel and meaningful for further investigations. This chapter introduces the origin of LBD, its major landmark studies, available tools, and resources. In particular, we explain the design and application of PKD4J to illustrate the principles and analytic decisions one typically needs to make. We highlight the recent developments in this area and outline remaining challenges.

Swanson's Pioneering Work

Swanson's work on Raynaud disease/fish-oil discovery exemplified the problem of mining undiscovered public knowledge from biomedical literature (Swanson 1986a). According to Swanson (1986a, b). LBD (a.k.a. UDPK) can be public, yet undiscovered, if independently created fragments of knowledge and information are logically related but never retrieved, interpreted, and studied together. In other words, when considered together, two complementary and non-interactive literature sets of articles (independently created fragments of knowledge) can reveal useful information of scientific interest not apparent in either of the two sets alone (Swanson 1986a, b).

Swanson formalizes the procedure to discover UPK from biomedical literatures as follows: Consider two separate literature sets, CL and AL, where the documents in CL discuss concept C and documents in AL discuss concept A. Both of these two literature sets discuss their relationship with some intermediate concepts B (also called bridge concepts). However, their possible connection via the concepts B is not discussed together in any of these two literature sets as shown in Fig. 7.1.

Swanson's UPK (or ABC) model can be described as the process to induce "A implies C", which is derived from both "A implies B" and "B implies C"; the

© Springer International Publishing AG 2017

C. Chen and M. Song, *Representing Scientific Knowledge*,
https://doi.org/10.1007/978-3-319-62543-0_7

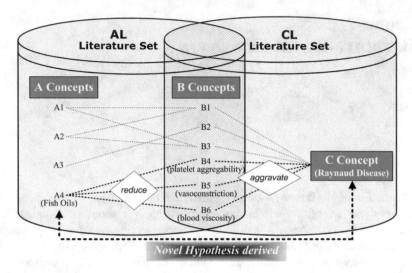

Fig. 7.1 Swanson's UPK model—the connection of fish oils and Raynaud disease

derived knowledge or relationship "A implies C" is not conclusive but hypothetical. For example, Swanson tried to uncover novel suggestions for what (B) causes Raynaud disease (C) or what (B) are the symptoms of the disease, and what (A) might treat the disease as shown in Fig. 7.1. Through analyzing the document set that discusses Raynaud disease he found that Raynaud disease (C) is a peripheral circulatory disorder aggravated by high platelet aggregation (B), high blood viscosity (B) and vasoconstriction (B). Then he searched these three concepts (B) against MEDLINE to collect a document set relevant to them. With the analysis on the document set he found out those articles show the ingestion of fish oils (A) can reduce these phenomena (B); however, no single article from both document sets mentions Raynaud disease (C) and fish oils (A) together. Putting these two separate literatures together, Swanson hypothesized that fish oils (A) may be beneficial to people suffering from Raynaud disease (C). This hypothesis that Raynaud disease might be treated by fish oil was hidden in the biomedical literature until Swanson uncovered through literature-based discovery. This novel hypothesis was later clinically confirmed by DiGiacomo et al. (1989). Later on, Swanson used the same approach to uncover 11 connections of migraine and magnesium (Swanson 1988).

One of the drawbacks of Swanson's method is that the method requires large amount of manual intervention and very strong domain knowledge, especially in the process of qualifying the intermediate concepts that Swanson names the "B" concepts. In order to reduce dependence on domain knowledge and human intervention and to automate the whole process as much as possible, several approaches have been developed to automate this discovery process based on Swanson's method (Lindsay and Gordon 1999; Pratt and Yetisgen-Yildiz 2003; Srinivasan 2004; Weeber et al. 2003). They have not only successfully replicated the Raynaud disease-fish-oil and migraine-magnesium discoveries, but also discovered new treatments for other diseases such as thalidomide (Weeber et al. 2003).

These research works have produced valuable insights into new hypothesis. On the other hand, substantial manual intervention is required to reduce the number of possible connections. We describe a fully automated approach for mining undiscovered public knowledge from biomedical literature. Our approach replaces ad hoc manual pruning with semantic knowledge from biomedical ontologies. We use semantic information to manage and filter the sizable branching factor in the potential connections among a huge number of medical concepts.

To efficiently find novel hypotheses efficiently and effectively from a huge search space of possible connections among the biomedical concepts, we need to first solve the problem of ambiguous biomedical terms. We utilize biomedical ontologies, namely UMLS and MeSH for this purpose. Our method requires minimal human intervention. Unlike other approaches (Hristovski et al. 2001; Pratt and Yetisgen-Yildiz 2003; Srinivasan 2004), our method only requires the user to specify the possible semantic relationships between the starting concept and the to-be-discovered target concepts rather than possible semantic types of the target concepts and the bridge concepts. Our method utilizes semantic knowledge (e.g., semantic types, semantic relations and semantic hierarchy) on bridge concepts and the target concepts to filter out irrelevant concepts and meaningless connections between concepts. Since there could be many plausible relationships between the bridge concepts and the target concepts, our method uses semantic relations to filter those relationships to identify desirable ones.

Major Trends of LBD

Swanson's pioneering work provides the framework on which almost all subsequent research in LBD is based (Cameron et al. 2013, Cohen et al. 2010, Malhotra et al. 2013, Spangler et al. 2014). The initial approach proposed by Swanson requires a laborious, time-intensive, manual process. The follow-up studies attempted to overcome these challenges by developing processes to make LBD easier and faster to perform and more automatic overall. Those studies proposed different techniques for concept extraction, computation of results, and sizes and types of input data. In LBD, human experts continue to play a significant role. New systems essentially follow Swanson's ABC model of discovery.

A recent trend in LBD is that more works has focused specifically on, and provided advancements in, automation of the LBD process. Using more advanced Natural Language Processing (NLP) techniques while at the same time exploiting metadata (e.g., from UMLS) has led to a reduction in the role of human experts (Wilkowski et al. 2011). Another trend is to use more advanced methods to capture important correlations between concepts. Hristovski et al. (2001) and Pratt and Yetisgen-Yildiz (2003) used an unsupervised machine learning algorithm (association rule mining) along with support and confidence metrics. In contrast,

Fig. 7.2 An example of Brat visualization of entity and relation

Wren et al. (2004) used statistical techniques to distinguish significant correlations. A related trend is the application of visualization. van der Eijk et al. (2004) differs from other work by giving a visual output directly to the user without the intermediate steps requiring human expert guidance. Overall, reducing reliance on human experts by increasing the degree of automation is an important recent trend in LBD research. The development of web-based visualization such as D3.js[1] and Brat[2] makes visualization of LBD scalable and accessible via web. The example of visualization with a PubMed sentence by Brat is shown in Fig. 7.2.

LBD Systems

We outline the design and functionality of three examples of LBD systems, namely the ArrowSmith developed in late 1990s, the BITOLA systems in mid 2000s, and the more recent Hypothesis Generator in 2015.

ArrowSmith

ArrowSmith is the very first LBD tool introduced by Swanson and Smalheiser (1997), which is publicly avaliable.[3] ArrowSmith provides a two-mode discovery method. The simple PubMed search function is available for the users to input two PubMed queries in order to define the two sets of articles A and C (Fig. 7.3).

To retrieve MEDLINE records corresponding to user queries in a fast mode, a local MEDLINE database was created. When a query is entered, the article ID numbers are downloaded from PubMed and the full MEDLINE records are retrieved from the local database, including a tokenized result of each article title after stopwords were removed. If articles are not found in the local database, then they are downloaded from PubMed as XML files, processed and stored in the local database. B-terms and their feature values are computed in a parallel mode by processing the sets of tokenized titles in chunks, and merging the results later on when each process is done. B-term features were pre-computed and stored in the term database for fast look-up.

---

[1]https://d3js.org/.

[2]http://brat.nlplab.org/features.html.

[3]http://arrowsmith.psych.uic.edu.

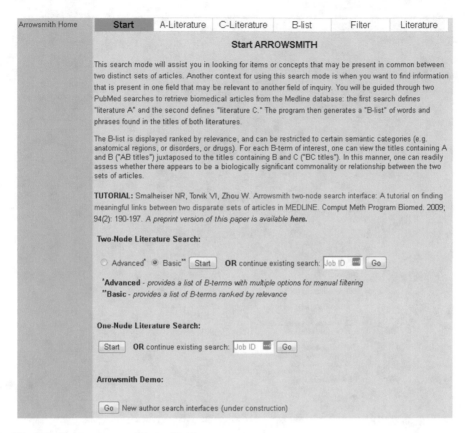

Fig. 7.3 The homepage of ArrowSmith

For instance, if we choose "Raynaud's disease" as the A-literature term and "Fish Oil" as the C-literature term, ArrowSmith returns the list of B terms after couple of minutes' execution time. With "Raynaud's disease" and "Fish oil" as A and C, ArrowSmith generates a total of 7093 B-terms that do not appear in both A and C literature (six articles that appeared in both A and C were excluded in the resulting b-term list). The list of B-terms is shown in the inner box of Fig. 7.4, which is sorted in order of predicted relevance score of a B-term that indicates a biological significance between the AB and BC literatures.

We can filter out the resulting B-term list by semantic types provided in UMLS. For instance, if we want to restrict the B-terms to the two semantic types, Activities & Behaviors and Anatomy, you can simply select the check box next to those two types once you click on "Restrict by semantic categories" button. It will result in the 730 B-terms that passed the filtering criteria (Fig. 7.5). Before clicking the button, you may want to scroll down the list to see if there are any non-highlighted B-terms that you want to keep. Use Ctrl to select additional B-terms.

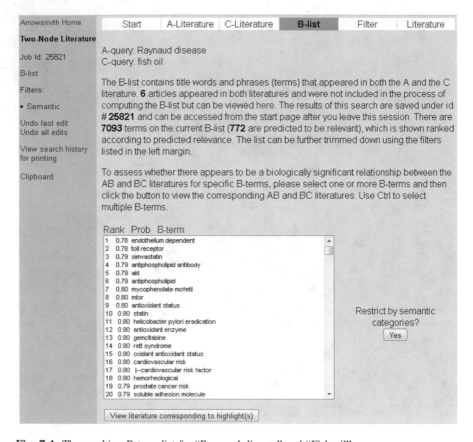

Fig. 7.4 The resulting B-term list for "Raynaud disease" and "Fish oil"

BITOLA

BITOLA is an web-based LBD system that has been around for about a decade (Hristovski et al. 2003), which is publicly available at.[4] The purpose of BITOLA is to help the biomedical researchers make new discoveries by discovering potentially new relations between biomedical concepts. The set of concepts contains MeSH and human genes from HUGO. BITOLA provides two discovery options: closed and open.

Open discovery allows the input of a single concept, then categories for first-order relatives of that concept, then categories for relatives of those first order concepts. Discovery algorithm for discovering new relations between medical concepts consists of the following five steps (Hristovski et al. 2001):

---

[4]http://arnika.mf.uni-lj.si/pls/bitola2/bitola.

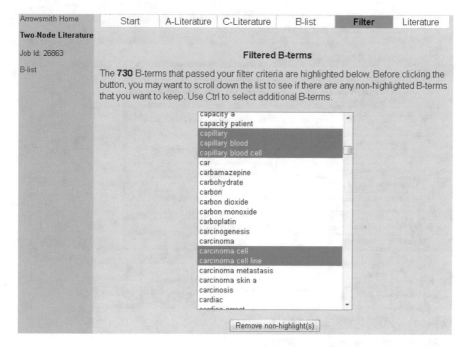

Fig. 7.5 Filtered B-terms

1. Given a starting concept of interest X
2. Find all concepts Y such that there is an association rule X → Y
3. Find all concepts Y such that there is an association rule Y → Z
4. Eliminate those Z for which an association X → Z already exists
5. The remaining concepts Z are candidates for an new relation between X and Z.

Because in MEDLINE each concept can be associated with many other concepts, the possible number of X → Z combinations can be extremely large. In order to deal this combinatorial problem, BITOLA applies filtering (limiting) and ordering functions to the discovery algorithm. The related concepts can be limited by the semantic type to which they belong and final possibility for limiting the number of related concepts or false related concepts is by setting thresholds on the support and confidence measures of the association rules. The main goal of the ordering is to present best candidates first to make human review as easy as possible (Hristovski et al. 2001).

For example, if Magnesium is the interest of search, type Magnesium and click on Find Starting Concept X in the BITOLA system, which will return a list of terms relevant to the query. As shown in Fig. 7.6, the query found 13 terms.

From the generated list, choose the very top one Magnesium, and BITOLA will fill in CUI (C0024467), the semantic type, and the chromosomal location automatically (if exists). Click on the button Find Related Zs, BITOLA will generate the

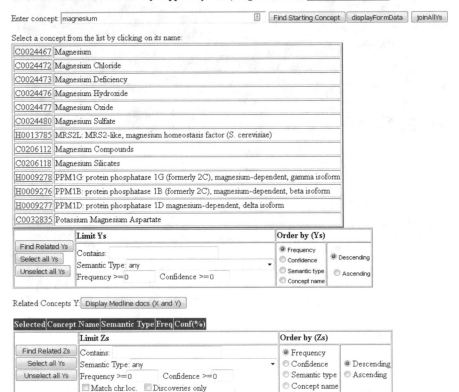

Fig. 7.6 The search results for the query *magnesium*

results, containing concept name, semantic type, frequency, confidence level, discovery, and chromosomal location (see Fig. 7.7).

   Once a list of related concepts Zs is displayed, click the button Find Intermediate
Ys, which will generate a list of substance terms that have been linked to
Magnesium in some articles. See Fig. 7.8.

   From this list of related concepts Ys, selecting the term Potassium with the
semantic type of Pharmacologic Substance and clicking on the button Display
Medline docs (X and Y) will display the two articles in PubMed about both
Magnesium (X) and Potassium (Y). The user can explore other links, or re-run the
query with other categories, so as to explore domains and chemicals that are linked
to both Magnesium and Potassium.

   In addition to the Closed Discovery option of BITOLA, the Open Discovery
option of BITOLA allows the users to expand their inquiry into one node basis
discovery. The Open Discovery option works quite similarly as the Closed
Discovery one. The only difference is the structure. With closed discovery the user
nominates X and Z then search for Y (limiting categories, if desired). With open

BITOLA - **Biomedical Discovery Support System (Program author: Dimitar Hristovski)**

Enter concept: [   ] [⊡] [Find Starting Concept] [displayFormData] [joinAllYs]

Starting Concept X

Concept: Magnesium   **Semantic Types:**
CUI: C0024467   Biologically Active Substance/ Element, Ion, or Isotope/

	Limit Ys		**Order by (Ys)**	
[Find Related Ys]	Contains: [ ]		⦿ Frequency	
[Select all Ys]	Semantic Type: any ▾		○ Confidence ⦿ Descending	
[Unselect all Ys]	Frequency >=0 Confidence >=0		○ Semantic type ○ Ascending	
			○ Concept name	

Related Concepts Y: [Display Medline docs (X and Y)]

Selected	Concept Name	Semantic Type	Freq	Conf(%)

	Limit Zs		**Order by (Zs)**	
[Find Related Zs]	Contains: [ ]		⦿ Frequency	
[Select all Ys]	Semantic Type: any ▾		○ Confidence ⦿ Descending	
[Unselect all Ys]	Frequency >=0 Confidence >=0		○ Semantic type ○ Ascending	
	☐ Match chr.loc. ☐ Discoveries only		○ Concept name	

Related Concepts Z:

Concept Name	Semantic Type	Freq	Conf(%)	"Discovery?"	Chr.Loc.
Rats	Mammal	42047	25.96		
Kinetics	Idea or Concept	18679	11.53		
Magnesium	Element, Ion, or Isotope	17916	11.06		
Magnesium	Biologically Active Substance	17916	11.06		
Potassium	Element, Ion, or Isotope	16743	10.34		
Potassium	Biologically Active Substance	16743	10.34		
Potassium	Pharmacologic Substance	16743	10.34		
Sodium	Biologically Active Substance	14659	9.05		
Sodium	Element, Ion, or Isotope	14659	9.05		
Sodium	Pharmacologic Substance	14659	9.05		
Cells, Cultured	Cell	13831	8.54		
Time Factors	Temporal Concept	11106	6.86		

Fig. 7.7 The results of the related concepts Z to "Magnesium"

discovery, the user nominates X, then search for Y (limiting categories, if desired), then search for Z (limiting categories, if desired).

Hypothesis Generator

Hypothesis Generator is a recently developed LBD system that is based on PKDE4J (Song et al. 2015) for entity and relation extraction (Baek et al. 2017). Hypothesis Generator was originally developed to examine how lactosylceramide is associated with arterial stiffness. However, due to the flexibility of the system, hypothesis generator can serve as the general LBD system.

A brief instruction for hypothesis generator is as follows. First, the user types in one or more search terms, for example, "Raynaud disease" (Fig. 7.9).

The search function is backed by the Apache Lucene information retrieval system. Hypothesis generator indexed the 2015 version of MEDLINE records with

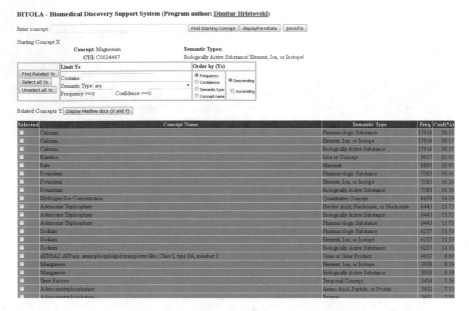

Fig. 7.8 The list of related concepts Y to the target term "Magnesium"

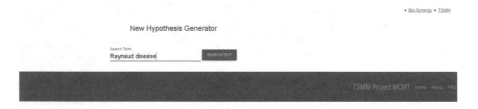

Fig. 7.9 The search homepage of the hypothesis generator

Lucene. The search term is highlighted in either the title or the abstract field (see Fig. 7.10).

PubMed ID for each result will be shown on the left and a direct link to the article is given on the right. The user can choose the number of PubMed records to be included for generating the paths.

On the search result page, the user can choose the number of PubMed records to extract entities from. This step is necessary since the current version of hypothesis generator extracts entities on the fly. In the future, extraction of entities will be done offline and stored in the database. If that is in place, this step will be eliminated. Once the number of records is chosen, you can click on the "generate paths" button, which will result in the follow result (Fig. 7.11).

The left panel shows the list of extracted entities and you can pick any two entities that you are interested into see the relation between two. Type in the entities that you want to conduct path analysis from the list of entity names. The left will be

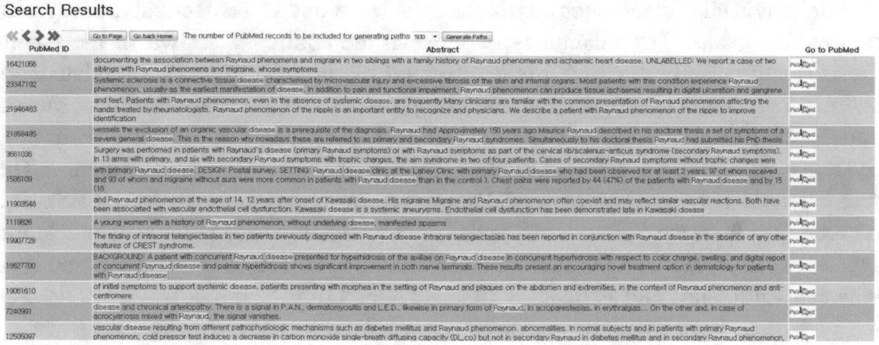

Search Results

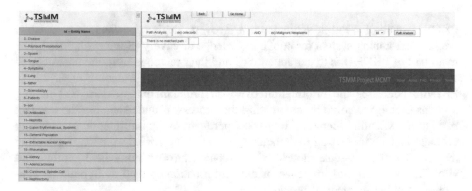

Fig. 7.10 The search result page for the query "Raynaud disease"

Fig. 7.11 The results of extracted entities (left) and the path analysis start page (right)

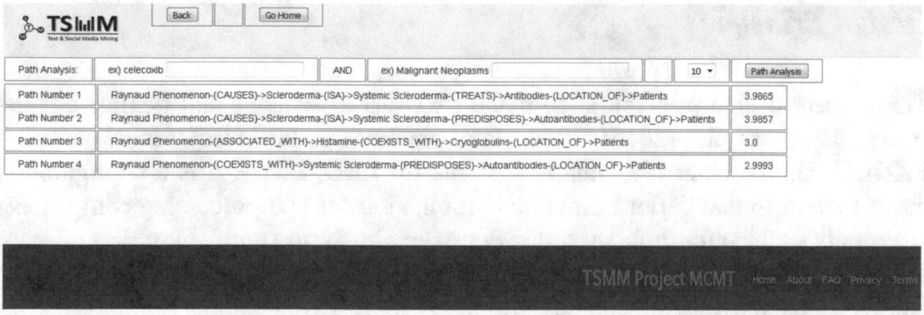

Path Analysis:	ex) celecoxib	AND	ex) Malignant Neoplasms	10 ▾	Path Analysis
Path Number 1	Raynaud Phenomenon-(CAUSES)->Scleroderma-(ISA)->Systemic Scleroderma-(TREATS)->Antibodies-(LOCATION_OF)->Patients				3.9865
Path Number 2	Raynaud Phenomenon-(CAUSES)->Scleroderma-(ISA)->Systemic Scleroderma-(PREDISPOSES)->Autoantibodies-(LOCATION_OF)->Patients				3.9857
Path Number 3	Raynaud Phenomenon-(ASSOCIATED_WITH)->Histamine-(COEXISTS_WITH)->Cryoglobulins-(LOCATION_OF)->Patients				3.0
Path Number 4	Raynaud Phenomenon-(COEXISTS_WITH)->Systemic Scleroderma-(PREDISPOSES)->Autoantibodies-(LOCATION_OF)->Patients				2.9993

Fig. 7.12 The path analysis result

the 'A-term' and the right will be the 'C-term' of your path. The user can choose the number of path you want to analysis as shown. For instance, if "Raynaud Phenomenon" is chosen as the A-term and "Patients" as the C-term, then the 'Path Analysis' will generate the results as shown in Fig. 7.12.

For "Raynaud Phenomenon" and "Patients" as A and C, respectively, the system returns four paths. The relation type between the entities is shown in the parenthesis. Importance of each path is determined by the overall semantic relatedness score. The overall relatedness score is computed by the average of a Phenomenon and Scleroderma. Pair 2 is Scleroderma and Systemic Scleroderma. Pair 3 is Systemic Scleroderma and Antibodies. Pair 4 is Antibodies and Patients. The relation type between Systemic Scleroderma and Antibodies is CAUSES. The relation type between Scleroderma and Systemic Scleroderma is IS-A. The relation type between Systemic Scleroderma and Antibodies is TREATS. The relation type between Antibodies and Patients is LOCATION_OF.

PKD4J: A Scalable and Flexible Engine

PKDE4J stands for Public Knowledge Discovery Engine for Java, is a scalable, flexible text mining system for public knowledge discovery (Song et al. 2015). The main task of PKDE4J is to extract entities and their relations from the unstructured text. PKDE4J extends Stanford CoreNLP written in Java (Manning et al. 2014). PKDE4J addresses the information overload problem that modern text mining systems promise to solve by automating the process of understanding the relevant parts of the scientific literature. Key tasks pertinent to the information overloading problem include increasing the efficiency of searching for information, facilitating the creation of large-scale models of the relationships of biomedical entities, and allowing for automated inference of new information as well as hypothesis generation to guide biomedical research.

Design Principle

The primary design principle is to make PKDE4J as scalable and flexible as possible. Song et al. (2015) used the pipeline architecture for developing PKDE4J. Unlike other text mining systems for LBD, PKDE4J is a configuration based system so that various different combinations of text processing components are readily enabled for different tasks. For example, for the problem of drug-disease interaction, we can use SIDER (http://sideeffects.embl.de/) for drug dictionary and KEGG (http://www.genome.jp/kegg/disease/) for disease dictionary. Another layer of flexibility is that entities can be extracted either by exact or approximate match. On top of the exact matching based entity extraction, bio entity can be extracted either by approximate matching-based, supervised-learning only, or a mixture of supervised-learning and dictionary. PKDE4J overcomes the problems of the dictionary-based approach by applying regular expression rules and N-gram to extract entities. Second, PKDE4J is a flexible extraction system that can be applied to different extraction tasks such as multi-class entity extraction, Protein-Protein Interaction (PPI), trigger extraction, etc.

Most of the current approaches are focused heavily on a specific application to solve a specific kind of problem. PKDE4J is designed to address the aforementioned issue by developing an extensible rule engine based on dependency parsing for relation extraction. It provides a rule configuration file that contains 17 rules to identify whether relation exists in a sentence and determine its relation type. Since a relation extraction task requires an unique set of extraction rules, one single optimized prediction model is only effective in a certain condition. For instance, a different model is required for the task of whether a sentence contains relation or not from the task of event extraction. In such scenario, supervised learning may not be the best option since for each task, a different classification model needs to be built. Thus, a flexible, plug-and-play module for a rule engine is the best option for different extraction tasks in an efficient manner.

Architecture

PKDE4J consists of four major components. The overall architecture of PKDE4J illustrates the connections between these components (Fig. 7.13).

The first component is preprocessing of input text. PKDE4J supports a verity of text formats, which includes PubMed in XML, PubMed Central in XML, ClincalTrials.gov in XML, and text data in CSV. The second component is entity extraction, including dictionary-based, supervised learning-based, a combination of dictionary with ontology like UMLS, and a combination of supervised learning-based with UMLS. The third component is relation extraction, which is based on a dependency tree-based rules. The fourth component is the storage and retrieval of the results from the entity and relation extraction components. The results are stored in a relational database in the format that can be used for visualization.

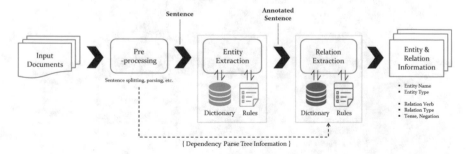

Fig. 7.13 The overall architecture of PKDE4J. *Source* Song et al. (2015)

Preprocessing

The preprocessing component covers various text processing tasks. The first one is tokenization. PKDE4J uses the Penn Treebank 3 (PTB) tokenization implemented in Stanford CoreNLP. PTBTokenizer is based on JFlex for an efficient, fast, and deterministic tokenization.

The second preprocessing task is sentence boundary detection. PKDE4J uses a Maximum Entropy model trained with the GENIA corpus for sentence splitting.

The third task is Part-Of-Speech (POS) tagging. PKDE4J uses the Stanford POS tagging algorithm for this task. The Stanford POS tagging algorithm is based on a flexible statistical CRF model.

The fourth task is lemmatization aided by Stanford CoreNLP. The fifth task is normalization of tokens. Token normalization is required since text contains various non-alphanumeric characters which may hinder the quality of entity extraction. The sixth task is n-gram matching. PKDE4J adopts the Apache Lucene ShingleWrapper algorithm, which constructs n-gram tokens from a token stream. The seventh task is approximate string matching. Approximate string matching may be needed when input text contains many spelling variations for the same entity name. PKDE4J extends the Soft-TFIDF algorithm that is a hybrid similarity measure introduced by Cohen et al. (2003).

Entity Extraction

Figure 7.14 shows the overall architecture of entity extraction component that consists of several steps. Step 1 is to load dictionaries. Dictionary loading is required when you choose the dictionary-based approach for entity extraction over other approaches. Depending on the target entities to be extracted, a list of

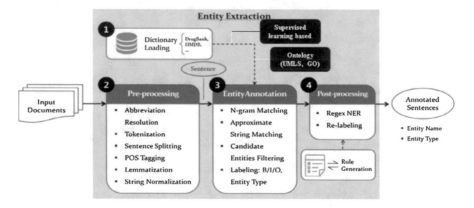

Fig. 7.14 Entity extraction component. An extended version of Song et al. (2015)

dictionaries are determined. Step 2 is preprocessing, which was described in the preprocessing component. Step 3 is entity annotation where the entity matching takes place between tokenized n-grams and dictionary entries. In entity annotation, there are four different options: (1) dictionary only, (2) a combination of dictionary with ontology, (3) supervised learning only, and (4) a combination of supervised learning with ontology. Step 4 is post-matching. For further improvement of extraction quality, PKDE4J uses the regular expressions to match the entities that are not found by dictionary. The regular expression rules define cascaded patterns over token sequences, which provides a flexible extension of the traditional regular expression language defined over strings.

Relation Extraction

The relation extraction component relies heavily on a set of dependency parsing based rules. Dependency parse trees provide a useful structure for the sentences by annotating edges with dependency types, e.g. subject, auxiliary, modifier. Dependency parse trees embed various information of dependencies within sentences, i.e. between words that are far apart in a sentence. The relation extraction module consists largely of three steps (See Fig. 7.15).

Step 1 is loading couple of dictionaries that contain biologically meaning verbs such as up-regulate, down-regulate, simptomize, etc. and nominalized terms like expression. Biologically meaningful verbs are classified into several categories and each category may have a few types (Table 7.1). The relation extraction component detects biologically meaningful verbs from sentences and map them to either categories or types, depending on the configuration setting.

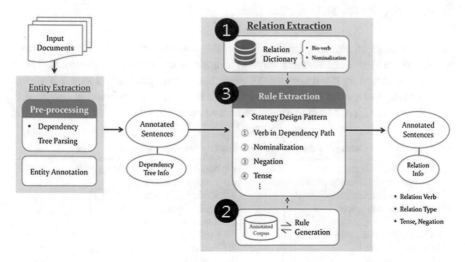

Fig. 7.15 Relation extraction component. *Source* Song et al. (2015)

Table 7.1 Classification of the biologically meaningful verb list

Category	Type	Verb example
Positive	Increase	Activate, promote, stimulate
	Transmit	Transport, link
	Substitute	Replace
Negative	Decrease	Inactivate, inhibit, block, arrest
	Remove	Breakbond, release, omit
Neutral	Contain	Embed, include, constitute
	Modify	Reconstitute, mutate, oxidize
	Method	Bleach, precipitate, coprecipitate
	Report	Prove, suggest, compare
Plain	Plain	Acquire, underlie, fix

Step 3 applies a set of relation rules to parsed dependency trees. After preprocessing, PKDE4J traverses the resulting dependency tree in postorder to find the relation triplets by using predefined set of relation rules for a dependency tree. In PKDE4J, each rule is called a strategy, which echoes the strategy design pattern adopted from Object-oriented system development. A strategy design pattern is particularly useful for creating objects which represent various strategies and a context object whose behavior varies as per its strategy object. In PKDE4J, a strategy represents a dependency tree-based relation rule. By applying a predefined set of strategies to each sentence, PKDE4J applies 17 predefined rules to the sentence, which generates a set of relation features such as relation type, tense, and negation for any given two entities located in the sentence (See Table 7.2).

Storing the Results of Extraction

At the last stage of pipeline, PKDE4J generates two major outputs. The first output is the extracted entities and the second output is the extracted relations. These outputs are stored in the relational database for further analysis. Table 7.3 shows the example of extracted entities. The example is a simplified version of output that only show PMID, entity name, entity type, and sentence where the entity is located

Table 7.2 A list of strategies that characterize relation between two entities

① Verb in dependency path	⑩ Number entities between entities
② No verb in dependency path	⑪ Entities in between
③ Detect nominalization	⑫ Surface distance
④ Weak nominalization	⑬ Entity counts
⑤ Negation	⑭ Same head
⑥ Tense (active/passive)	⑮ Entity order
⑦ Contain clause	⑯ Full tree path
⑧ Clause distance	⑰ Path length
⑨ Negation clause	

Table 7.3 Example of output of extracted entities

PMID	Entity	Type	Sentence
28482223	Phentolamine	DRUG	Phentolamine is one of the most representative nonselective αadrenoreceptor blocking agents, which have been proved to be owned various pharmacological actions
28482223	protein	FOOD	With the aid of multiple biophysical techniques, this scenario was to detailed explore the potential biorecognition between phentolamine and the hemeprotein in the cytosol of erythrocytes, and the influences of dynamic characters of protein during the bioreaction
28482223	protein	FOOD	Biorecognition can induce fairly structural transformation (selfregulation) of protein conformation

in. In addition to those four attributes, there are other attributes available such as beginning and ending position of entity as the results of entity extraction.

The second output is the relation extraction result shown in Table 7.4. The output consists of PMID, relation type, left entity name, left entity type, right entity name, right entity type, verb, voice, negation, and sentence where two entities are located in.

Table 7.4 Example of output of extracted relations

Field	Value 1	Value 2
PMID	8447197	27983686
Relation Type	PLAIN	RESULT_OF
Entity 1	Alcohol	Dairy
Entity 1 Type	FOOD	FOOD
Entity 2	Alcoholic	Drink
Entity 2 Type	FOOD	FOOD
Verb	Play	Containing
Tense	ACTIVE	ACTIVE
Negation	POSITIVE	POSITIVE
Sentence	Many variables, aside from the amount and duration of alcohol consumption, play a role in the development and progression of alcoholic liver disease (ALD)	In a placebo controlled, randomized, crossover study, 35 healthy males received either six placebo gelatin capsules consumed with 200 mL of water, six capsules with 800 mg polyphenols derived from red wine and grape extracts, or the same dose of polyphenols incorporated into 200 mL of either pasteurized dairy drink, soy drink (both containing 3.4% proteins) or fruit flavored protein free drink

Recent Developments and Remaining Challenges

Recently, LBD research has paid attention to deep learning as an effort to improve the quality of discovery. Rather et al. (2017) applied a word embedding technique called Word2Vec to the LBD problem. They used the MRDEF subset of UMLS Metathesaurus to train the Word2Vec model and reported a 23% overlap between their approach and MRREL. Deep learning has also been applied to the task of phenotyping (Che et al. 2015) used to identify patient subgroups based on individual clinical markers. Žitnik et al. (2013) conducted a study on non-negative matrix factorization techniques for fusing various molecular data to uncover disease-disease associations and show that available domain knowledge can help reconstruct known and obtain novel associations. Despite the recent interests in deep learning, it is still premature. More advanced studies of the applications of deep learning to the LBD problems are needed to evaluate how deep learning can advance LBD research.

There are several remaining challenges in LBD. The first challenge is how to implement a comprehensive procedure to obtain manually labeled samples. Although state-of-the-art machine learning methods have been utilized to automate the process, current approaches still observe degraded performance in the face of limited availability of labeled samples that are manually annotated by medical experts. Another major challenge is the convergence of multi-disciplinary teams that are pertinent to LBD. Although collaboration among researchers from various different fields is prevalent in LBD, it is often observed that development is separated from evaluation and end-usage of the tool developed. The third challenge is the standardization of evaluation. Evaluation in LBD is often ad hoc based and no general guidelines are established for LBD researchers to follow. Although there is a movement of standardization such as PubAnnotation,[5] we still need to put much effort into setting up the guidelines for LBD research.

References

Baek SH, Lee D, Kim M, Lee JH, Song M (2017) Enriching plausible new hypothesis generation in PubMed. PLoS ONE 12(7):e0180539

Cameron D, Bodenreider O, Yalamanchili H, Danh T, Vallabhaneni S, Thirunarayan K, Sheth AP, Rindflesch TC (2013) A graph-based recovery and decomposition of Swanson's hypothesis using semantic predications. J Biomed Inform 46:238–251. doi:10.1016/j.jbi.2012.09.004

Che Z, Kale D, Li W, Bahadori MT, Liu Y (2015) Deep computational phenotyping. In: Proceedings of the 21th ACM SIGKDD International Conference on Knowledge Discovery and Data Mining, pp 507–516 (ACM, 2015)

Cohen T, Whitefield GK, Schvaneveldt RW, Mukund K, Rindflesch T (2010) EpiphaNet: an interactive tool to support biomedical discoveries. J Biomed Discov Collab 5:21–49

---

[5]http://pubannotation.org/.

Cohen WW, Ravikumar P, Fienberg SE (2003) A comparison of string metrics for matching names and records. In: Paper Presented at the International Conference on Knowledge Discovery and Data Mining (KDD) 09, Workshop on Data Cleaning, Record Linkage, and Object Consolidation

DiGiacomo RA, Kremer JM, Shah DM (1989) Fish-oil dietary supplementation in patients with Raynaud's pheomenon: a double blind, controlled, prospective study. Am J Med 86:158–164

Hristovski D, Peterlin B, Džeroski S, Stare J (2001) Literature based discovery support system and its application to disease gene identification. In: Proceeding AMIA Symposium 928

Hristovski D, Peterlin B, Mitchell JA, Humphrey SM (2003) Improving literature based discovery support by genetic knowledge integration. Stud Health Technol Inform 95:68–73

Lindsay RK, Gordon MD (1999) Literature-based discovery by lexical statistics. J Am Soc Inf Sci 50(7):574–587

Malhotra A, Younesi E, Gurulingappa H, Hofmann-Apitius M (2013) 'HypothesisFinder:' a strategy for the detection of speculative statements in scientific text. PLoS Comput Biol 9(7): e1003117. doi:10.1371/journal.pcbi.1003117

Manning CD, Surdeanu M, Bauer J, Finkel J, Bethard SJ, McClosky D (2014) The stanford CoreNLP natural language processing toolkit. In: Proceedings of 52nd Annual Meeting of the Association for Computational Linguistics: System Demonstrations, pp 55–60

Pratt W, Yetisgen-Yildiz M (2003) LitLinker: capturing connections across the biomedical literature, K-CAP'03, pp 105–112, Sanibel Island, FL, 23–25 Oct 2003

Rather NN, Patel CO, Khan SA (2017) Using deep learning towards biomedical knowledge discovery. Int J Math Sci Comput (IJMSC) 3(2):1–10. doi:10.5815/ijmsc.2017.02.01

Song M, Kim WC, Lee DH, Heo GE, Kang KY (2015) PKDE4J: entity and relation extraction for public knowledge discovery. J Biomed Inform 57:320–332

Spangler S, Wilkins AD, Bachman BJ, Nagarajan M, Dayaram T, Haas P, Regenbogen S, Pickering CR, Comer A, Myers JN, Stanoi I, Kato L, Lelescu A, Labrie JJ, Parikh N, Lisewski AM, Donehower L, Chen Y, Lichtarge O (2014) Automated hypothesis generation based on mining scientific literature. In: Paper Presented at the Proceedings of the 20th ACM SIGKDD International Conference on Knowledge discovery and data mining, New York, NY, USA

Srinivasan P (2004) Text mining: generating hypotheses from MEDLINE. J Am Soc Inf Sci 55 (4):396–413

Swanson DR (1986a) Fish oil, Raynaud's syndrome, and undiscovered public knowledge. Perspect Biol Med 30:7–18

Swanson DR (1986b) Undiscovered public knowledge. Libr Q 56(2):103–118

Swanson DR (1988) Migraine and magnesium: eleven neglected connections. Perspect Biol Med 31(4):526–557

Swanson DR, Smalheiser NR (1997) An interactive system for finding complementary literatures: a stimulus to scientific discovery. Artif Intell 91:183–203

van der Eijk C, Van Mulligen E, Kors JA, Mons B, Van den Berg J (2004) Constructing an associative concept space for literature-based discovery. J Am Soc Inf Sci Technol 55(5):436–444

Weeber M, Vos R, Klein H, de Jong-Van den Berg LT, Aronson AR, Molema G (2003) Generating hypotheses by discovering implicit associations in the literature: a case report for new potential therapeutic uses for Thalidomide. J Am Med Inf Assoc 10(3):252–259

Wilkowski B, Fiszman M, Miller CM, Hristovski D, Arabandi S, Rosemblat G, Rindflesh TC (2011) Graph-based methods for discovery browsing with semantic predications. In: AMIA Annual Symposium Proceedings, pp 1514–1523

Wren JD, Bekeredjian R, Stewart JA, Shohet RV, Garner HR (2004) Knowledge discovery by automated identification and ranking of implicit relationships. Bioinformatics 20(3):389–398

Žitnik M, Janjić V, Larminie C, Zupan B, Pržulj N (2013) Discovering disease-disease associations by fusing systems-level molecular data. Sci Rep 3

Chapter 8
Patterns and Trends in Semantic Predications

Abstract We demonstrate a series of studies of semantic predications from Semantic MEDLINE, including the detection of semantic predications with burstness and in association with conflict, contradictory, or other sources of uncertainties of scientific knowledge. Semantic networks of predications are analyzed within the framework of structural variations. Examples in this chapter represent scientific knowledge at a level of granularity that differs from those studies of scientific knowledge at the level of articles or journals of scholarly communication.

Semantic MEDLINE Database

The backend of Semantic MEDLINE is the Semantic MEDLINE Database (SemMedDB) (Kilicoglu et al. 2012). As of December 31, 2016, SemMedDB contains about 89.2 million predications from 26.7 million bibliographic records from MEDLINE. Its primary coverage is the biomedical literature. These predications are extracted by SemRep. The current version of SemMedDB is semmever30.

Representing Semantic Predications as a Graph

SemMedDB contains several tables of citations (in the MEDLINE sense of the term), i.e. the metadata of a published article, original sentences, and predications. For example, the SENTENCE table contains information on individual sentences such as SENTENCE_ID, PubMed ID (PMID), and the sentence. The PREDICATION table contains various information about predications such as PREDICATION_ID, a SENTENCE_ID, PMID (PubMed ID), PREDICATE, SUBJECT_CUI, SUBJECT_NAME (preferred name of the subject of the predication), and similar fields for the object of the predication. We loaded SemMedDB version 24 to a MySQL database. The examples explained below are based on this version. Figure 8.1 shows a

© Springer International Publishing AG 2017
C. Chen and M. Song, *Representing Scientific Knowledge*,
https://doi.org/10.1007/978-3-319-62543-0_8

visualization of a network of semantic predications in Neo4j, a graph database. The visualization shows that the semantic connections are unevenly distributed. Some entities are connected by a lot of semantic relations, whereas some are connected by few connections. The unevenness implies a level of uncertainty.

A distinct advantage of a graph database over the traditional relational database is a reduced complexity of queries. As illustrated in Table 8.1, a complex and time-consuming query with multiple table joins in a relational database can be reduced to a simple and efficient query in a graph database in Neo4j with the Cypher query language. The query in the graph database is in Cypher, a powerful query language supported by Neo4j. The query is to find paths that start with a doctor node and connect to a therapy node through at least four other types of nodes in between. A Cypher query shares some similarities with MySQL queries in terms of their style.

A study of scientific claims often need to address a series of questions based on the current results of a search. Graph databases such as Neo4j provide the desirable flexibility. Consider the following questions concerning the interest in scientific knowledge relevant to a body of scientific publications. These questions may remind you Heilmeier's Catechism we discussed in Chap. 1.

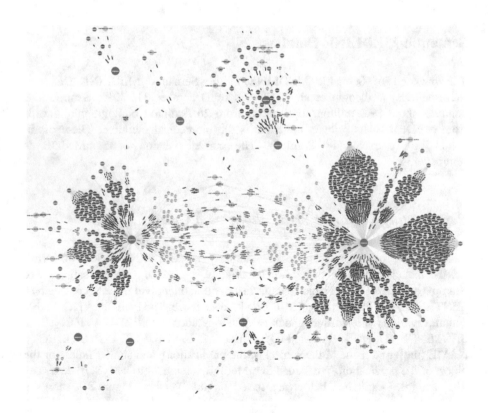

Fig. 8.1 A network of semantic predications visualized in Neo4j

Table 8.1 The complexity of a query can be reduced in a graph database

Database Type	Query

Relational	SELECT C.DiagnosisID as diagnosis, C.TherapyID as therapy
	FROM prescriptions as C JOIN
	  (SELECT
	    DiagnosisID as ID, A.ICD10_CODE as code, A.ICD10_TEXT as text
	  FROM diagnosis as A JOIN
	  (SELECT
	    PracticeID, DoctorID, PatientID, Age, DiagnosisID, Action
	  FROM actions
	  ON A.DiagnosisID=B.DiagnosisID) as D
	ON C.DiagnosisID=D.ID;

Graph	MATCH path=(d:DOCTOR)-[r*4..]- (t:THERAPHY) RETURN path LIMIT 100;

1. When did a claim (hypothesis, assertion, conjecture, statement) appear for the first time in a body of source information, e.g. the literature? In which paper/source?
2. How certain was the assertion?
3. How many papers made the same claim subsequently?
4. How many papers made claims that contradict a given claim subsequently?
5. What are the closely related but distinct claims for a given claim?
6. Given a claim, which reference is most frequently cited along with the claim's citation context?
7. Which references are frequently co-cited in relation to a given claim?
8. How many dimensions (clusters or eigenvectors or topics) are associated with the citation contexts of a claim?

There are several advantages of addressing these questions in a graph database. In particular,

1. Much faster responses than using relational databases such as MySQL
2. Much more flexible to formulate complex queries for complex questions
3. Much easier to incrementally update the database
4. Particularly suitable for detecting emerging trends in research.

We illustrate the flexibility of the generic approach with an example of 13 full-text publications of our own. The small graph contains 12 authors, 48 cited references, and 36 sentences that contain citations. The approach is applicable to a wide variety of subject areas regardless their overall uncertainty levels because the mechanisms for differentiating uncertainties from claims will be in place as a unique feature of the approach. The Cypher query below is equivalent to the question: who are the authors that have published papers containing sentences that cited references in this dataset?

```
MATCH (a:Author)-[w:PUBLISHES]->
        (p:Paper)-[c:CONTAINS]->
        (s:Sentence)-[d:CITES]-(r:Ref)
RETURN a, w, p, c, s, d
LIMIT 200;
```

Figure 8.2 depicts the result of the Cypher query. The red node is an author. The orange nodes are articles published by the author. The green nodes are sentences in a published article. The purple nodes are references cited by the sentence of the green node connected to them. For instance, the green node at the upper right corner of Fig. 8.2 corresponds to the sentence: "CiteSpace follows a simple model of the dynamics of scholarly communication …" The subject of the sentence is CiteSpace.

More specific questions can be formulated with a Cypher query. For instance, we are interested in authors who published articles that contain statements, or claims, with CiteSpace as the subject. The question is formulated in the Cypher query below.

```
MATCH      (c:Claim)<-[r:MAKES]-(s:Sentence)<-[i:CONTAINS]-(p:Paper)<-
[j:PUBLISHES]-(a:Author)  where  c.subject  =~  '(?i).*CiteSpace.*'  return
c,r,s,i,p,j,a;
```

Figure 8.3 shows a visualization of the query result in Neo4j. The visualization shows that the author published two papers that satisfied the criteria. The author is a red node in the graph, connecting to two yellow nodes of papers, which in turn contain sentences in green. The purple nodes are references cited by the green sentences. In other words, the set of green nodes represent citation contexts.

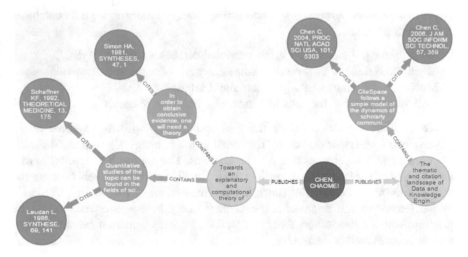

Fig. 8.2 A graphical answer to the question: who has published what paper containing sentences that cited which references?

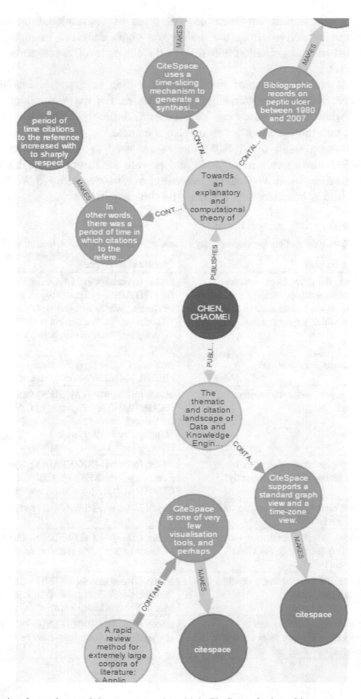

Fig. 8.3 A sub-graph containing sentences in which CiteSpace is the subject

The representation is very flexible. All types of conventional bibliographic networks can be derived from the underlying graph database, including author collaboration networks, citation networks at the author level, article level, and the sentence level.

Table 8.2 includes a list of questions concerning scientific claims made in published articles and corresponding Cypher queries. For example, claims that are associated with hedging words can be identified. Similarly, claims that are associated with uncertain cue words can be also identified.

The following series of MySQL queries are used to collect sentences from MEDLINE articles in the virus dataset and prepare for them to be uploaded to a graph database in Neo4j. We include the completion time on Lenovo W530 so that the reader who is interested can estimate the time required.

Table 8.2 Questions and corresponding queries in Cypher to a graph of scientific publications

Questions in English	Questions in Cypher
When did a claim (hypothesis, assertion, conjecture, statement) appear for the first time in a body of source information, e.g. the literature? In which paper/source?	match (c:Claim)<-[r:MAKES]-(s:Sentence)<-[i:CONTAINS]-(p:Paper)<-[j:PUBLISHES]-(a:Author) where c.object =~ '(?i).*CiteSpace.*' OR c.subject =~ "(?i).*CiteSpace.*' return min(p.year), max(p.year);
How certain was the assertion?	match (c:Claim)-[r]-(s:Sentence)-[:HEDGES]-(w:HedgeTerm) return w.weight;
How many papers made the same claim subsequently?	match (c:Claim)<-[r:MAKES]-(s:Sentence)<-[i:CONTAINS]-(p:Paper)<-[j:PUBLISHES]-(a:Author) where (c.object =~ '(?i).*CiteSpace.*' OR c.subject =~ '(?i).*CiteSpace.*') return count(p);
How many papers made claims that contradict a given claim subsequently?	match (p:Paper)-[:CONTAINS]->(s:Sentence)-[:MAKES]->(d:Claim)-[:CONTRADICTS]-(c:Claim) return count(p);
What are the closely related but distinct claims for a given claim?	match (d:Claim)-[r]-(c:Claim) return d;
Given a claim, which reference is most frequently cited along with the claim's citation context?	match (r:Ref)<-[:CITES]-(s:Sentence)-[:MAKES]->(c:Claim) return r, max(sum(r));
Which references are frequently co-cited in relation to a given claim?	match (s1:Sentence)-[:CITES]->(r1:Ref), (s1:Sentence)-[:CITES]->(r2:Ref), (s1:Sentence)-[:MAKES]->(c:Claim) return r1, r2;
How many dimensions (clusters or eigenvectors or topics) are associated with the citation contexts of a claim?	match (s:Sentence)-[:CITES]->(r:Ref) return s; hierarchical clustering(sentence by word)

```
SELECT *
FROM sentence
INTO OUTFILE 'D:/temp/sentences.csv'
FIELDS ENCLOSED BY '"'
TERMINATED BY ','
LINES TERMINATED BY '\r\n';

Query OK, 143,045,997 rows affected (35 min 39.65 sec)
```

```
CREATE TABLE _virus_sentence
SELECT s.sentence_id AS sid, s.pmid AS pmid,  s.type AS type,
       s.number AS number, s.sentence AS sentence
FROM _virus_year_text AS v, sentence AS s
WHERE v.sid=s.sentence_id;

Query OK, 662132 rows affected (1 hour 59 min 32.65 sec)
Records: 662132  Duplicates: 0  Warnings: 0
```

```
SELECT *
FROM _virus_sentence
INTO OUTFILE 'D:/temp/virus.sentences.csv'
FIELDS ENCLOSED BY '"'
TERMINATED BY ','
LINES TERMINATED BY '\r\n';

Query OK, 662132 rows affected (1.60 sec)
```

The sentences in CSV are loaded to a graph database in Neo4j with the following Cypher queries:

```
CREATE CONSTRAINT ON (s:Sentence) ASSERT s.sid IS UNIQUE;
USING PERIODIC COMMIT 500
LOAD CSV
FROM  'file:///D:/temp/virus.sentences.csv' AS line
WITH LINE LIMIT 662132
MERGE (sentence:Sentence {sid:line[0]})
SET sentence.type=line[2],
    sentence.number=toInt(line[3]),
    sentence.sentence=line[4]
DROP CONSTRAINT ON (s:Sentence) ASSERT s.sid IS UNIQUE;

Added 997459 labels, created 997459 nodes, set 3997459 properties,
statement executed in 449713 ms.
```

The next step is to export the virus dataset of semantic predications from MySQL and then load them to Neo4j. Although it is possible to transfer the data directly from MySQL to Neo4j, it is efficient and reliable to divide the conversion into several smaller steps due to the size of Semantic MEDLINE. The sentence table alone contains 140 million rows.

```
SELECT *
FROM _virus_year_text
INTO OUTFILE 'D:/temp/virus.csv'
FIELDS ENCLOSED BY '"'
TERMINATED BY ','
LINES TERMINATED BY '\r\n';

Query OK, 662132 rows affected (1.60 sec)
```

The semantic predications on virus are uploaded to the Neo4j server with the following Cypher queries. Five types of nodes are added to the graph, namely, paper, sentence, claim, concept, and text. A paper node contains properties such as PMID and the year of publication. A claim corresponds to a semantic predication in Semantic MEDLINE. The subject and object concepts are mapped to concept nodes in the graph. Nodes are connected accordingly based on their types. For example, a paper node CONTAINS a sentence. A sentence MAKES a claim. A subject node as a UMLS concept REPRESENTS its original text.

```
CREATE CONSTRAINT ON (p:Paper) ASSERT p.pmid IS UNIQUE;
CREATE CONSTRAINT ON (s:Sentence) ASSERT s.sid IS UNIQUE;
CREATE CONSTRAINT ON (c:Claim) ASSERT c.pid IS UNIQUE;
CREATE CONSTRAINT ON (a:Concept) ASSERT a.name IS UNIQUE;
CREATE CONSTRAINT ON (t:Text) ASSERT t.text IS UNIQUE;
USING PERIODIC COMMIT 500
LOAD CSV FROM 'file:///D:/temp/virus.csv' AS line
WITH LINE LIMIT 1000000
MERGE (paper:Paper {pmid:line[2], year:toInt(line[8])})
MERGE (sentence:Sentence {sid:line[0]})
MERGE (claim:Claim {pid:line[1], link:line[3]})
        SET claim.subject=line[4], claim.object=line[6]
MERGE (subject:Concept {name:line[4]}) SET subject.type=line[5]
MERGE (object:Concept{name:line[6]}) SET object.type=line[7]
MERGE (s_text:Text {text:line[9]})
MERGE (o_text:Text {text:line[10]})
MERGE (paper)-[:CONTAINS]->(sentence)
MERGE (sentence)-[:MAKES]->(claim)
MERGE (subject)-[r:CONNECTS {type:line[3]}]->(object)
MERGE (subject)-[:REPRESENTS]->(s_text)
MERGE (object)-[:REPRESENTS]->(o_text)
DROP CONSTRAINT ON (p:Paper) ASSERT p.pmid IS UNIQUE;
DROP CONSTRAINT ON (s:Sentence) ASSERT s.sid IS UNIQUE;
DROP CONSTRAINT ON (c:Claim) ASSERT q.pid IS UNIQUE;
DROP CONSTRAINT ON (a:Concept) ASSERT a.name IS UNIQUE;
DROP CONSTRAINT ON (t:Text) ASSERT t.text IS UNIQUE;
```

The process took 748,162 ms to complete. The resultant graph contains 1.07 million nodes, 4.29 million properties, and 1.39 million relationships. It takes up 1.8 GB of disk space on the computer. As shown in Table 8.3, the semantic graph on virus features 553,169 sentences from 320,818 MEDLINE articles. These sentences collectively make 136,209 claims, i.e., semantic predications, involving 18,723 UMLS concepts, which in turn represent 66,584 words or phrases in the original unstructured text.

We will illustrate the usage of the virus graph with a few examples. One is to find claims of causality concerning Ebola.

Table 8.3 The graph constructed from the semantic predications on virus

Node type	Node count	Link type	Link count
Paper	320,818	CONTAINS	1,106,338
Sentence	553,169	MAKES	1,255,560
Claim	136,209		
Concept	18,723	CONNECTS	271,703
Text	66,584	REPRESENTS	140,858

Causality Claims on Ebola

Claims that identify how one entity may influence another are defined as causality claims, for example, as in the claim that Ebola virus ~ CAUSES ~ Hermorrhagic fever. The response time of the following Cypher query is less than 0.5 s. The Cypher query is asking for claim nodes such that the Ebola concept is either the subject or the object. The resultant claims are shown in Table 8.4, where c.pid is the semantic predication ID in Semantic MEDLINE.

```
match (c:Claim) where c.subject =~ '(?i).*ebola.*' or c.object =~
'(?i).*ebola.*' return c.pid, c.subject, c.link, c.object order by c.link limit
10;

Returned 10 rows in 477 ms.
```

If we want to dig deeper, we can ask which papers made these claims with the following Cypher query. This query takes slightly over 1 s to complete (1026 ms to be precise).

```
match (p:Paper)-[r*1..3]->(c:Claim)
where c.subject =~ '(?i).*ebola.*' or c.object =~ '(?i).*ebola.*'
return p.pmid, p.year, c.pid, c.subject, c.link, c.object order by c.link limit 10;
```

Table 8.4 Causality claims concerning Ebola

c.pid	c.subject	c.link	c.object
14126539	Mutation	AFFECTS	Ebola virus
1705776	Ebola virus	CAUSES	Hemorrhagic Fevers, Viral
6248254	Ebola virus	CAUSES	Disease
7722924	Ebola virus	CAUSES	Hemorrhagic Fever, Ebola
7481712	Ebola virus	CAUSES	Hemorrhagic Disorders
7351555	Ebola virus	CAUSES	Acute Disease
9991484	Ebola Virus, Sudan	CAUSES	Communicable Diseases
9991327	Ebola Virus, Zaire	CAUSES	Disease
9501704	Ebola virus	CAUSES	Infection
9050015	Ebola virus	CAUSES	Laboratory Infection

A question that is probably more relevant to the purpose of identifying emerging trends would be: who and which paper was the first to make claims on Ebola? As shown in Table 8.5, the earliest causality claim involving Ebola first appeared in 1978. The claim that Ebola virus causes viral hemorrhagic fevers appeared first in a 1978 MEDLINE article (PMID 352653).

```
match (p:Paper)-[r*1..3]->(c:Claim)
where c.subject =~ '(?i).*ebola.*' or c.object =~ '(?i).*ebola.*'
return p.pmid, p.year, c.pid, c.subject, c.link, c.object order by p.year limit 5;
```

The multi-type relationships among articles, claims, and semantic types are visualized in Fig. 8.4 through a built-in visualization function in Neo4j. To generate the visualization, the user just needs to formulate a Cypher query to specify the conditions to be satisfied.

Table 8.5 Papers that made the earliest causality claims on Ebola

p.pmid	p. year	c.pid	c.subject	c.link	c.object
307455	1978	1650932	Immunofluorescence	DIAGNOSES	Ebola virus
352653	1978	1705776	Ebola virus	CAUSES	Hemorrhagic Fevers, Viral
503930	1979	1705776	Ebola virus	CAUSES	Hemorrhagic Fevers, Viral
119829	1979	2294309	Vero Cells	LOCATION_OF	Ebola virus
94087	1979	2294309	Vero Cells	LOCATION_OF	Ebola virus

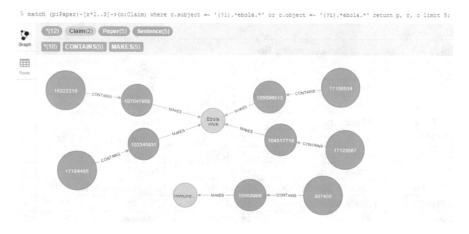

Fig. 8.4 The earliest causality claims involving Ebola

Conflicting Claims

Conflicting claims are a major source of uncertainty in scientific literature. The hallmark of a domain expert is the ability to differentiate conflicting information and contradictory claims. In addition to the positive causality claims shown in Table 8.5, there are claims that negate the causality associated with Ebola (Table 8.6). The claim that virus NEG_CAUSES Ebola Hemorrhagic Fever is extracted from the sentence: "These results suggest that the different clinical outcomes of EBOV infection do not result from virus mutations." By contrasting contradictory claims, one may identify provenance of evidence associated with such contradictions and track the developments that may lead to a reconciliation. A valuable practical potential of this capability is to monitor claims made by retracted papers and to detect the consistency between a given paper and the rest of the body of knowledge.

When Was a Causal Relationship Initially Hypothesized?

Researchers often need to trace to the origin of a hypothesis in the literature. The following example illustrates how we can query when and which paper hypothesized a causal relationship based on the semantic graph. The Cypher query below retrieves chains of papers ~ sentences ~ claims such that the sentences contain the word 'hypothesi,' and the semantic links are one of the casual relations. The partial word 'hypothesi' is used to catch variations such as hypothesis and hypothesized,

```
match (p:Paper)-[]-(s:Sentence)-[]-(c:Claim)
where s.sentence<>" and s.sentence =~ '(?i).*hypothesi.*' and c.link =~
    '(?i).*CAUSES|AFFECTS|TREATS|INHIBITS|DISRUPTS|PREVENTS|PRED
    ISPOSES|CONVERTS_TO'
return p.pmid, p.year, s.type,s.sentence, c.subject,c.link, c.object
LIMIT 15;

Returned 9 rows in 2698 ms.
```

Table 8.6 Negations of causality claims on Ebola

p.pmid	p.year	c.subject	c.link	c.object
11752702	2002	Virus	NEG_CAUSES	Hemorrhagic Fever, Ebola
17940961	2007	Ebola Virus, Zaire	NEG_CAUSES	Disease

Table 8.7 illustrates two examples from titles, whereas Table 8.8 shows two examples from abstracts. For example, the idea that a virus can cause cancer first appeared in a 1967 article (PMID 5596354).

Now we know the existence of two claims, or semantic predications, that virus CAUSES Malignant Neoplasms, and that virus CAUSES Neoplasm. Were these predications discussed in other MEDLINE articles? Given a semantic predication, show us the articles associated with the predication. Suppose we are interested in the latter predication, which has a predication ID of 544471 in Semantic MEDLINE. The Cypher query below retrieves all the articles with sentences connected to the predication.

```
match (p:Paper)-[]-(s:Sentence)-[]-(c:Claim{pid:'544471'})
where s.sentence =~ '(?i).*' and
      c.link =~ '(?i)CAUSES|INTERACTS_WITH|AFFECTS|PREVENTS|TREATS'
return count(distinct(p)), count(distinct(s)), count(distinct(c));
```

Table 8.9 shows the results of the Cypher query. A total of 118 articles are connected to the predication 544,471. The earliest appearance was 1926.

Measuring the Importance of Semantic Predications

One way to identify important semantic predications is based on whether they have attracted a lot more attention than their peers from the scientific community. From a sociological perspective of scientific change (Fuchs 1993), researchers are driven by their competitions for recognition and reputation. A topic that attracts much attention from researchers is apparently considered important. The uncertainty associated with a high-attention topic must be also high—there must be a lot of potential to make high-reward discoveries or something that can dramatically boost one's recognition or reputation. Thus, competitions in high-profile and high-risk areas of research tend to be more intensive than other areas. In contrast, research areas with a low level of uncertainty are unlikely to sustain intensive competitions.

Table 8.7 MEDLINE articles that hypothesized causal relations in titles

PMID	Year	Sentence	Subject	Relation	Object
5596354	1967	[Hypothesis that cancer can be caused by a virus].	Virus	CAUSES	Malignant Neoplasms
1435387	1992	Human cancers and viruses: a hypothesis for immune destruction of tumours caused by certain enveloped viruses using modified viral antigens.	Virus	CAUSES	Neoplasm

Table 8.8 MEDLINE articles that hypothesized causal relations in abstracts

PMID	Year	Sentence	Subject	Relation	Object
8780661	1996	It was hypothesized that if the transmission of CMV through transfusion causes CMV disease in human immunodeficiency virus-positive hemophiliacs, then hemophiliacs with CMV AIDS would be more likely to have received transfusions than those with AIDS-defining disease not caused by CMV (non-CMV AIDS)	Cytomegalovirus	CAUSES	Disease
17429926	2007	The objective of this study was to investigate the association hypothesis that outcome following respiratory syncytial virus (RSV) induced bronchiolitis (RSVB) and RSV induced wheeze (RSVW) are different	Respiratory syncytial virus	CAUSES	Bronchiolitis

The reasoning concerning attention, uncertainty, risk, potential reward, and ultimate recognition suggests that we may learn valuable insights from semantic predications that stand out in how long and how much of attention they have ever generated in scientific articles published in the past. Burst detection (Kleinberg 2002) is a generic algorithm that can be used to identify the level of attention as a type of burstness over time. If a semantic predication appears at a much higher level of frequency than other semantic predications within the same research domain, then intuitively, the semantic predication is having a burst. The duration in which a relatively high-level frequency is observed defines a duration of burstness. If multiple levels of frequency are observed in association with a semantic predication, then the predication may experience hierarchically related bursts. For example, a semantic predication concerning the semantic relation between virus and infection

Table 8.9 Earliest sentences concerning the predication: Virus CAUSES Neoplasm (PID: 544471)

PMID	Year	SID	Sentence
19869151	1926	6276721	Approximately 1 c.mm. of spleen tissue in 3,000 c.mm. of medium may on occasion maintain a concentration of Rous virus in this fluid sufficient to produce a tumor upon inoculation into chickens
19870262	1934	5867204	It is often impossible to determine whether the neoplasms caused by the virus of Strain 2 are of endothelial or mesenchymal origin, and it is possible that both types of cells may be stimulated by the same virus
19870455	1936	5451445	This virus produces neoplasms only when brought in contact with bone or cartilage
21001044	1946	5828267	Title: Induction of neoplasia in vitro with a virus
13185069	1953	6779315	Title: [Cellular multiplication and tumors induced by virus; cancer as an infection]

may have a level of frequency higher than some relatively rare diseases in the Semantic MEDLINE, thus, the predication may have a long period of relatively low-level burstness. From time to time, the semantic predication may have an even higher level of frequency, e.g., when particular types of virus, such as HIV, H1N1, and SARS are involved.

CiteSpace provides a simple user interface for applying the burst detection to semantic predications (Fig. 8.5). The user can specify the semantic types of semantic predications. In the following example, causal relations such as CAUSES, AFFECTS, INHIBITES, DISRUPTS, PRODUCES, and PREVENTS are specified. Parameters for the burst detection model are set so that a burst must last at least for three consecutive years between 1914 and 2014, a 101-year time span.

Kleinberg's algorithm found 167 qualified semantic predications. Figure 8.6 illustrates some of the earliest ones. These predications are considered particularly important by scientists who publish in biomedicine. The darker blue bars depict the period in which a semantic predication appears in MEDLINE for the first time. Each red bar depicts the duration of a burst, which is a period in which the frequency of the occurrences of the predication is considerably higher relative to that of other predications at the same time. The overall profile of the red lines suggests that (1) the burstness of a predication tends to shift over time and (2) the burstness of a predication becomes relatively short except the first few long-lasting ones. Thus, the immediate conclusion is that researchers' focus changes over time. This observation can be used as the basis of detecting emerging topics at a large scale.

Table 8.10 shows a more selective group of semantic predications that have the strongest bursts (burst strength > 10.0) among the 167 predications with bursts. The earliest predication, Virus CAUSES Influenza, first appeared in 1918, which would immediately remind us the 1918–1919 Spanish Flu. A particularly strong burst, highlighted in Table 8.2, is found with the *HIV => Acquired Immunodeficiency*

Fig. 8.5 Detecting bursts in semantic predications on causal relations in research on virus

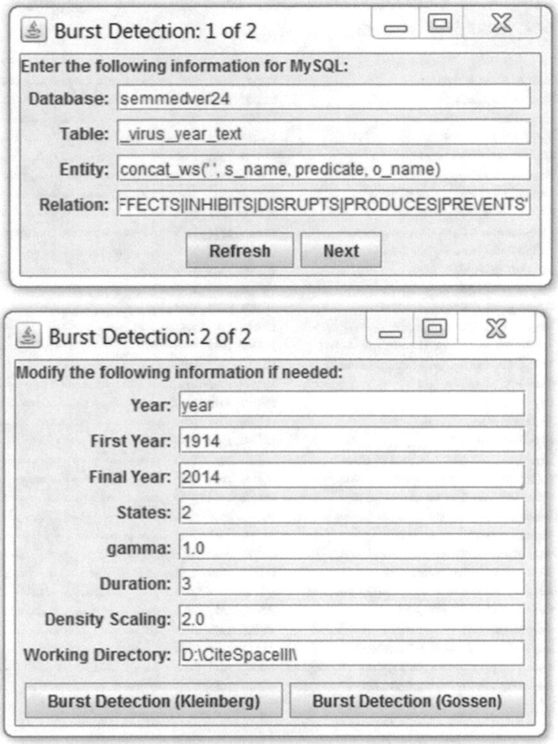

Burst Detection: 1 of 2

Enter the following information for MySQL:

Database: semmedver24

Table: _virus_year_text

Entity: concat_ws(' ', s_name, predicate, o_name)

Relation: FFECTS|INHIBITS|DISRUPTS|PRODUCES|PREVENTS'

Refresh   Next

Burst Detection: 2 of 2

Modify the following information if needed:

Year: year

First Year: 1914

Final Year: 2014

States: 2

gamma: 1.0

Duration: 3

Density Scaling: 2.0

Working Directory: D:\CiteSpaceIII\

Burst Detection (Kleinberg)   Burst Detection (Gossen)

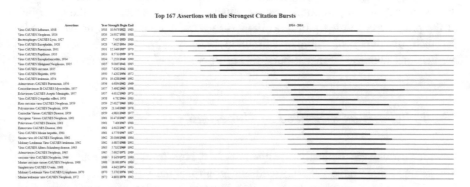

Fig. 8.6 Some of the earliest semantic predications found with bursts

Syndrome predication, which first appeared in 1984 and started to burst for 10 years from 1991 to 2000. The strongest burst belongs to the predication *Human Papillomavirus => Malignant neoplasm of cervix uteri*, which first appeared in 1986. It didn't burst until 26 years later in 2011. It is worth noting that the Nobel

Table 8.10 Semantic predications with burstness strengths > 10.0 from top 167 ones with bursts. '=>' denotes CAUSES

Assertions	Strength	Begin	End	1914 - 2014
Virus => Influenza, 1918	10.9479	1922	1983	
Virus => Neoplasm, 1926	24.9327	1931	1988	
Virus => Pneumonia, 1931	12.3469	1937	1970	
Virus => leukemia, 1954	19.4288	1960	1992	
Rous sarcoma virus => Neoplasm, 1959	25.6127	1965	1993	
Polyomavirus => Neoplasm, 1959	21.148	1965	1978	
Oncogenic Viruses => Neoplasm, 1961	10.4718	1967	1995	
Simian virus 40 => Neoplasm, 1962	20.0066	1968	1994	
Murine sarcoma viruses => Neoplasm, 1968	18.661	1974	1986	
Avian Sarcoma Viruses => Neoplasm, 1975	10.6576	1981	1991	
Virus => Disease, 1922	11.5348	1981	1984	
Virus => Acute Erythroblastic Leukemia, 1977	10.9621	1983	1998	
Cytomegalovirus => Pneumonia, 1975	11.1421	1985	1997	
Virus => Acquired Immunodeficiency Syndrome, 1983	16.5985	1989	1995	
Retroviridae => Acquired Immunodeficiency Syndrome, 1984	23.2685	1990	1996	
HIV => Acquired Immunodeficiency Syndrome, 1984	**38.7063**	**1991**	**2000**	
Retroviridae => Disease, 1982	11.4849	1991	1998	
HIV => Infection, 1985	24.7191	1993	2001	
Hepatitis C virus => Hepatitis C, 1989	13.3138	1996	2005	
HIV-1 => Infection, 1989	12.5993	1998	2004	
Kaposi Sarcoma AFFECTS human herpesvirus 8, 1996	17.1303	2003	2010	
Hepatitis C virus => Chronic liver disease NOS, 1989	14.1651	2004	2009	
human herpesvirus 8 => Kaposi Sarcoma, 1996	24.5536	2004	2009	
Respiratory syncytial virus => Lower respiratory tract infection, 1986	14.6341	2005	2014	
Hepatitis C virus => Fibrosis, 1989	14.974	2007	2014	
Hepatitis C virus => Primary carcinoma of the liver cells, 1990	12.7617	2007	2014	
West Nile virus => Encephalitis, 1969	11.8932	2009	2014	
SARS coronavirus => Severe Acute Respiratory Syndrome, 2003	17.8423	2009	2014	
Norovirus => Gastroenteritis, 1999	25.8451	2010	2014	
Human Metapneumovirus => Respiratory Tract Infections, 2003	12.9203	2010	2014	
Norovirus => Acute gastroenteritis, 2000	19.819	2010	2014	
Human Papillomavirus=>Malignant neoplasm of cervix uteri, 1986	**42.2078**	**2011**	**2014**	
Human Papillomavirus => Anogenital venereal warts, 1983	16.2154	2012	2014	

Prize in Physiology or Medicine in 2008[1] was awarded to Harald zur Hausen for his discovery of "human papilloma viruses causing cervical cancer" and the other half of the prize was awarded jointly to Françoise Barré-Sinoussi and Luc Montagnier for their discovery of "human immunodeficiency virus"—the HIV. The two predications with the strongest burstness are both Nobel Prize winning topics!

Contradictions as a Source of Uncertainty

Research fronts are typically involved with a high-level of uncertainty, where research questions have yet answered and controversial findings have yet settled. Scientific claims surrounded by cues that indicate the involvement of incomplete, conflicting, and contradictory information point to areas of research where the uncertainty is high and the competition is likely worthwhile.

Funding agencies' decisions on high-risk and high-reward research are often revealing in terms of how people make decisions involving a significant degree of

---

[1]https://www.nobelprize.org/nobel_prizes/medicine/laureates/2008/press.html.

uncertainty. Wagner and Alexander (2013) evaluated the Small Grants for Exploratory Research (SGER) program of the U.S. National Science Foundation (NSF). SGER was a 16-year program operated from 1990 until 2006. The program was designed to serve as a special funding device to support high-risk and high-reward research that is unlikely to get funded through the traditional evaluation system. Citation counts, expert interviews, and the results of a survey all indicate a successful SGER. On the other hand, the evaluation of the program reveals that the NSF program directors were perhaps overly conservative—they spent far less than the allowable funds that were ear-marked for exploratory research. The program directors remained risk averse even with a program particularly designed to encourage transformative research. Similarly, Laudel and Glaser (2014) studied links between epistemic properties and institutional conditions for research based on projects funded by the European Research Council (ERC). They found that research that is important for the progress of a field is difficult to fund with common project grants. The conventional funding mechanisms appear to discourage unconventional research across all disciplines.

The perceived risk is in part due to the epistemic uncertainty—the scientific community is simply lacking the knowledge to remove the uncertainty or the controversies. Semantic predications and their original text in Semantic MEDLINE provide a useful resource for the study of scientific claims along with the extent to which conflicting or contradictory information is involved.

Each semantic predication consists of a subject, a predicate, and an object. If we can identify those predicates that are particularly associated with sentences in scientific articles that contain indicators of conflicting or contradictory information, then we may reach a better understanding of their patterns.

We constructed two sets of sentences: one set contains sentences that include terms such as conflicting, contradictory, and inconsistent; the other set contains sentences that do not include such terms. We call the former set the conflict set and the latter the no conflict set. Next, we compare the semantic types of the semantic predications associated with each of the two sets.

Figure 8.7 is a log-log plot of predicates found in the no conflict set (x-axes) against their frequencies in the conflict set. The size of a node represents the frequency of the corresponding predicate. Predicates in dark red are those involved in causal relations, such as TREATS, AFFECTS, and CAUSES, whereas those in light red are involved in structural or ontological relations, such as PROCESS OF, IS A, and PART OF. The line divides predicates into two parts. Predicates above the line, such as, TREATS, AFFECTS, and PREVENTS, appeared more often in the conflict set, whereas predicates below the line appeared more often in the no-conflict set, such as CAUSES, INHIBITS, and DISRUPTS. In both groups, predicates of causal relations are dominating the overall semantic predications. The conflict set appears to have more active or positive predications, i.e., the A causes B pattern. In contrast, the no-conflict set appears to have more passive or negative predications, i.e., the A suppresses B pattern.

Figure 8.7 also shows that the conflict set contains relatively more semantic predications with the semantic types such as PROCESS_OF, IS_A,

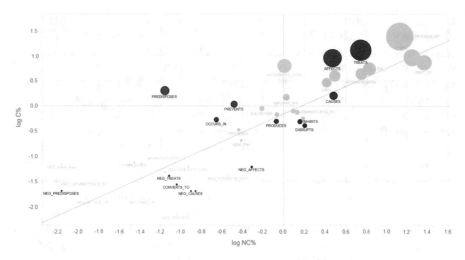

Fig. 8.7 Major semantic relations in the conflict versus no conflict sets of articles from Semantic Medline

ASSOCIATED_WITH, and USES as well as TREATS, AFFECTSS, and other types of relations. The division suggests the richness of the conflict set and the overall significance of the research topics that are intrinsically more essential to research.

Semantic Predications on Virus Research (1914–2014)

We constructed a subset of semantic predications on virus from Semantic MEDLINE (version: semmedver24) so that we can explore semantic predications closely in terms of their various properties associated with representing scientific knowledge, including advantages and weaknesses.

The virus subset contains semantic predications and their original text as long as either the subject or the object of a predication has the semantic type of virus, which is the name used in MEDLINE.

The MySQL query to extract the subset and the status report are as follows:

```
CREATE TABLE _virus
SELECT count(*)
FROM predication_aggregate
WHERE s_type = 'virs' OR o_type='virs';

Query OK, 638792 rows affected (54.92 sec)
```

A total of 638,792 qualified predications were selected from the table predication_aggregate. The citation table contains 23.7 million MEDLINE articles

(23,657,386 for sememdver24). Linking the two large tables with a MySQL Join took 52 min on a Lenovo Workstation W530. Among the 638,792 predications, about 21.3% (136,214) are unique predications, which involve 28 types of relations.

To reduce the need for possibly time-consuming table joins in the future, we created a temporary table that contains all the major information about predications along with the year of publication for the original article. Adding the time, i.e. the year of publication, allows us to perform algorithms such as burst detection, which identifies which semantic predications are attracting the attention of researchers in terms of how fast their frequencies increase in the literature. The following query joined two tables so that each predication can be timestamped by the year of the original article's publication. The new table contained 638,780 predications.

```
CREATE TABLE _virus_year
SELECT pid, sid, a.pmid as pmid,
    predicate, s_name, s_type, o_name,
    o_type, b.pyear as year
FROM
    _virus AS a,
    citations AS b
WHERE a.pmid=b.pmid;

Query OK, 638780 rows affected (52 min 33.63 sec)
Records: 638780  Duplicates: 0  Warnings: 0
```

Figure 8.8 depicts the distribution of the number of semantic predications per year. Since 1975, there are 5000 predications each year. The 5-year moving average is closely tracking the top of the bars. In 2013, the number of the predications is over 30,000.

The distribution of the semantic predications on virus research shows that the majority of the predications appeared after 1975 and there were more predications in early 1990s than the value of a 5-year moving average.

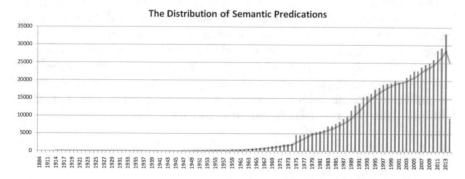

Fig. 8.8 The distribution of semantic predications over time. The red line is the 5-year moving average

The third query aims to match specific information from a sentence with a given predication. On a Lenovo W530 workstation, it took 2 h to complete the query.

```
CREATE TABLE _virus_year_text
SELECT a.sid as sid, a.pid as pid, a.pmid as pmid, a.predicate as
       predicate, a.s_name, a.s_type, a.o_name, a.o_type, a.year,
       b.subject_text, b.object_text
FROM _virus_year AS a, sentence_predication AS b
WHERE a.sid=b.sentence_id AND a.pid=b.predication_id;

Query OK, 662132 rows affected (2 hours 2 min 8.92 sec)
Records: 662132  Duplicates: 0  Warnings: 0
```

There are a total of 136,209 distinct semantic predications in the _virus_year_text table. The following query can be used to find predications that appeared most frequently in the subset of virus:

```
SELECT
    b.c, -log2(b.c/136209),
    b.pid, b.predicate,
    concat(b.s_name, ' ', b.predicate, ' ', b.o_name)
FROM (
        SELECT count(*) AS c, pid,
           s_name,
           predicate,
           o_name
        FROM _virus_year_text
  GROUP BY pid
  ORDER BY count(*) DESC
) AS b
LIMIT 10;
```

Table 8.11 lists the top 10 most frequently appeared semantic predications in the virus subset. The top 10 predications' predicates include three PROCESS OF, six PART OF, and one LOCATION OF.

The information content of most frequently appeared semantic predications are lower than low-frequency predications. Predications that are relative rare have high information contents. Table 8.12 is generated with the following query, which searches for predications that have a particular number of appearances, e.g. b. c = 10 for 10 appearances, in the virus dataset. The query also calculates the IC of each predication based on the total of 136,209 predications. In addition, predicates are limited to predicates with implications of causal relations such as AFFECTS, CAUSES, TREATS, INHIBITES, and DISRUPTS.

Table 8.11 Top 10 most frequently appeared semantic predications in the virus subset

Count	IC	PID	Predicate	Predication
26077	2.3850	2634059	PROCESS_OF	Communicable Diseases PROCESS OF Hepatitis C virus
6076	4.4866	1666937	PART_OF	DNA PART OF Human Papillomavirus
5456	4.6418	946718	PROCESS_OF	Disease PROCESS OF Cytomegalovirus
5111	4.7361	2383142	PROCESS_OF	Communicable Diseases PROCESS OF HIV-1
4474	4.9281	2407960	PART_OF	RNA PART OF HIV
3764	5.1774	541931	LOCATION_OF	Cells LOCATION OF Virus
3331	5.3537	3027212	PART_OF	Vaccines PART OF Human Papillomavirus
3318	5.3594	4640493	PART_OF	RNA PART OF Hepatitis C Virus
2578	5.7234	1467311	PART_OF	Large T-Antigen PART OF Simian virus 40
2221	5.9385	663163	PART_OF	DNA PART OF Simian virus 40

Table 8.12 Some examples of rare predications with 1, 5, or 10 appearances in total

Count	IC	PID	Predicate	Predication
10	13.7335	608940	CAUSES	Enterovirus CAUSES Syndrome
10	13.7335	764171	AFFECTS	Infection AFFECTS Virus
10	13.7335	1076906	CAUSES	Virus CAUSES Latent Infection
10	13.7335	1138218	CAUSES	Papillomavirus, Cottontail Rabbit CAUSES Papilloma bit CAUSES
10	13.7335	1238699	CAUSES	Enterovirus CAUSES Conjunctivitis, Acute Hemorrhagic
5	14.7335	588468	CAUSES	Adenoviruses CAUSES Pharyngo-Conjunctival Fever
5	14.7335	652125	CAUSES	ECHOVIRUS 11 CAUSES Disease
5	14.7335	695371	CAUSES	Simplexvirus CAUSES Primary infection NOS
5	14.7335	740895	CAUSES	Fibroma Virus, Rabbit CAUSES Neoplasm
5	14.7335	771239	CAUSES	Virus CAUSES Tick-Borne Encephalitis
1	17.0556	541848	AFFECTS	Carcinoma AFFECTS Tick-Borne Encephalitis Virus
1	17.0556	543509	CAUSES	Rabies virus CAUSES Multiple Sclerosis
1	17.0556	547396	CAUSES	Echoviruses CAUSES Meningococcal meningitis
1	17.0556	550507	CAUSES	Mumps virus CAUSES comatose
1	17.0556	558749	CAUSES	sarcoma virus CAUSES Malignant Neoplasms

```
SELECT
    b.c,
    -log2(b.c/136209),
    b.pid,
    b.predicate,
    concat(b.s_name, ' ', b.predicate, ' ', b.o_name)
FROM (
        SELECT
            count(*) AS c,
            pid, s_name, predicate, o_name
        FROM _virus_year_text
        WHERE
            predicate regexp
                'AFFECTS|CAUSES|TREATS|INHIBES|DISRUPTS'
        GROUP BY pid
        ORDER BY count(*) DESC
    ) AS b
WHERE b.c=10
LIMIT 5;
```

We can export the content of the table to a comma separated values (CSV) file with the following query:

```
SELECT count(*), pid, predicate
FROM _virus_year_text
GROUP BY pid
INTO OUTFILE 'D:/temp/frequencies_of_pids.csv'
FIELDS TERMINATED BY ','
LINES TERMINATED BY '\r\n';
```

Exploring a Semantic Network of Predications in CiteSpace

A collection of semantic predications forms a network with UMLS concepts as nodes and semantic types as relations. Given the variety of visual analytic functions provided by CiteSpace, structural and temporal patterns in a set of semantic predications can be studied as an associative network. For example, a semantic network of UMLS concepts and their semantic relations can be constructed from a given set of semantic predications. Similarly, as shown at the beginning of the chapter, one can also construct a graph database and explore various graph-theoretical questions in graph database query languages such as Cypher in Neo4J.

Causal Relations in Virus Research

As summarized in Table 8.13, the total of 662,132 instances of semantic predications concerning a virus in one way or another came from 320,818 MEDLINE articles. The number of unique predications is 136,209. On average, each

Table 8.13 Statistics of semantic predications concerning viruses

	Semantic predications		MEDLINE articles
	Total	Unique	Unique
Virus	662,132	136,209	320,818
Causal relations in virus	50,861	15,902	38,256

predication is expected to appear five times, although we know its distribution is skewed. There are 15,902 unique semantic predications are related to assertions on causal relations, such as HIV CAUSES AIDS. There are a total of 50,861 instances of these causal predications from 38,256 MEDLINE articles.

Table 8.14 shows top 20 most popular types of predicates in the set of predications on virus. Predicates such as PART_OF, PROCESS_OF, LOCATION_OF, and IS_A are essential to ontological structures, whereas predicates such as CAUSES, INTERACTS_WITH, AFFECTS, and PREVENTS are assertions concerning the impact of one concept on another or changes that one may cause in the other. Predicates on the second half of the table are a series of predication types that negate those in the first half. NEG_CAUSES, for example, negates the predicate CAUSES as in HIV NEG_CAUSES AIDS, which would be equivalent to the assertion that HIV does not cause AIDS.

Table 8.14 Top 20 most popular types of predicates in the virus dataset

Count	Predicate
248756	PART_OF
163969	PROCESS_OF
111078	LOCATION_OF
41860	CAUSES
27942	ISA
24062	INTERACTS_WITH
19361	DIAGNOSES
12807	COEXISTS_WITH
6283	AFFECTS
1838	PREVENTS
910	NEG_LOCATION_OF
783	NEG_INTERACTS_WITH
632	NEG_PART_OF
611	NEG_CAUSES
386	NEG_PROCESS_OF
293	NEG_COEXISTS_WITH
239	NEG_DIAGNOSES
204	NEG_AFFECTS
53	NEG_PREVENTS
35	compared_with

In CiteSpace, under the Data menu, there is an item Semantic MEDLINE > Semmed2WoS. This function executes the following query to retrieve predications in which the subject causes changes in the object. In particular, several relations meet this condition, namely CAUSES, TREATS, AFFECTS, PREVENTS, INHIBITS, and INTERRUPTS. Each of the relations specifies a change induced by the subject of the predication. In addition, we are also interested in the negation of such relations, for example, NEG_CAUSES and NEG_AFFECTS because of the importance of knowledge concerning the causality.

```
SELECT *
FROM _virus_year_text
WHERE predicate REGEXP
    'CAUSES|TREATS|AFFECTS|PREVENTS|
     NEG_CAUSES|NEG_AFFECTS|NEG_PREVENTS|
     INHIBITS|INTERRUPTS';
```

The above query found 38,256 MEDLINE articles containing 50,861 semantic predications on causal relations, which represent 15,902 unique semantic predications. CiteSpace converts these MEDLINE records to a format similar to the Web of Science such that the user can use CiteSpace's visual analytic functions to explore the structure and dynamics of these predications over time (Chen 2017). The user can simplify the network with functions such as Pathfinder network scaling and analyze transformative potentials of MEDLINE articles through Structural Variation Analysis (Chen 2012).

Semantic predications of a MEDLINE article are converted to a format that extends the standard Web of Science format (Fig. 8.9). For example, an article (PMID: 24099575) published in 2013 contains four semantic predications. These predications are mapped to an extended field XX, which can be recognized by CiteSpace to visualize such records as part of a network of concepts linked by corresponding semantic predications. The number of predications is set as the value of the TC field, which can be used as basis for selecting articles based on how many distinct predications they have.

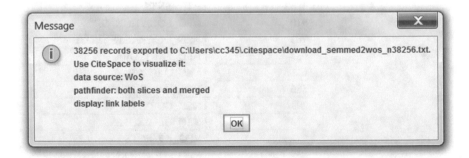

Fig. 8.9 A total of 38,256 MEDLINE records are converted to a data file for subsequent analysis with CiteSpace

```
PT J
TI Rabbit Hemorrhagic Disease Virus CAUSES RHD
SO Medline PMID 24099575
DT Article
DE Rabbit_Hemorrhagic_Disease_Virus~CAUSES~Rheumatic_Heart_Disease;
   European_brown_hare_syndrome_virus~CAUSES~Hepatitis;
   Rabbit_Hemorrhagic_Disease_Virus~CAUSES~Hepatitis; Lagovirus~CAUSES~Hepatitis
NR 0
TC 4
PD JUN-15
PY 2013
PM 24099575
XX rabbit hemorrhagic disease virus     rhd      0.03886223059344139     CAUSES
   european brown hare syndrome virus  hepatitis 0.03886223059344139     CAUSES
   rabbit hemorrhagic disease virus      hepatitis 0.03886223059344139     CAUSES
   lagovirus    hepatitis 0.03886223059344139     CAUSES
ER
```

Visual Analysis of Semantic Predications

In a network of co-cited references, each node is a scientific publication. Two nodes are connected if a subsequently published article cited both of them within the article. The network can be divided into clusters, or groups of references, such that each group can be characterized by some themes. Furthermore, these themes are typically shared by references within the same cluster, but it is less likely to be shared with references in other clusters (Chen et al. 2010; Chen 2017). Each cluster is resultant from the work of a specialty, i.e. a community of researchers who tend to publish in a set of interrelated journals or bump into each other at conferences they regularly attend. We can identify an important article by its citation burst and we will attribute the most significant contributions made by the article to the article as a whole. Thus each article serves a role that is much like a concept. Small (1978), who pioneered much of the co-citation analysis methodology, coined the term concept symbols.

The resolution of a GPS device determines the extent to which it can locate a position with confidence. It becomes helpless if the precision required to accomplish a task is below the finest level of granularity the GPS can reach. Scientometrics at the granularity of an article level can answer many of our questions. However, with the resolution at the article level, it is difficult for us to address many more specific questions. A semantic predication represents a semantic relation between two concepts. Usually, one of the concepts is called the subject and the other is called the object. The semantic relation represents how the subject and the object are connected semantically. For example, "HIV causes AIDS" is semantic predication. HIV is the subject, whereas AIDS is the object. The verb causes is the semantic relation.

There are distinct advantages of representing the knowledge of a scientific domain in terms of semantic predications. Semantic predications provide more precise representations of knowledge than using articles as a whole.

To what extent is the methodology that we have demonstrated at the article level applicable to the study of a scientific domain at the level of semantic predications? We will adapt the methodology and apply it to the study of virus research in the following example.

The source of input data is MEDLINE. Similar to a bibliographic record in sources such as the Web of Science, a MEDLINE record includes the meta-data of a scientific publication, including the title, the abstract, and a list of keywords. Unlike a record in the Web of Science, a MEDLINE record is indexed by a number of MeSH terms—Medical Subject Headings. MeSH terms are from the controlled vocabulary thesaurus compiled by the U.S. National Library of Medicine. MeSH terms are organized in a hierarchy. Unlike the Web of Science records, MEDLINE records do not include information on cited references. There are two ways to obtain semantic predications from scientific publications, primarily within the scope of biomedicine research. One is to extract semantic predications by using SemRep and the other is to use semantic predications extracted by the SemMed project. We have introduced both SemRep and SemMed in Chap. 5.

Each MEDLINE record may have one or more semantic predications. Semantic predications from the same MEDLINE record are co-occurring predications. As we have seen, a set of co-occurring entities can be represented as a network of inter-connecting entities. We assign a timestamp to each predication. The timestamp registers the time when the semantic predication appears in our dataset for the first time. Thus we can treat the collection of semantic predications in the same way as we treat co-occurring keywords in the Web of Science records. For example, we can run a burst detection to see which semantic predications have abrupt increases in their occurrences. We can divide a network of semantic predications into clusters so that we can see which semantic predications tend to be discussed together. We can generate timeline visualizations and see how they evolve over time. We can perform a structural variation analysis and identify novel connections between semantic predications. In other words, we can apply many analytic techniques developed for document co-citation analysis to semantic predications.

Constructing a Semantic Network

Unlike bibliography records in the Web of Science or Scopus, a MEDLINE article does not include references cited by the article. When we construct a network of cited references in CiteSpace, a common strategy is to select articles that have been cited to an extent themselves and build the network of references cited by these elite articles. The principle is to emphasize the input from established sources.

When we converted the 38,256 MEADLINE articles to analyze the structural and temporal patterns of semantic predications, there is no information about either the references they cited or how frequently they have been cited by other articles. It is possible to collect an equivalent dataset from sources such as the Web of Science and then extract semantic predications, which is in fact what we are currently working on. In the examples to follow, we construct semantic networks by selecting MEDLINE articles from this dataset with two options.

The first option is to select MEDLINE articles that are the top N articles from each year in terms of the number of semantic predications. For instance, the user can select top 50 or top 100 MEDLINE articles each year in terms of their values in the TC field, which is the number of semantic predications in each article. Semantic predications from the selected MEDLINE articles will be used to construct a semantic network of concepts and their relationships defined by predicates in associated semantic predications.

The second option is to select MEDLINE articles based on a generalized g-index (Egghe 2006). The number g is defined as the average of the first g occurrences of semantic predications per MEDLINE article. Using the g-index has an advantage over the first option. The selection of the top N in the first option is arbitrary because it does not take into account the distribution of the occurrences of semantic predications in MEDLINE articles. In contrast, the second option is based on the g-index, which provides a less arbitrary cutoff point.

Option 1: Top N MEDLINE Articles

A semantic network was generated from top 50 MEDLINE articles between 1980 and 2016 in terms of the number of semantic predications per article. The resultant network consists of 338 UMLS concepts that appeared either as the subject or the object of a semantic predication. The largest connected component (LCC) of the network contains 331 concepts, or 92% of the entire network. The modularity of the network with respect to the partition by its clusters is 0.4125, which is in the middle of the range. The average silhouette of the network is relatively low at 0.267, which means the heterogeneity of a cluster is generally high. In other words, the diversity of predications in a given cluster is relatively high.

Figure 8.10 depicts a visualization of the largest connected component of the network without applying any link filters. Each semantic predication consists of a subject, a predicate, and an object. The subject and the object are represented by concepts defined in the UMLS metathesaurus. An UMLS concept is a term that represents a group of semantically equivalent terms. Each UMLS concept has a unique identifier CUI. In the virus example, HIV is an UMLS concept (CUI = C0019682). The HIV concept is used as the representative of 101 various kinds of semantically equivalent phrases found in text. Table 8.15 shows some of the most commonly occurred terms in text. All of these terms are mapped to the UMLS concept HIV. In addition to the term HIV itself, terms such as Human immunodeficiency virus, HTLV-III, LAV, and HIV-1 are unified under the same UMLS concept HIV. There are 101 such terms identified as the subject of a

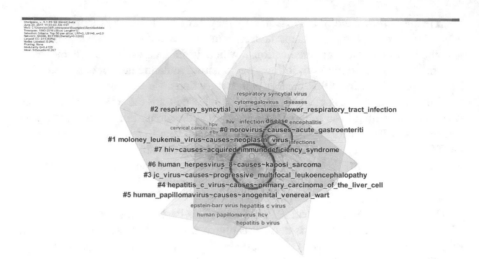

Fig. 8.10 A semantic network of 338 UMLS concepts connected by 1158 semantic predications of causality relations (1980–2016). CiteSpace: Top 50. Largest CC: 331 (92%). Q: 0.4125. S: 0.267

Table 8.15 Most frequent terms mapped to the UMLS Concept HIV (CUI = C0019682) as subjects and objects

Count	As subject	Count	As object
4892	HIV	18841	HIV
1562	Human immunodeficiency virus	4041	Human immunodeficiency virus
138	HTLV-III	460	HIV-1
98	LAV	237	HTLV-III
66	HIV-1	172	LAV
34	Human immunodeficiency viruses	132	AIDS virus
30	HIV-1LAI	90	HIVDR
23	AIDS virus	74	HIV-1IIIB
22	LAV/HTLV-III	43	TDR
19	lymphadenopathy-associated virus	41	lymphadenopathy-associated virus

predication and 268 such terms identified as the object of a predication. The mapping is done by SemRep, which we have introduced earlier.

Each node in the network is an UMLS concept, such as HIV. Two concepts are connected by corresponding predicates through semantic predications. For example, concepts of HIV and AIDS can be connected by the predicate CAUSES through the semantic predication HIV CAUSES AIDS. Individual semantic predications serve as local constraints on UMLS concepts and connect them based on their roles in semantic predications. The network therefore is a semantic network because the connections are defined by the predicates in their relationship.

CiteSpace divides the network into clusters of nodes that are tightly connected. Nodes within the same cluster appeared more often in the same predications than nodes between different clusters. Each cluster is labeled by the most representative semantic predication that is responsible for the linkage within the cluster. Cluster labels are displayed as strings of text starting with cluster IDs #0, #1, and so on. The size of a cluster is in descending order of its ID. Cluster #0 is the largest one, followed by Cluster #1. As shown in Fig. 8.10, we can see some of the nodes are labeled such as respiratory syncytial virus, hiv, and cervical cancer. Labels of the majority of the nodes in the network are not shown because they have lower frequencies than the ones that are shown. The largest cluster of UMLS concepts are labeled as #0 norovirus CAUSES acute gastroenteritis. Cluster #7 is labeled by the predication HIV CAUSES Acquired immunodeficiency syndrome.

Once we constructed a semantic network based on the semantic predications, many visual analytic functions in CiteSpace can be readily applied to the study of these predications. Figure 8.11 shows a timeline view of the network. Each line represents a cluster. Clusters are arranged in descending order from the top of the display downwards. Figure 8.12 zooms into make the fine-grain details more readable. Large circles on the left are concepts that appeared earlier on. They are connected with subsequently appeared concepts in their own clusters through the reinforcement of semantic predications. A purple rim of a circle indicates its high betweenness centrality in the network. A red ring indicates a detected period of burstness.

Figure 8.13 shows the same network such that we can identify the most frequently appeared concepts in semantic predications in Semantic MEDLINE. The size of a node represents the frequency of the concept in the virus dataset. The color of a node denotes its cluster membership. Salient concepts include virus, disease, and infection based on their size. Connections between concepts represent causal relations linked by predicates such as CAUSES.

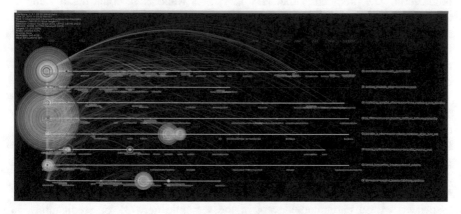

Fig. 8.11 A timeline view of the semantic predications on causality relations

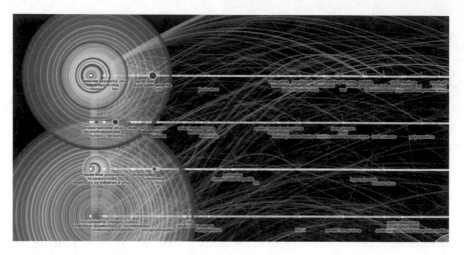

Fig. 8.12 A close-up to the timeline view of the four largest clusters of predications

Figure 8.14 shows the neighboring concepts of the concept HIV. The arrow of a link points from the subject of a predication to the object. For example, the predication HIV CAUSES Kaposis sarcoma is represented as a directed link from the concept HIV to the concept Kaposis sarcoma. Similarly, several predications are conveyed:

HIV CAUSES acquired immune deficiency syndrome.
HIV CAUSES diseases
HIV CAUSES disorder
HIV CAUSES AIDS
HIV CAUSES cytopathic effect.

The number shown on a link is the relative frequency of the particular semantic predication. The predication that HIV CAUSES Kaposis sarcoma has a value of 0.04, which is the probability of seeing the particular predication. It is based on the ratio of the number of instances of this particular predication over the number of instances of all the predications that connect the two concepts. One can also normalize the prevalence of the predication with reference to the total number of links connecting any two concepts through all semantic types.

As shown in a historical view in Fig. 8.15, the concept of HIV first appeared in 1987. Its burst was detected in 1990 and it lasted for one year, but its frequency continued to increase and peaked in 2012 involved in 94 predications that year. Figure 8.16 shows the history of the concept Virus since 1980. The concept has a period of burst that lasted for 9 years from 1980 till 1988. The concept appeared in 3481 MEDLINE articles.

From the citation history view in CiteSpace, one can look up articles that are associated with a particular concept in the semantic network of predications. In Fig. 8.17, the predication of interest is shown at point 1, cytomegalovirus CAUSES

Fig. 8.13 Most frequently appeared concepts in the virus dataset

colitis. The value of 4 at point 2 indicates there are 4 MEDLINE articles containing this predication. The metadata of one of the articles is shown in the figure. In particular, the location of the predication in the abstract is underlined.

The network visualized in Fig. 8.10 is rather crowded. Clusters are overlapping with one another considerably, which affect the clarity of the view. The visualization shown in Fig. 8.18 has improved the clarity by pruning the excessive links from the network and preserving only the salient links through an algorithm called Pathfinder network scaling. The result of Pathfinder network scaling is called a Pathfinder network. Links in a Pathfinder network must meet a triangular inequality condition. Otherwise, links that fail to meet the conditions are removed from the network. In this way, the number of links is reduced while the integrity of the network is maintained by the condition.

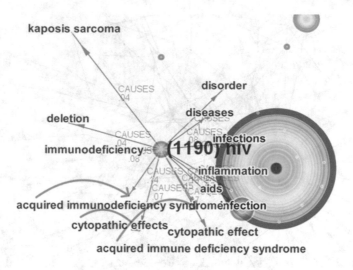

Fig. 8.14 The concept hiv and its neighbouring concepts connected through causal connections

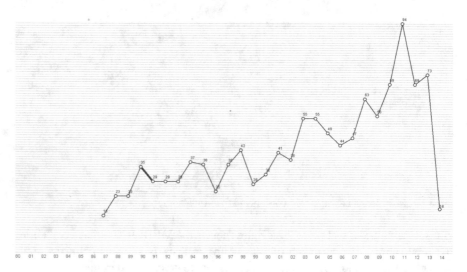

Fig. 8.15 The HIV concept has a burst of 6.7642 between 1990 and 1992. It appeared in 1190 PubMed records

The triangular inequality condition requires that the cost or weight of a direct link between nodes n_i and n_j should not be greater than the total cost of an alternative path that connects the two nodes. Otherwise, the alternative path provides more insightful connections than the direct link. Therefore, it is justifiable to eliminate the direct link from the network. In our everyday life, similar principles apply to many situations when we need to choose from multiple routes between two

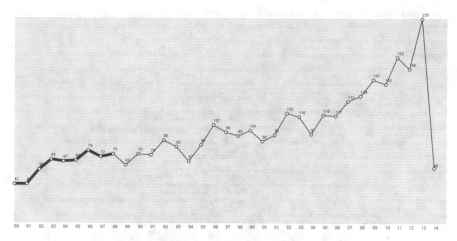

Fig. 8.16 The burstness of the concept Virus (Strength: 110.9355, duration 1980–1988). The concept appeared in 3481 PubMed records

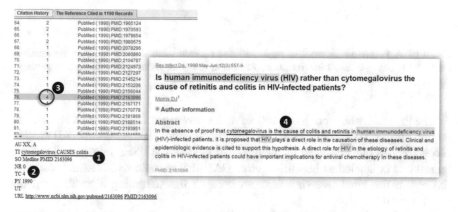

Fig. 8.17 Explore the source of a semantic predication

locations, for instance, choosing between a non-stop flight from Philadelphia to London Heathrow and a flight that makes one or two stops before London. The cost of a path could be either the door-to-door time or the price of the ticket plus the extra meals on a longer flight.

Pathfinder network scaling was originally developed by psychologists to identify major connections out of a potentially complex network. Sometimes when we compare two concepts directly, their similarity may seem low. However, once we insert the third concept in between, it may suddenly become clear how the two concepts are indeed connected through some profound ways. As soon as we see an example that can justify the closer-than-I-thought proximity, we would be more willing to revise our estimate of the similarity. The previously thought less similar concepts may appear to be more closely related.

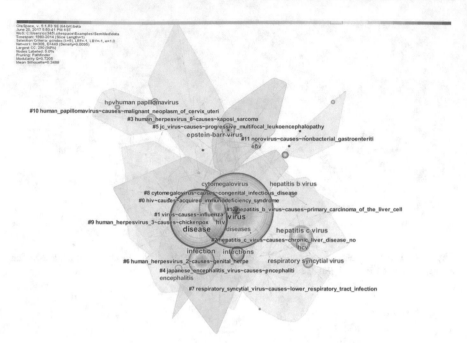

Fig. 8.18 A Pathfinder network of predications. 31 clusters labelled by LLR on predications. Node selection: g-index; Link retention: Pathfinder on time-sliced networks and the merged network

The Pathfinder network has a noticeable improvement in terms of its clarity. The largest cluster #0 is represented by the predication HIV CAUSES AIDS. The second largest one is labeled by the predication virus CAUSES influenza. The third largest one is labeled as hepatitis c virus CAUSES chronic liver disease nos. Note the nos in the label was shown as no because the stemming algorithm did not recognize NOS.

Figure 8.19 shows how the user can interact with the visualized semantic network. Upon clicking on the concept node HIV, its neighboring concepts will be highlighted while other concepts will be suppressed. An arrow from HIV points to the disease concept, representing the predication HIV CAUSES Disease. Similarly, an arrow points to the concept infections with a probability of 0.12 and a link for HIV CAUSES AIDS (0.15).

Figure 8.20 depicts a timeline view of the Pathfinder network. The nodes are selected based on their g-index. Several large clusters have high-frequency concepts, which are shown as large circles.

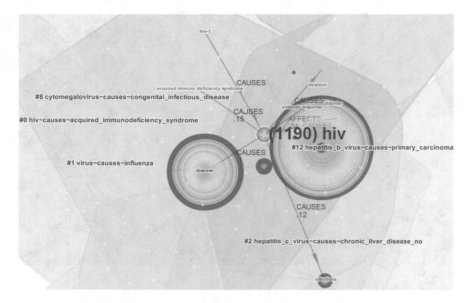

Fig. 8.19 The concept HIV and neighboring concepts. For example, HIV causes AIDS (0.15)

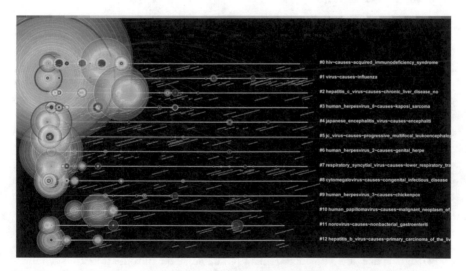

Fig. 8.20 A timeline view of the Pathfinder network. Nodes are selected by their g-index scores

Figure 8.21 shows the same timeline after the user zoomed in. The first line has several nodes with red rings. These red rings depict the durations of detected burst. The purple rims indicate concepts with high betweenness centrality scores in the network. The slightly slanted labels identify the three most frequent concepts each year in their corresponding clusters. For instance, the rightmost node on the second lowest line is labeled as human papilloma virus, which has a period of burst.

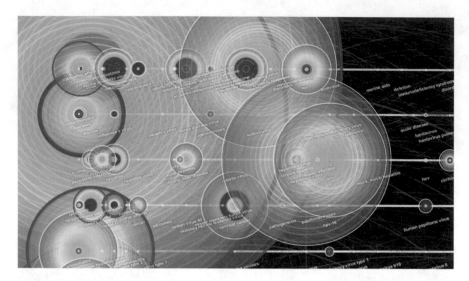

Fig. 8.21 Zooming in

Structural Variations

Analyzing semantic predications in their semantic network allows the analyst to examine novel connections added to the network. More importantly, sometimes a local structure's variation may lead to a change of the global structure. Semantic predications that have the potential to induce such global structural changes are considered important. In CiteSpace, global changes induced by local links are measured in terms of the rate of modularity change, the rate of inter-cluster linkage change, and the distribution of betweenness centrality change.

The theory of structural variation has intuitive interpretations of scientific change. Profound global changes of scientific knowledge may be caused by local changes, which serve as perturbations to a complex adaptive system. The process takes two stages. At the first stage, a novel connection needs to be made. Semantic connections, especially causality, that were previously thought unlikely or even never thought of, are proposed. Proposers are usually researchers who are visionary and creative. Novel connections that have the potential to broaden the current knowledge space are considered most valuable. The introduction of such new connections is likely to transform the state of the art of a scientific field. This stage may correspond to the first stage of Shneider's four-stage evolution model, i.e. the conceptualization stage. The key to the next stage is whether the novel connection can establish itself and attract enough followers to contribute towards the further development of the idea, including applying the original idea to a broad range of domains. This stage may partially correspond to the third stage of Shneider's model —the application stage. The structural variation functions in CiteSpace provide a concrete tool to identify the early sign of a potentially new conceptualization. It is critical whether the process can reach the second stage, i.e. whether it can attract enough followers to keep the original idea alive.

The dashed lines in Fig. 8.22 are novel connections. An article published in 1983 (PubMed ID: 6870184) is responsible for these potentially transformative links. Adding these novel links induced the largest modularity change rate of 15.25. The global structure of the new network is significantly different from the network prior to the addition of these links. This is very useful information for the conceptualization stage. One can generate creative hypotheses that have not been considered in the scientific literature. Then the new hypotheses must be examined and attract enough researchers to make the new ideas sustainable.

In this case, the article published in 1983 (PubMed ID: 6870184) induced the largest modularity change rate of 15.25 (PubMed ID: 6870184). The dashed links represent unprecedented links connecting distinct clusters for the first time (within the scope of the dataset analyzed). Table 8.16 lists the semantic predications made by the article along with the year of their first appearance and corresponding PMIDs. In this case, these predications are not entirely new. They appeared prior to the publication of the 1983 article, but they did not meet the network modeling criteria to become part of the Pathfinder network.

Sometimes emerging patterns are more apparent if trajectories of novel links added by multiple articles are shown simultaneously. Figure 8.23 shows the trajectories of novel semantic predications made by the top 10 MEDLINE articles that are responsible for the strongest modularity change rates. Given that cluster labels are centered at the weight center of each cluster, the concentrations of the dashed lines suggest that novel predications are connecting Clusters #3, #1, #0, and #2. In particular, there are many novel inter-cluster links between Clusters #0 and #1.

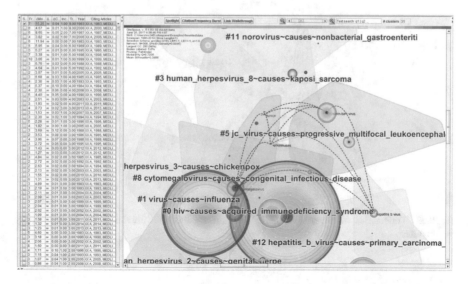

Fig. 8.22 Novel connections in dashed lines are made by a 1983 article (PubMed ID: 6870184)

Table 8.16 Semantic predications on causal relations from the 1983 article (PubMed ID: 6870184)

PID	Subject	Predicate	Object	First appear	PMID
1174925	Primary carcinoma of the liver cells	AFFECTS	Hepatitis B Virus	1978	680585
1398038	Carcinoma of Nasopharynx	AFFECTS	Herpesvirus 4, Human	1977	199059
1428074	Burkitt Lymphoma	AFFECTS	Herpesvirus 4, Human	1975	200925
1686084	Kaposi Sarcoma	AFFECTS	Cytomegalovirus	1978	212367
1840068	Retroviridae	CAUSES	Neoplasm	1979	85722
3589577	Malignant Neoplasms	AFFECTS	Human virus	1983	6870184
3589638	Carcinoma	AFFECTS	Herpesvirus 2, Human	1983	6870184

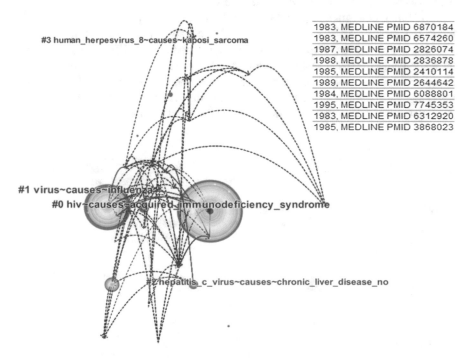

Fig. 8.23 Trajectories of novel links added by top 10 articles with the strongest modularity change rates

CiteSpace supports several ways to build an associative network. Apart from qualifying nodes based on the g-index, TopN is an alternative option. Its main advantage is its simplicity. The TopN node selection criterion selects the top N articles by citations or frequencies from each time slice to form the network. Figure 8.24 is a Pathfinder network of semantic predications between 1990 and 2014, generated with CiteSpace based on 3-year intervals and top 100 most common predications per interval.

Figure 8.25 shows the result of a structural variation analysis (Chen 2012). The semantic predications are selected from those appeared between 1990 and 2014 in 3-year intervals. Top 100 most popular semantic predications per time interval are included. The nodes of the network consists of UMLS concepts that appeared as the subject or the object of a semantic predication, such as HIV and AIDS. Connections between concept nodes are determined by semantic predications. For instance, given the predication HIV CAUSES AIDS, the concept nodes HIV and AIDS are connected in the network with a semantic link CAUSES.

The network is then divided into clusters of sub-networks based on the connectivity in the network such that concepts within the same cluster are tightly connected by semantic predications, whereas concepts between distinct clusters are loosely connected at most. In addition, Pathfinder network scaling is applied to the network, which means links that do not meet the triangle inequality condition imposed by the Pathfinder network scaling algorithm will be removed. The resultant Pathfinder network preserves the links that satisfy the triangle inequality

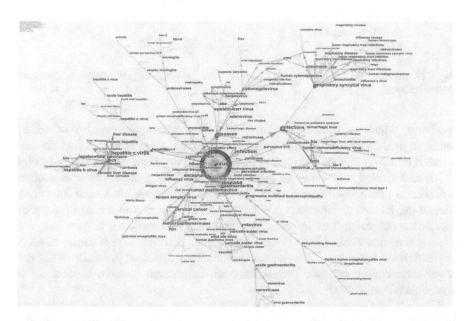

Fig. 8.24 A Pathfinder network of semantic predications generated in CiteSpace. Time slicing: 3; TopN: 100; Range: 1990–2014

Fig. 8.25 Structural variation analysis of the semantic predications (1990–2014) in CiteSpace (3-year intervals)

condition, which make the remaining structure more representative of salient semantic predications. Each cluster is labeled by the most representative concept selected by log-likelihood ratio tests in CiteSpace. For instance, Cluster #3, labeled as hiv and located near the upper right of the diagram, contains concepts such as hiv, human immunodeficiency virus, cmv, aids, retrovirus, hiv-1, and infections. Cluster #4, labeled as infectious mononucleosis and located in the mid-right area, contains concepts such as Epstein-barr virus (ebv), adenovirus, herpesvirus, and kaposis sarcoma. Dashed lines linking concepts in distinct clusters depict novel or unprecedented semantic connections at the time they appeared in MEDLINE. Such novel cross-cluster semantic connections are considered to have transformative potentials (Chen 2012).

Figure 8.26 shows novel semantic links between distinct clusters made by a MEDLINE article published in 2004 (PMID: 14766405). This article yielded a modularity change rate of 7.63, which is significantly high. It has 14 transformative

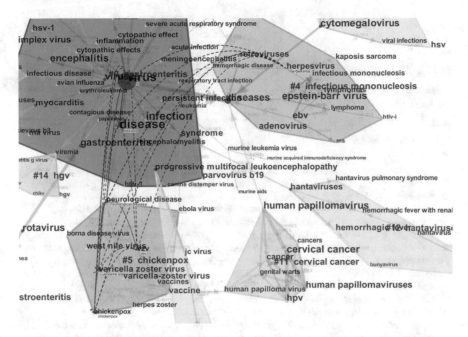

Fig. 8.26 A closer view of novel semantic links between distinct clusters made by a MEDLINE article (PMID: 14766405)

links and a centrality divergence of 0.04. These structural variation metrics indicate that since the semantic predications made by the article connect distinct clusters, there is a significant boundary-spanning potential. A boundary-spanning mechanism is considered as one of the most common types of scientific discoveries (Chen 2011).

Figure 8.27 shows annotations of the six semantic predications extracted from the MEDLINE article (PMID: 14766405). The first five predications are positive causal relations. The last one is a negation; however, the assertion is in the context of animals rather than humans, which is an important distinction that should have been preserved to avoid potential contradictions if one works on the level of extracted predications only. Furthermore, the strength of a semantic connection can be estimated based on how often the particular semantic type appears out of all the possible types connecting the two concepts. For example, the strongest link among the six relations is the first one (0.3707), followed by the third (0.1602) and the fifth (0.1454). The negation is the weakest (0.0550).

The visualized network is based on causal relations only. To obtain all semantic predications associated with the article, one can use the following query. The result is listed in Table 8.17.

```
SELECT distinct(pid), year, s_name, predicate, o_name
FROM _virus_year_text
WHERE pmid=14766405;
```

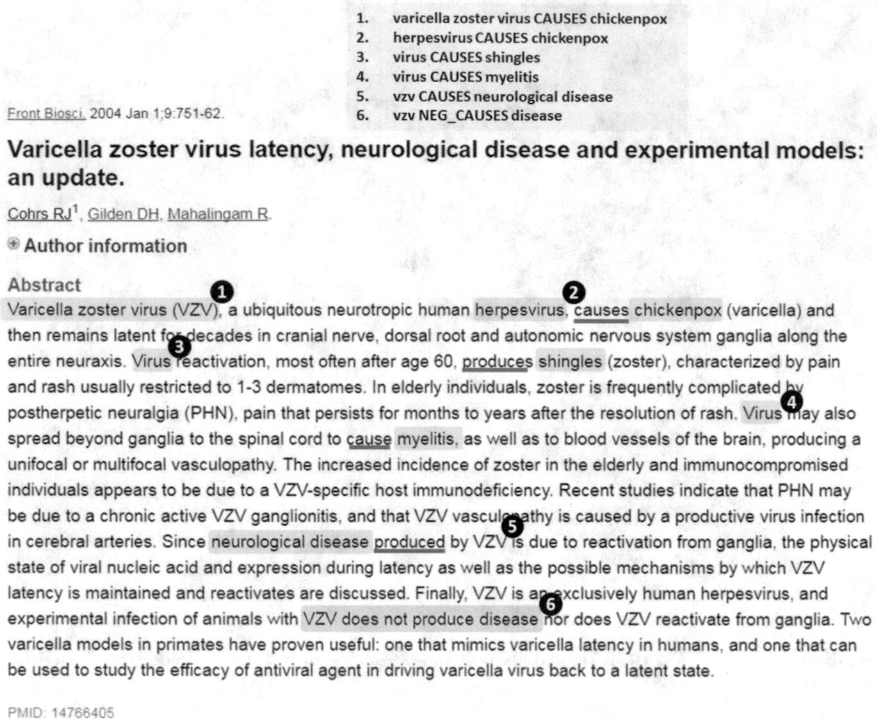

1.   varicella zoster virus CAUSES chickenpox
2.   herpesvirus CAUSES chickenpox
3.   virus CAUSES shingles
4.   virus CAUSES myelitis
5.   vzv CAUSES neurological disease
6.   vzv NEG_CAUSES disease

Front Biosci. 2004 Jan 1;9:751-62.

Varicella zoster virus latency, neurological disease and experimental models: an update.

Cohrs RJ[1], Gilden DH, Mahalingam R.

⊕ Author information

Abstract

Varicella zoster virus (VZV), a ubiquitous neurotropic human herpesvirus, causes chickenpox (varicella) and then remains latent for decades in cranial nerve, dorsal root and autonomic nervous system ganglia along the entire neuraxis. Virus reactivation, most often after age 60, produces shingles (zoster), characterized by pain and rash usually restricted to 1-3 dermatomes. In elderly individuals, zoster is frequently complicated by postherpetic neuralgia (PHN), pain that persists for months to years after the resolution of rash. Virus may also spread beyond ganglia to the spinal cord to cause myelitis, as well as to blood vessels of the brain, producing a unifocal or multifocal vasculopathy. The increased incidence of zoster in the elderly and immunocompromised individuals appears to be due to a VZV-specific host immunodeficiency. Recent studies indicate that PHN may be due to a chronic active VZV ganglionitis, and that VZV vasculopathy is caused by a productive virus infection in cerebral arteries. Since neurological disease produced by VZV is due to reactivation from ganglia, the physical state of viral nucleic acid and expression during latency as well as the possible mechanisms by which VZV latency is maintained and reactivates are discussed. Finally, VZV is an exclusively human herpesvirus, and experimental infection of animals with VZV does not produce disease nor does VZV reactivate from ganglia. Two varicella models in primates have proven useful: one that mimics varicella latency in humans, and one that can be used to study the efficacy of antiviral agent in driving varicella virus back to a latent state.

PMID: 14766405

Fig. 8.27 The semantic predications extracted from the article (PMID: 14766405)

Table 8.17 All the semantic predications associated with the MEDLINE article (PMID: 14766405)

PID	First	Current	Subject	Predicate	Object
818714	1950	2004	Virus	CAUSES	Herpes zoster disease
1655451	1973	2004	Human herpesvirus 3	ISA	Herpesviridae
1968010	1977	2004	Ganglia	LOCATION_OF	Human herpesvirus 3
2680886	1986	2004	Herpesviridae	CAUSES	Chickenpox
3067201	1983	2004	Human herpesvirus 3	CAUSES	Chickenpox
3405215	1986	2004	Virus	CAUSES	Myelitis
3415496	1989	2004	Human herpesvirus 3	CAUSES	nervous system disorder
5873382	1996	2004	Posterior root of spinal nerve	LOCATION_OF	Human herpesvirus 3
10852298	2003	2004	Cranial Nerves	LOCATION_OF	Human herpesvirus 3
10852497	2004	2004	Human herpesvirus 3	NEG_CAUSES	Disease

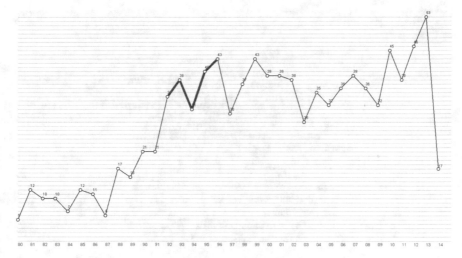

Fig. 8.28 The burstness of the concept Epstein-Barr virus between 1992 and 1995

Burst detection is a generic method. It is applicable to identify the areas of intensive interest. In this case, burst detection can be used to identify highly attractive concepts as well as semantic predications. Figure 8.28 shows the burstness of the concept Epstein-Barr Virus (EBV) in Cluster #4 of the network. Prior to the burst that started in 1992, the concept already appeared in the dataset with an average of 10 appearances each year in the first few years of the 1980s. The annual frequency jumped from 21 to 34 when its burst was detected. Since the concept is a major node in Cluster #4 given the font size of its label, the burst of the concept may indicate the emergence of Cluster #4 to a new level. Combining with the burst of semantic predications, one can explore the dynamics of research from different perspectives at multiple levels of granularity.

Option 2: MEDLINE Articles by g-Index

In addition to select MEDLINE articles based on the number of semantic predications per time slice, CiteSpace also allows the user to construct a semantic network based on the g-index. The original g-index is defined based on citations. However, given a set of predications, the user can select qualified MEDLINE articles based on the g-index of the number of semantic predications.

The following example is based on 38,256 MEDLINE articles on virus research. The relevance of each record is determined based on whether it is indexed by the MeSH term virus. Figure 8.29 shows the largest connected component of a network of co-occurring semantic predications on causal relations over a 101-year period of time (1914–2014) on virus research. In order to be included in the network, a

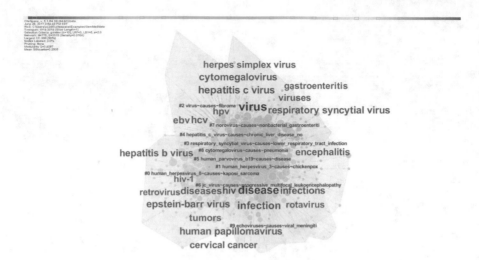

Fig. 8.29 A network of co-occurring semantic predications extracted from MEDLINE articles on virus research over 101 years (1914–2014). Node selection was based on the g-index (k = 10). Clusters of semantic predications are labelled by semantic predications with all the citing articles

semantic predication must appear twice or more in at least one year. The network contains 775 semantic predications. The largest connected component contains 699 (90%). The modularity of the network is relatively low (0.4087) as well as a low silhouette score of 0.2505, suggesting that these semantic predications are highly interrelated but the heterogeneity of each group is low.

Figure 8.30 depicts a timeline visualization of the 10 clusters contained in the largest connected component. The size of a node represents the occurrences of the corresponding semantic predication. The rings in red indicate the detected period of burst. The first 8 clusters run up all the way to 2014. The timelines of clusters #8 and #9 stopped earlier.

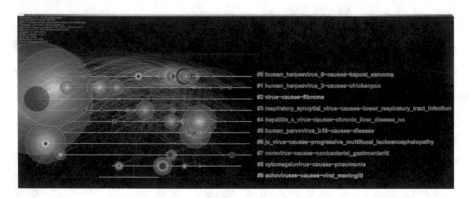

Fig. 8.30 A timeline visualization of the semantic predications on causal relations

Semantic predications in the largest cluster #0 are essentially concentrated between 1974 and 1986. The most recent predication with a burst of occurrences appeared in 1995. The cluster is labeled with the predication "human herpesvirus 8 causes Kaposi sarcoma," suggesting that the predication is the most representative for the cluster. The most representative predication for the next cluster #1 is human herpesvirus 3 causes chickenpox. Cluster #1 includes a few very popular predications in 1920s, 1940s, and 1950s. It also has a string of more recent predications with bursts.

The distribution of the circles in the timeline view indicates the activity level of each cluster (see Fig. 8.31). For example, Cluster #0 and Cluster #1 have different patterns of the distributions. Cluster #0 has predications concentrated between mid-1970s and mid-1980s, which correspond to the most active period of research in AIDS.

Figure 8.32 shows a list of 25 UMLS concepts that have a period of burst for 25 years or longer. There are many more concepts that have shorter periods of burst. These concepts may serve as the subject or the object of a semantic predication. The one with the longest period of burst is virus, which has a 74-year long lasting burst period between 1914 and 1987. Given that we are dealing with a collection of semantic predications on virus, this is hardly surprising.

The concept with the second long lasting burst is herpes virus for 67 years between 1925 and 1991. The timeline visualization shows that human herpesvirus was the subject of both Cluster #0 and Cluster #1. In Cluster #0, the most representative predication is that human herpesvirus 8 causes Kaposi sarcoma, whereas in Cluster #1, the leading predication is that human herpesvirus 3 causes chickenpox.

The third one is phage (1927–1989). The concept of phage was a focus in Cluster #5, which is labeled by the leading predication "human parvovirus b19 causes disease."

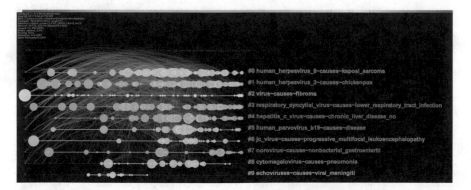

Fig. 8.31 A slightly different view of the timeline visualization with an emphasis on the distribution of predications over time

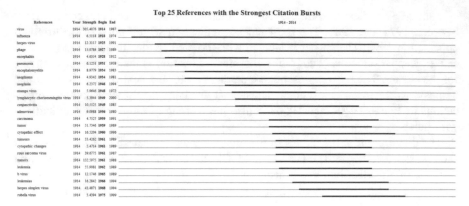

Fig. 8.32 Semantic predications with a period of burst for 25 years or longer

Influenza has the fourth longest period of burst (1918–1974). The year 1918 reminds us the 1918 flu pandemic, or the Spanish flu, which killed 50–100 million people, or 3–5% of the world's population. It was one of the deadliest natural disasters in human history.

Using the same methodology as we have applied to the study of the scientific literature of terrorism research, we generated a hierarchical structure of the semantic predications in the largest cluster (#0). The hierarchy has two branches (Fig. 8.33). The upper branch includes two major semantic predications shown in Table 8.18. We will refer them by their predication IDs in the following discussion.

Predication 7581872 on primary effusion lymphoma has two children nodes, including a branch led by predication 5292122 on Kaposi sarcoma. The former predication first appeared in the virus dataset in 1998 and first appeared in the cluster #0 in 1999 (see Fig. 8.34).

Fig. 8.33 The ontological tree of semantic predications in the largest cluster (#0)

Table 8.18 Two major semantic predications in cluster #0

Predication ID	Subject	Concept ID	Predicate	Object	Concept ID
7581872	Human herpesvirus 8	C0376526	CAUSES	Primary effusion lymphoma	C1292753
5292122	Human herpesvirus 8	C0376526	CAUSES	Kaposi sarcoma	C0036220

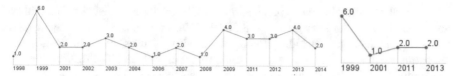

Fig. 8.34 The distributions of predication 7581872 in the collection of predications on virus (left) and within cluster #0 (right)

Table 8.19 presents a few examples of source sentences from which the predication 7581872 was extracted. These sentences referred to the discovery of HHV8, for example, "HHV-8 was discovered in 1994…" and the "recently identified Kaposi's sarcoma-associated herpesvirus (KSHV) and it is now "formally called human herpesvirus 8 (HHV8). Each article's PubMed ID is listed in the table.

Cluster #0 is labeled by the predication 5292122. Its distributions are shown in Fig. 8.35. It first appeared in the virus dataset in 1996 and first appeared in this cluster in 1999.

Table 8.19 Source sentences of the HHV8 and KS predication in articles published in 1996

PubMed ID	Year	Source sentences of the predication (ID: 7581872)
8627015	1996	Human herpesvirus 8 is present in the lymphoid system of healthy persons and can reactivate in the course of AIDS
8640314	1996	Human herpesvirus 8 (HHV-8, KSHV) was discovered in 1994 by means of a molecular biology approach which permitted to characterize fragments of its genomic sequence
8684008	1996	In addition, HBL-6 harbors DNA sequences of the recently identified Kaposi's sarcoma-associated herpesvirus (KSHV), now formally called human herpesvirus 8 (HHV8)
8692871	1996	Recently, DNA sequences from a novel herpesvirus, termed KS-associated herpesvirus (KSHV), or human herpesvirus 8 (HHV-8) have been identified within KS tissue from both HIV-positive and HIV-negative cases
8866603	1996	Recently, herpesvirus-like deoxyribonucleic acid (DNA) sequences, defining a new herpesvirus termed "human herpesvirus 8" (HHV8) or "Kaposi's sarcoma-associated herpesvirus" (KSHV), were detected in Kaposi's sarcoma of acquired immune deficiency syndrome (AIDS) and non-AIDS patients

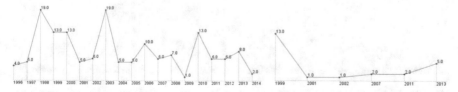

Fig. 8.35 Distributions of the predication 5292122 within cluster #0 (left) and in the entire virus dataset (right)

As another example, major predications in Cluster #4 are related to hepatitis c and liver disease. As shown in Fig. 8.36, the cluster is characterized by several paths of predications, including "hepatitis c virus causes liver cirrhosis" and "hepatitis c virus causes primary carcinoma of the liver cell." The distributions of the leading predication show a steady increase since 1990 (Fig. 8.37).

Analyzing the structure and dynamics of semantic predications enables us to study the knowledge of a domain at a finer level of granularity than the conventional article-level citation or co-citation analysis. The visual analytic framework that we have developed for exploring the abstract landscape of a knowledge main provides an extensible platform for us to examine various aspects of the knowledge domain as a complex adaptive system. Each time when a new article is published, semantic predications introduced by the article serve as a source of perturbation to the current organization of semantic predications. Although perturbations act directly on local structures of the existing knowledge organization, sometimes local changes may have global and system-wide consequences. Information that can cause global changes is certainly of our interest. The following example illustrates a structural variation analysis of the high-dimensional space of relevant semantic predications on virus research. The primary goal is to demonstrate that the structural variation theory is applicable to the new level of granularity.

Structural Variations

Computing structural variation rates for the dataset is a computationally expensive. It took 14,661.465 s on a Lenovo W530 to complete the numerous but necessary comparisons required, which is just over 4 h.

Fig. 8.36 A hierarchy of major semantic predications in Cluster #4 on relations between hepatitis c virus and the liver disease

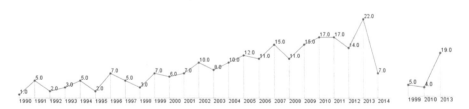

Fig. 8.37 Distributions of the leading predication in Cluster #4 (left: the entire virus dataset and right: Cluster #4)

Figure 8.38 depicts an interaction with the visualized network of concepts connected by various semantic predications after the structural variation model is completed. The five dashed lines are novel links introduced by a 1999 MEDLINE article authored by Ellen Feigal (PMID: 9989205), entitled "AIDS-associated malignancies: research perspectives." These novel connections are derived from the following statement in the article's abstract: "The appearance in 1981 of a usually rare malignancy, Kaposi's sarcoma, in homosexual men [1] was one of the first harbingers of an epidemic caused by a retrovirus, human immunodeficiency virus (HIV), which causes the acquired immunodeficiency syndrome (AIDS)." More interestingly, these concepts belong to different clusters in the network. Linking concepts across different clusters draws our attention to this article's transformative potential. FEIGAL1999 is in fact a review article. It highlights some recent findings from the vantage point of the year 1999, including

- "discovery of a new gamma-herpes virus, human herpes virus 8 (HHV8) or Kaposi's sarcoma herpes virus (KSHV), in 1994 which led to a rapid series of investigations strengthening links of this virus in the pathogenesis of all forms of Kaposi's sarcoma (KS) [2]"
- "association of a rare type of B cell tumor called primary effusion lymphoma with HHV8 [3];"

Obviously, the two findings mentioned above are indeed the two concepts in the predication hierarchy of the largest cluster #0,—Kaposi sarcoma and primary effusion lymphoma. The FEIGAL1999 article has been cited 40 times on Google

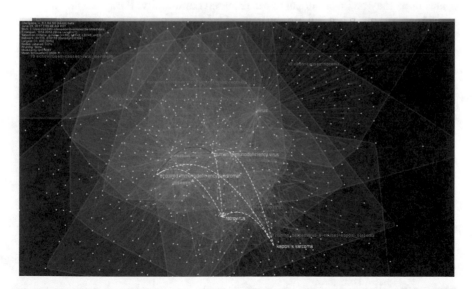

Fig. 8.38 Structural Variation Analysis: the five dashed lines are novel links introduced by a 1999 MEDLINE article (PMID: 9989205)

Scholar. As we will see shortly, this article could be a very good starting point to understand the largest cluster of the virus research.

Table 8.20 lists the semantic predications extracted from the FEIGAL1999 article that made five transformative links, i.e. novel inter-cluster links that are unprecedented in the dataset we have examined. These semantic predications represent three types of semantic relations: ISA, PROCESS_OF, and CAUSES. For example, HIV is a retroviridae. Kaposi sarcoma is a cancer—malignant neoplasms. Retroviridae causes Kaposi sarcoma. And, HIV causes AIDS. This is a highly informative set of predications. This is part of the domain knowledge.

Kaposi sarcoma is a rare type of tumor prior to the AIDS era. It is primarily found in elderly men of Mediterranean descent and in patients on immunosuppressive therapy. In individuals with HIV positive, the incidence of Kaposi sarcoma is 75,000-fold greater and about sevenfold higher in homesexual or bisexual men than other HIV risk groups. The FEIGAL1999 review article introduced five transformative links because it focuses on infectious agents that share common etiological roles in viral infection, immune dysregulation, and cancer pathogenesis. Since the review pulls together the existing knowledge and current advances from distinct research communities such as molecular biology, immunology, virology, and anti-viral therapy, the FEIGAL1999 review is essentially serving the role of a broker of intellectual ideas originated from different disciplinary blocks. The brokerage role is likely to transform the organizational structure of the underlying domain.

It is obvious from the timeline view shown in Fig. 8.39 that dashed lines of transformative links connect concepts in different clusters. These newly added connections strengthen the tie between Cluster #0 and Cluster #5. The cross-cluster connections may be inspirational to the research community. For example, one may ask what the new relationship implies and what new discoveries would become logical. How are human herpesvirus 8 and human parvovirus b19 related? What do they have in common? If more and more articles follow up and reinforce this

Table 8.20 Semantic predications extracted from the article with five transformative links (PMID: 9989205)

PID	Subject	Predicate	Object
2383214	HIV	CAUSES	Acquired Immunodeficiency Syndrome
7348435	HIV	CAUSES	Malignant Neoplasms
2383195	HIV	ISA	Retroviridae
2050367	Kaposi Sarcoma	ISA	Malignant Neoplasms
2310015	Kaposi Sarcoma	PROCESS_OF	Male population group
1081869	Lymphoma, Non-Hodgkin's	ISA	Malignant Neoplasms
4762282	Primary central nervous system lymphoma	ISA	Malignant Neoplasms
9848528	Retroviridae	CAUSES	Kaposi Sarcoma
3723926	Retroviridae	CAUSES	Malignant Neoplasms

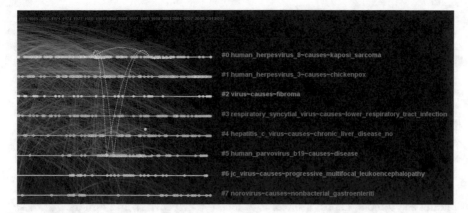

Fig. 8.39 FEIGAL1999 made transformative links across different clusters

pattern, one can imagine that the separation between the two clusters would diminish and eventually the two currently distinct clusters may merge into a single cluster.

Structural variation analysis in CiteSpace provides several metrics of the global changes induced by a particular article. In addition to count the number of transformative predications, one can inspect transformative changes measured by metrics such as the modularity change rate, cluster linkage change rate, and the relative entropy of the distribution of betweenness centrality. Different metrics are sensitive to different types of global structural variation. Figure 8.40 shows the footprint of

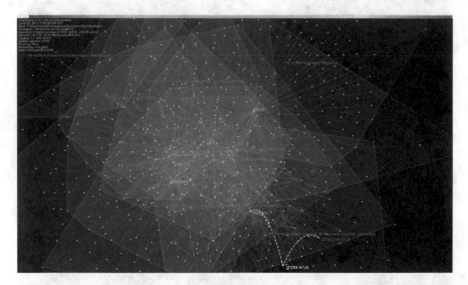

Fig. 8.40 A 1972 MEDLINE article (PMID: 4340152) added two novel predications and reinforced one existing predication. The modularity change rate induced by the article is 7.83. It also shifted the distribution of betweenness centrality scores of the nodes by a degree of 0.05

an article published in 1972 (PMID: 4340152). This article has the highest modularity change rate of 7.83. It added two novel links—one is between gross virus and tumors and the other between gross virus and polyoma virus (dashed white lines) and reinforced the existing link between polyoma virus and tumor (solid purple line).

Figure 8.41 shows the footprint of a 2001 article (PMID: 1134302), which has the largest number of incremental links of five. Unlike a transformative link, an incremental link connects concepts that belong to the same cluster. According to the structural variation theory, an article that essentially contributes incremental links is more likely to have its focus on some established research topics than articles that contribute transformative links. The article contributed semantic predications that connect concepts such as Kaposi's sarcoma and tumors, primary effusion lymphoma and tumors, kaposi's sarcoma-associated herpesvirus and tumors. All these concepts belong to the largest cluster #0. The label of the cluster is centered on the centroid of the cluster near the lower right corner of the network.

The timeline view shown in Fig. 8.42 makes it obvious—all the semantic relations contributed by the article PMID: 1134302 are within Cluster #0. The

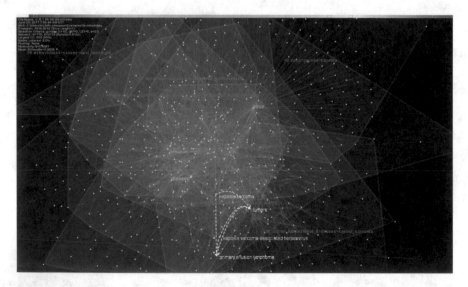

Fig. 8.41 The footprint of a 2001 article (PMID: 1134302), which has the largest number of incremental links

Fig. 8.42 Incremental links made by article PMID: 1134302 are all within Cluster #0

Table 8.21 Semantic predications extracted from article PMID: 1134302

PID	Subject	Predicate	Object
2805744	Homologous Gene	AFFECTS	Cell cycle
5807171	Kaposi Sarcoma	AFFECTS	human herpesvirus 8
5979163	multicentric Castleman's disease	AFFECTS	human herpesvirus 8
7351621	Neoplasm	AFFECTS	human herpesvirus 8
5979188	Primary Effusion Lymphoma	AFFECTS	human herpesvirus 8
5643730	Signal Transduction	AFFECTS	Immune response
7730985	Basal Cell	PART_OF	human herpesvirus 8
1356169	Complement System Proteins	PART_OF	Virus
540196	Neoplasm	PROCESS_OF	Human

specific semantic predications extracted from the article are listed in Table 8.21, involving three types of semantic relations, namely AFFECTS, PART_OF, and PROCESS_OF. Five of the nine predications involve the concept human herpesvirus 8, which is the key concept for the largest cluster. The predication that Kaposi Sarcoma AFFECTS human herpesvirus 8 is semantically equivalent to the predication that human herpesvirus CAUSES Kaposi Sarcoma, which is the most representative predication that characterizes the cluster.

Summary

With the structural variation analysis, we can inspect the potential of an article in terms of the degree to which its semantic predications conform to the existing structure. A departure from the norm is an early sign of a potentially transformative idea. Novelty is a necessary condition for a scientific breakthrough. However, it is not a sufficient condition. A novel idea may not materialize itself for a long time. A sufficient condition of transformative research is its acceptance by the relevant research community. A revolutionary idea is not part of the knowledge of the underlying domain until fellow researchers start to pay attention to it and start to do something about it.

Citations to an article, altmetrics on social media, and the strength or length of a period of citation burst are among some of the simplest indicators of a potential impact of scientific contributions. Since these indicators do not directly reflect the structure of a domain's knowledge, they are extrinsic measures of scientific change. In contrast, metrics derived from structural variations are intrinsic measures because they directly reflect the changes of the structure of a domain's knowledge.

In next chapter, we will address a fundamental concept concerning the meta-knowledge, i.e. the knowledge of knowledge—the uncertainty of a scientific claim at a particular point of time. Take the predication that HIV causes AIDS as example. Our knowledge of what we know today may change drastically tomorrow. This is one kind of uncertain associated with our knowledge. We will discuss relevant issues in more detail in next chapter.

References

Chen C (2011) Turning points: the nature of creativity. Springer, Berlin

Chen C (2012) Predictive effects of structural variation on citation counts. J Am Soc Inform Sci Technol 63(3):431–449. doi:10.1002/asi.21694

Chen C (2017) Science mapping: a systematic review of the literature. J Data Inf Sci 2(2):1–40

Chen C, Ibekwe-SanJuan F, Hou J (2010) The structure and dynamics of co-citation clusters: a multiple-perspective co-citation analysis. J Am Soc Inf Sci Technol 61(7):1386–1409. doi:10.1002/asi.21309

Egghe L (2006) Theory and practise of the g-index. Scientometrics 69(1):131–152. doi:10.1007/s11192-006-0144-7

Kilicoglu H, Shin D, Fiszman M, Rosemblat G, Rindflesch TC (2012) SemMedDB: a PubMed-scale repository of biomedical semantic predications. Bioinformatics 28(23):3158–3160. doi:10.1093/bioinformatics/bts591

Kleinberg J (2002) Bursty and hierarchical structure in streams. In: Proceedings of Proceedings of the 8th ACM SIGKDD International Conference on Knowledge Discovery and Data Mining pp 91–101

Laudel G, Glaser J (2014) Beyond breakthrough research: epistemic properties of research and their consequences for research funding. Res Policy 43(7):1204–1216. doi:10.1016/j.respol.2014.02.006

Small H (1978) Cited documents as concept symbols. Soc Stud Sci 8(3):327–340

Wagner CS, Alexander J (2013) Evaluating transformative research programmes: a case study of the NSF Small Grants for Exploratory Research programme. Res Evaluat 22(3):187–197. doi:10.1093/reseval/rvt006

Chapter 9
Visual Analytic Observatory of Scientific Knowledge

Abstract A conceptualization of research on uncertainties in scientific knowledge is presented. Several common sources of uncertainties in scientific literature are characterized, notably, retracted scientific publications, hedging, and conflicting findings. Semantically equivalent uncertainty cue words and their connections with semantic predications are identified and visualized as the first step towards a systematic study of uncertainties in accessing and communicating the status of scientific assertions.

Introduction

As new discoveries and advances are made, scientific knowledge, conveyed through the content of scientific literature, is subject to constant changes. These changes could be revolutionary as well as evolutionary (e.g., Kuhn 1962; Fuchs 1993; Shneider 2009). Despite the tremendous growth in terms of scholarly metrics to measure various aspects of scientific activities and the growing efficiency in retrieving relevant scientific publications in general, accessing scientific knowledge to meet our needs for assessing the state of the art of a research area and making various decisions remains a major challenge (Chen 2016).

Today, we still have to build our understanding of the state of the art of science through painstakingly time-consuming and cognitively demanding processes. We still have to piece together sporadically distributed information and transform it to a cohesive conceptualization of our own. The knowledge acquisition process from the vast volume of scientific literature remains the most challenging bottleneck not only for scientists and researchers, but for everyone seeking to obtain an accurate picture of the state of the art. Although increasingly sophisticated techniques

© Springer International Publishing AG 2017
C. Chen and M. Song, *Representing Scientific Knowledge*,
https://doi.org/10.1007/978-3-319-62543-0_9

emerge to address one or more specific aspects of the knowledge acquisition bottleneck, the scientific community as a whole is still limited by the lack of integrative and widely accessible options to increase the throughput of the bottleneck and in turn to increase the efficiency and effectiveness of the transformation from information to knowledge. Furthermore, the development and evaluation of such tools is hindered by the lack of accessible and persistently maintained resources such as classic cases and training materials of in-depth studies of representative high-impact research, contemporary and innovative metrics and analytic tools, metadata and gold standards for comparative and evaluative studies.

Visual Analytic Observatory of Scientific Knowledge

We envisage a widely accessible and persistently maintained community resource —a *visual analytic observatory of scientific knowledge* (VAO). The central idea of the VAO is that the essence of scientific knowledge can be captured by a set of semantically organized assertions along with their status of uncertainty and that knowledge represented in this way can fundamentally increase the efficiency and accuracy of our understanding of scientific knowledge. As a result, many existing analytic methods will be fruitfully extended to the new level of granularity. A sketch of the architecture is illustrated in Fig. 9.1. The development of the VAO[1] is supported by the Science of Science and Innovation Policy (SciSIP) program of the National Science Foundation.

In this ambitious framework, unstructured text in a scholarly publication will be transformed to a semantic network of assertions along with their epistemological status and the provenance of their evolution. A set of scientific articles will be represented by a more extensive but organizationally equivalent semantic network. Ultimately, the body of scientific literature of a scientific domain can be represented in this framework. This framework will eventually enable us to transform how we communicate and keep abreast with the advances of science.

A unique focus in the VAO development is the role of uncertainty in advances of scientific knowledge. The goal of the VAO is to improve the clarity of the representation of scientific knowledge substantially and especially improve the clarity of the uncertainties associated with particular areas of scientific knowledge. Ultimately, the VAO will make scientific knowledge easy to access with the level of clarity that one can communicate efficiently to address Heilmeier's series of questions regarding the planning, execution, or evaluation of scientific inquiries.

---

[1]https://www.researchgate.net/project/Research-A-Visual-Analytic-Observatory-of-Scientific-Knowledge-VAO.

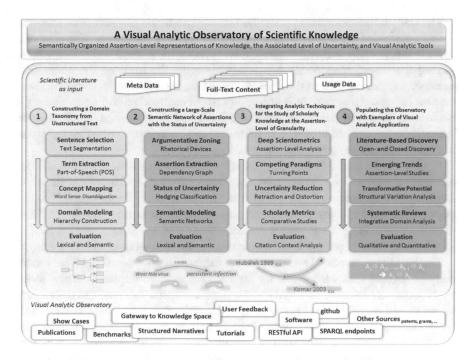

Fig. 9.1 Architecture of a visual analytic observatory of scientific knowledge

Many efforts in representing scientific knowledge attempt to reduce scientific knowledge to a set of propositions. For instance, in Semantic MEDLINE, scientific assertions are extracted from unstructured text of published articles and expressed as propositions in a generic form of (*Subject*)-(*Predicate*)-(*Object*) as in (*West Nile Virus*)-(*Causes*)-(*Persistent Infection*) or (HIV)-(Causes)-(AIDS). The negation of an assertion is also an assertion. An assertion itself can be embedded as the subject or the object of another assertion, e.g. (Assertion$_1$)-(is a)-(Assertion$_2$). On the one hand, scientific knowledge is represented by propositions that have been aggregated and mapped to standard vocabularies such as UMLS concepts and semantic types. The complexity of the diverse expressions in natural languages is considerably reduced and it is easier to handle the represented knowledge computationally. On the other hand, much of the meta-knowledge is lost, notably the epistemic status of a scientific assertion in terms of how scientists try to communicate the subtlety as precisely as possible through carefully chosen words, and, perhaps more importantly, how scientists handle conflicting and contradictory reports in the scientific literature on the exact same topic. We believe that scientific knowledge should be represented and communicated along with its epistemic status and the provenance of its status. Such meta-knowledge is an integral part of an expert's domain expertise. The knowledge of uncertainty is an expert's expertise!

The study of uncertainties in scientific knowledge should pay particular attention to two sources of uncertainty: (1) hedging and (2) contradictions. If a scientific claim is modified by hedging devices, then it indicates that the researcher who is making the assertion evidently has reservations to the truth of the claim. For instance, a statement that HIV causes AIDS leaves no doubt to its audience about the firm belief of its author. In contrast, one may become doubtful when reading a more carefully crafted statement: *a recently published study suggest that X might be responsible for Y if condition Z is met*. Hedging may become necessary when information is incomplete or entirely missing.

Intuitively, the level of uncertainty is higher when it is evident that contradictory information prevents scientists from making a positive and absolute assertion. Conflicting, contradictory, and controversial results must be reconciled before speculations and hypotheses can be accepted as part of scientific knowledge. How often do we come across topics or research areas that are puzzled by conflicting information? How important is it for scientists to reconcile contradictory findings?

The VAO aims to provide an integrative, extensible, and shared platform for the study of scientific knowledge and for the research and development of new tools. As a community resource, the VAO will enable scientists, analysts, and the general public to accomplish several types of analytic tasks that have been so far cognitively demanding and time consuming. It will enable the study of scientific knowledge to reach a deeper level of granularity and, more importantly, a potentially more efficient and effective way to understand critical information in scientific discovery and in the public understanding of science. It has the potential to increase the productivity of research at a reduced cost.

Types of Uncertainties in Scientific Literature

Scientific knowledge is never free of uncertainty. It is difficult to communicate uncertainty clearly, especially on issues with widespread concerns, such as climate change (Heffernan 2007) and Ebola (Johnson and Slovic 2015). The way in which the uncertainty of scientific knowledge is communicated to the public can influence the perceived level of risk and the trust (Johnson and Slovic 2015). A good understanding of the underlying landscape of uncertainty is essential, especially in areas where information is incomplete, contradictory, or completely missing. For instance, there is no information on how long the Ebola virus can survive in a water environment (Bibby et al. 2015). If surrogates with similar physiological characteristics can be found, then any knowledge of such surrogates would be valuable. Currently, finding such surrogates in the literature presents a real challenge (Bibby et al. 2015).

According to sociological views of scientific change, competition leads to scientific change (Fuchs 1993). Three types of scientific change are likely to emerge: permanent discovery, specialization, and fragmentation. The severity of competition is the strongest in settings that lead to permanent discovery. A lighter degree of competition is associated with specialization. The least competitive environment is associated with fragmentation.

Scientists compete for recognition and reputation. Many other tangible or intangible benefits may come with established reputation and authority. Publishing novel and interesting discoveries is one of the long established traditions in science. The threshold of publishing a scientific article has been lowered over the years. New journals are launched at a high speed.

From the competition point of view, novel, interesting, and controversial ideas are likely to attract more attention than commonly known, trivial, and expected results. Sociologists suggest that the interestingness of a topic depends on whether it challenges our current beliefs. If we know the information that we are about to learn is contradict to our current belief, then we can expect that the gain from understanding the new information is likely to be the highest.

Table 9.1 presents some examples of sentences from MEDLINE articles. These sentences indicate some common types of uncertainties in biomedicine. The first column is a list of terms that indicate some types of uncertainty—we call them cue words of uncertainty, for example, the term unknown in the first sentence "The mechanism is unknown." The uncertainty is high when the mechanism of a disease is unknown. Contradictions are another type of uncertainty. One must validate each of the contradicting components before making selections. Similarly, controversial and inconsistent results in published articles all represent a degree of uncertainty. In summary, if there are competing alternative interpretations, then we are dealing with uncertainty.

Hedging and Speculative Cues

Hedging is a particularly relevant concept for characterizing the tentative and context-dependent nature of scientific claims (Hyland 1996). Hedging is a rhetorical means, or a communicative technique, to convey the degree of uncertainty associated with a statement or an assertion (Behnam et al. 2012; Clark et al. 2011; Di Marco et al. 2006; Horn 2001; Kilicoglu and Bergler 2008). The presence of hedge words can mitigate an otherwise overstated scientific claim such that the status of the knowledge is documented more accurately. Reinstating hedging information surrounding an assertion can help us to understand precisely what is currently known about the assertion. Introducing hedging information provides an additional and important means to characterize the role of an assertion in the context of the domain knowledge as a whole. Furthermore, it will enable us to understand not only the current status of a scientific assertion, but also the trajectory of the evolution of its status over time. We will be able to better understand how the uncertainty associated with a scientific assertion changes as new information, e.g. new discoveries, becomes available. We will be able to better assess the potential of a research program in terms of the extent to which it reduces the uncertainty of the scientific knowledge of a particular area.

Table 9.1 Sentences that indicate uncertainties in scientific knowledge

Terms on uncertainty	Type	Instances	PMID	Sentence ID	Sentence
Unknown	ab	300800	165704	10667452	The mechanism is unknown
Suspect	ab	165545	12351994	77704397	An immunopathology is suspected
Unclear	ab	164034	7260869	10419608	The etiology is unclear
Unusual	ab	141237	3629081	33402065	Such cases are unusual
Controversial	ab	122406	2499131	34124598	The results are controversial
Consensus	ab	113464	23979725	152414767	There is no consensus on treatment
Incomplete	ab	95914	2419361	29557765	This association is incomplete
Conflicting	ab	91371	11433428	68757263	These reports are conflicting
Contrary	ab	68059	8324612	45236707	On the contrary it is increasing
Debatable	ab	64233	860951	13010435	Possible causes are being debated
Inconsistent	ab	53353	10434263	63377317	The results are inconsistent
Uncertain	ab	48831	3585876	3573539	The etiology is uncertain
Unexpected	ab	46336	2260033	53665387	This result is unexpected
Confusing	ab	39363	2250070	44822246	This was confusing and misleading
Paradoxical	ab	38218	7635297	51365510	This leads to a paradox

Uncertainty cues in scientific writing in general come from adjectives, adverbs, auxiliaries, verbs, conjunctions, and nouns. Szarvas et al. (2012) identified uncertainty cues in each of these categories. For instance, probably, likely, and possible are uncertainty cues in the adjective and adverb category. Examples of auxiliaries as uncertain cues include may, might, and could. Speculative verbs include suggest, seem, and appear. Nouns include speculation, proposal, and rumor.

Researchers have developed heuristics that can be used to detect propositions with uncertainty based on uncertainty cues. For example, based on the suggestions of Kilicoglu and Bergler (2008), one can derive the following heuristics to identify propositions that are likely to involve uncertainties:

- If a proposition has an uncertain verb, noun, preposition, or auxiliary as a parent in the dependency graph of the sentence, then the event is regarded as uncertain.
- If a proposition has an uncertain adverb or adjective as its child, then it is regarded as uncertain.

Figure 9.2 shows a log-log plot of hedging words appeared in the conflict (x) and non-conflict (y) axes. Words appear below the dashed line appeared more often in the conflicting set than in the non-conflicting set. The conflict set consists of MEDLINE articles with sentences containing words such as conflict, contradictory, and inconsistent, whereas the non-conflict set does not contain such cue words.

Uncertainty cue detection has mostly been developed in the biomedicine domains (Szarvas et al. 2012). Researchers studied the distribution of hedging cues in scientific writings of different domains. Rizomilioti (2006) studied publications from three domains, namely, archeology, literacy criticism, and biology and found that uncertainty cues were the highest in archeology and the fewest in literacy criticism. Hyland (1998) found that writers in humanities use hedging devices significantly more than writers in sciences. Falahati (2006) compared psychology, medicine, and chemistry and found that hedges are more often in psychology than in medicine and chemistry.

Table 9.2 shows the uncertainty levels of scientific disciplines in Elsevier's full text repository Consyn as of August 13, 2015. The uncertainty level of each scientific discipline, or subject area, in Consyn is estimated by the proportion of items containing any of the five words: *conflicting*, *contradictory*, *inconsistent*, *discrepant*, and *irreconcilable*. These five words are useful indicators of controversial and unresolved alternative interpretations. They indicate the lack of clarity of the status of scientific knowledge.

The top four disciplines with the highest rates of uncertainty items are all social sciences. Mathematics, physics and astronomy, chemical engineering, along with material science and chemistry are the five disciplines with the lowest level of uncertainty word use. Psychology has 32%—the highest rate of items characterized

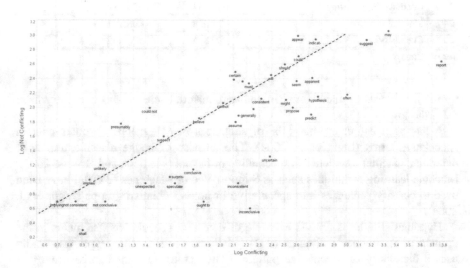

Fig. 9.2 Hedging words in the conflicting set versus the non-conflicting set

Table 9.2 The uncertainties of scientific disciplines

1	Subject area (as of 8/13/2015)	Journal items only	Subtotal items in area	Rate (%)
2	Psychology	70,096	220,250	32
3	Business, management and accounting	26,717	97,083	28
4	Social sciences	74,835	283,598	26
5	Economics, econometrics and finance	27,920	113,083	25
6	Neuroscience	99,908	434,270	23
7	Medicine and dentistry	423,391	2,093,102	20
8	Veterinary science and veterinary medicine	24,390	126,768	19
9	Pharmacology, toxicology and pharmaceutical science	56,441	305,601	18
10	Nursing and heal professionals	39,692	218,124	18
11	Arts and humanities	14,470	78,844	18
12	Environmental sciences	56,594	328,192	17
13	Immunology and microbiology	51,184	310,404	16
14	Agricultural and biological sciences	63,010	400,272	16
15	Biochemistry, genetics and molecular biology	120,012	800,766	15
16	Computer science	32,040	252,366	13
17	Decision sciences	17,500	144,119	12
18	Earth and planetary sciences	24,393	225,816	11
19	Engineering	45,281	510,624	9
20	Energy	18,253	235,489	8
21	Mathematics	17,737	239,676	7
22	Physics and astronomy	28,507	498,418	6
23	Chemical engineering	17,434	355,512	5
24	Material science	24,038	608,991	4
25	Chemistry	20,585	52,2442	4

Source Consyn

by the five words of uncertainty. In contrast, material science and chemistry have 4%—the lowest.

It has been estimated that 11% of sentences in MEDLINE abstracts contain speculative terms (Light et al. 2004). The purpose of hedge classification is to determine whether a sentence is speculative or factual (Medlock and Briscoe 2007). Machine learning techniques such as Support Vector Machines (SVMs) have been used to classify sentences into speculative or non-speculative groups (Light et al. 2004).

HypothesisFinder is a good example of detecting speculative statements in the domain of Alzheimer's diseases (AD) (Malhotra et al. 2013). HypothesisFinder uses a dictionary of speculative patterns. Their study identified three groups of speculative patterns and their ability to detect speculative sentences accurately. For example, the strongest signals are given by phrases such as "*might be involved*," "*hypothesized that*," and "*raising the possibility that*." The medium-strength signals include "*seems to*," "*appears to be*," and "*can be anticipated*." Weak patterns include "*presume*," "*suppose*," and "*would*." HypothesisFinder is available online[2] as part of the information retrieval system SCAIView Academia. A precision of 0.91 and a recall of 0.73 were reported for their evaluation based on the BioScope corpus (Szarvas et al. 2008).

Finding Semantically Equivalent Uncertainty Cues

We are developing a new method for uncertainty cue word recognition (Chen et al. 2017). Unlike earlier studies that commonly used hand-crafted rules and dependency graphs to identify cues of uncertainty, we found that recent advances in deep learning and distributional semantics have the potential to make substantial improvements (McDonald, and Ramscar 2001).

The distributional hypothesis is that words appearing in the same contexts tend to have similar meanings (Harris 1954). They are likely semantically equivalent. Word2vec (Mikolov et al 2013) is one of the most popular word embedding models in the recent years (see Chap. 6). Using a Word2vec model training on Google news, we expanded a list of hand-picked uncertainty cue words to obtain many more semantically equivalent uncertainty cue words.

The seed list is shown in Table 9.3. The selection of the initial uncertainty cue words was based on our own heuristics of how an uncertain can be directly characterized or indirectly inferred. For example, words in the original seed list include words such as inconsistent, ambiguous, debatable, bizarre, and surprising. When these words are found in a scientific publication, one can expect that the statement implies some degree of uncertainty. For example, inconsistent results may imply that a research question involves uncertainties because researchers cannot settle it yet and extra efforts are required to clarify the current inconsistency. Similarly, if a study has produced surprising results, then the underlying theory is questionable because it was not capable of predicating the results correctly.

The word2vec expansion increased the number of semantically equivalent uncertainty cue words by almost 10 times with a total of 469 words combined. The expanded words represent 83.37% of the combined set. The original seed list represents 16.63 of the combined set. Figure 9.3 visualizes the combined set of uncertainty cue words. Words from the original seed list are shown in red, including prominent words such as inconsistent, contrary, ambiguous, bizarre, and debatable.

---

[2]http://www.scaiview.com/scaiview-academia.html.

Table 9.3 A seed list of uncertainty cue words

Ambiguity or -ous	Irreconcilable
Baffling	Misbelief
Bizarre	Misconception
Conflicting	Misleading
Confusing	Mystery or -ious or -ies
Consensus	Paradox or -ical
Contentious	Perplexity
Contradictory	Puzzling
Contrary	Skeptic
Controversial	Surprising or surprise
Debatable	Suspect
Deceptive	Suspicion
Dispute	Unanticipated
Doubtful	Uncertain
Dubious	Uncertainty
Fallacy	Uncharted
Flaw	Unclear
Implausible	Unconvincing
Impossible	Undetermined
Improbable	Undiscovered
Incoherent	Unexpected
Incompatible	Unexplained
Incomplete	Unidentified
Incomprehensible	Unknown
Inconceivable	Unpredictable
Inconclusive	Unrecognized
Incongruity	Unreliable
Inconsistent	Unusual

In contrast, words expanded from the word2vec model are shown in green, including words such as misguided, inaccurate, tricky, muddled, and contradictive.

Figure 9.4 shows the network of 469 uncertainty cue words colored in 11 communities, i.e. semantically equivalent classes. The size of a label is proportional to the eigenvector centrality of the corresponding node in the network. For instance, inconsistent has the highest eigenvector centrality, followed by contrary and ambiguous, all of which belong to the same class.

Uncertainty cue words can be used to select sentences that may involve a degree of uncertainty. Furthermore, uncertainties surrounding semantic predications can be identified.

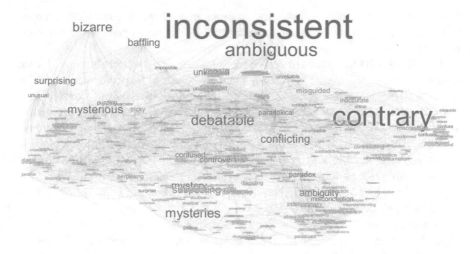

Fig. 9.3 Uncertainty cue words from the original seed list (in red) and expanded (in green)

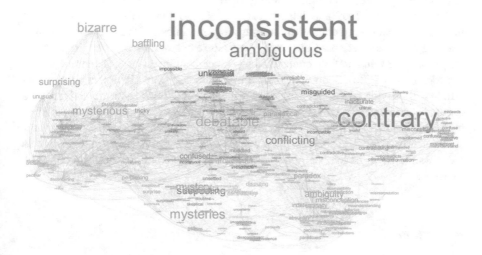

Fig. 9.4 Uncertainty cue words colored in 11 semantically equivalent classes. The size of a label is proportional to the eigenvector centrality of the corresponding word

Citation Distortion and Provenance of Evidence

Citation distortion is a significant source of uncertainty. The scholarly impact of a scientific discovery is often considered in terms of how it is cited in subsequent publications. The strength of the influence of a single published statement on subsequent interpretation reveals interferences in chains of collective reasoning (Rzhetsky et al. 2006). However, as authors of citing articles commonly paraphrase the original work in citation contexts, citations may distort the intended interpretation of the original source. Such distortions deviate the true epistemic status of the original finding.

Greenberg (2009) demonstrated how citations are overwhelmingly biased towards citing supportive as opposed to refuting papers of a specific claim and the important role of review papers in directing the flow of citations concerning a scientific claim. His study shows that primary data that weakened or refuted claims were ignored and citations exponentially amplified supportive claims over time. Greenberg's analysis also found evidence of how the status of a scientific hypothesis may be distorted in subsequent citations and a hypothesis was incorrectly referred to as a matter of fact—"*This subclaim had transformed from hypothesis to "fact" through citation alone, a process that might be called citation transmutation*" on page 5 of (Greenberg 2009).

Greenberg found that citation biases, amplifications, and citation diversions appeared not only in scientific literature, but also in nine grant proposals funded by the NIH. His investigation raises an important question that the science of science and innovation needs to address concerning the trustworthiness of how scientific knowledge is stated, paraphrased, and quoted.

Citation contexts of a published article refer to the sentence containing an instance of citation along with surrounding sentences. Citation contexts and hedging are connected in an interesting way. Research shows that hedging is more frequently seen in citation contexts than other sentences of a scientific article (Di Marco et al 2006). Semantic predications extracted from citation contexts of an article will provide additional insights into the semantic relations extracted from the original statements. These semantic predications and corresponding information of their uncertainties form a chain of evidence of how the original work impacts subsequently published studies. Taken together, the provenance of evidence is valuable for us to develop a good understanding of scientific knowledge and its dynamics.

Retraction

If hedging and citation distortions indirectly indicate the possibility of uncertainties involved in scientific knowledge, the retraction of a published article sends direct signals that some claimed scientific knowledge must be re-examined and re-validated (Chen et al. 2013). In such situations, the uncertainty should increase and scientific knowledge as a collective belief system should be rolled back to the point prior to the publication of the retracted article. Notorious examples of retracted studies include the highly controversial study on the connection between MMR vaccines and autism by Wakefield et al. (1998), the Bell Lab physicist's forging data (see Service RF 2002), the high-profile retraction of Hwang (Kakuk 2009), and the rise and fall of STAP.[3]

We may all have heard a variant version of the same story. A hen lays a golden egg every day. However, that is not good enough for its owner, who would rather to have all the golden eggs all at once. So the owner killed the hen to retrieve the golden eggs, but to his surprise, he ended up with no golden eggs not only for that day but forever. To a scientist, a high-profile breakthrough would be a golden egg. Under the intensive competition, the more golden eggs he could produce, the better. Unlike the hen that can produce a golden egg every day, a researcher may not guarantee when he can deliver a golden egg. In fact, no one can plan for the delivery of a golden egg in his entire scholarly career.

Imagine two scientists are competing for recognition in a high-profile area of research. The one who makes a breakthrough first is likely to receive all the attention and all the resources. In contrast, his competitor is likely to suffer a great deal of loss in terms of attention and resources. The two scientists not only have to publish, but also maximize the chance that what they publish will attract the attention of the field.

A retraction is a step that can undo the process of a publication. Retractions are most common in areas that are advancing very fast. Publications in such areas have a relatively low half-time expectation. Chen et al. (2013) found that the most active and fast-moving areas of research have a higher rate of retraced articles. This is the type of scientific change that is resulted from the highest degree of competition, namely permanent discoveries. As Fuchs explained, scientists with high mutual dependence in the research fronts and working on research with a high degree of uncertainty have the highest stake.

Severe competition and pressure is not an excuse for compromising the integrity of one's scholarship. It is, however, something that one can anticipate as a result of the interplay of a broad spectrum of social, psychological, and behavioral factors.

The retraction of a published article is a mechanism for restoring compromised scientific knowledge. Figure 9.5 shows numerous highly cited retracted articles. The fact that a retracted article has been highly cited requires investigations at a deeper level. Why has it been cited? Did authors cite the article before its retraction or afterwards? If they cited after the retraction, are they aware of the fact that is has been retracted? What difference will a retraction make as far as the contemporary scientific knowledge is concerned? Each node labeled in the visualization represents a retracted publication. The size of a node is proportional to its citation counts. The larger size a node, the more citations it has. In Fig. 9.6, the node labeled as Nakao N has a large size. In fact, the retracted article (Nakao et al. 2003) is among the top 10 most cited retracted articles in the Web of Science.

Figure 9.7 illustrates various information of the retracted article by Nakao et al. The title indicates that this article is about a randomized controlled trial of a combination treatment in non-diabetic renal disease. The sentences highlighted in

---

[3]http://www.nature.com/news/stap-1.15332.

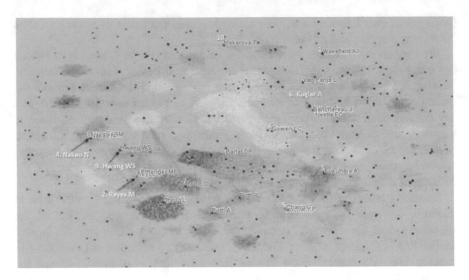

Fig. 9.5 Retracted articles (red dots) in a co-citation network. *Source* Chen et al. (2013)

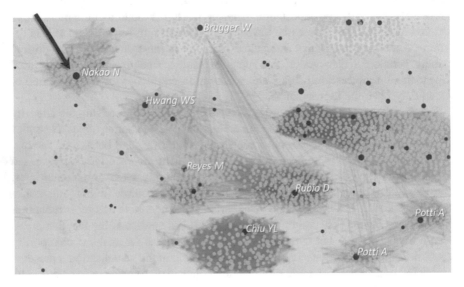

Fig. 9.6 A retracted article by Nakao et al

yellow are sentences from which semantic predications are extracted. For example, one semantic predication is extracted from the title of the article (the first row in Table 9.4). The subject of the predication is Angiotensin-Converting Enzyme Inhibitors. The object is Diabetic Nephropathy. The predicate is TREATS. The diagram shown in the lower right corner of Fig. 9.7 depicts how these predications are connected. The publication of the article imposes this small network of

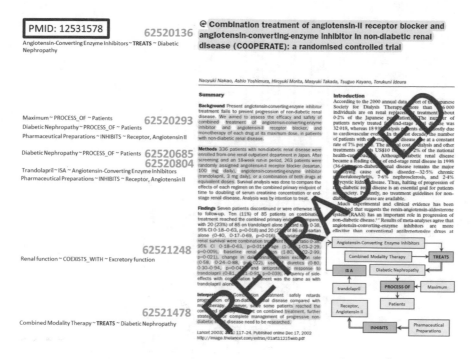

Fig. 9.7 The first page of the retracted article by Nakao et al. (2003) along with semantic predications extracted from the article

predications over the existing scientific knowledge, which can be represented by semantic predications from all published articles. The retraction removes such a network of predications from the current global network of predications. Thus publications and retractions both alter the structure of scientific knowledge. More specifically, the publication of Nakao et al. establishes or strengthens the claim that combined modality therapy TREATS diabetic nephropathy (Predication: 7762151). The retraction of the article weakens the semantic relation.

A retraction increases the uncertainty of scientific claims. If the affected predications have a unique position, it may in turn influence the uncertainty of a much larger area in scientific knowledge. We will introduce other types of uncertainties associated with scientific knowledge.

Distributions of Uncertainty Cues

The advances of science, engineering, and technology have considerably pushed the boundaries of scientific knowledge. At the same time, the integral role of uncertainty in science has been very much overlooked.

Table 9.4 Semantic predications extracted from Nakao et al. (PMID: 12531578)

Sentence	Predication	Subject	Predicate	Object
62520136	3281525	Angiotensin-converting enzyme inhibitors	TREATS	Diabetic nephropathy
62520293	878505	Maximum	PROCESS_OF	Patients
62520293	1111567	Diabetic nephropathy	PROCESS_OF	Patients
62520293	2931514	Pharmaceutical preparations	INHIBITS	Angiotensin-converting enzyme inhibitors
62520293	4958352	Pharmaceutical preparations	INHIBITS	Receptor, angiotensin II
62520685	1111567	Diabetic nephropathy	PROCESS_OF	Patients
62520804	2653028	Trandolapril	ISA	Angiotensin-converting enzyme inhibitors
62520804	4958352	Pharmaceutical preparations	INHIBITS	Receptor, angiotensin II
62521248	581854	Renal function	COEXISTS_WITH	Excretory function
62521478	7762151	Combined modality therapy	TREATS	Diabetic nephropathy

Table 9.5 shows distributions of uncertainty cue words in the most representa-tive and most comprehensive sources of scientific publications. Google Scholar, the Web of Science, PubMed contain meta-data of scientific publications, whereas ScienceDirect, Springer, Mendeley, and Core are sources of full text articles. For each word, its frequency in a data source is compared with the frequency of the word knowledge. For example, the frequency of the word unknown in Google Scholar is 99% of the frequency of the word knowledge, whereas the frequency of the word contrary only appears 52.59% of the frequency of the word knowledge. Within the same data source, we can compare the popularity of an uncertainty cue word. Between different data sources, we are able to compare the relative fre-quency. For example, the word unknown is relatively more popular in the Web of Science (132.94% of knowledge). The term uncertainty is most frequently found in Google Scholar (69.52%).

Table 9.6 lists distributions of uncertainty cue words in non-scientific publica-tions. Non-scientific sources including U.S. Supreme Courts, patents and applica-tions, New York, Google, and NSF.

The leading cue words in the U.S. Supreme courts include words such as con-trary and controversial. Interestingly, both USPTO and the New York Times are led by the word impossible. The NSF award abstracts are led by the word uncertainty.

Contradictory Claims

These observations have two implications: one on the interestingness and the other on the uncertainty. The interestingness explains the motivations behind the dynamics of the discourse of the argumentation. According to a theory proposed by sociologist Murray Davis (1971), the best way to attract people's attention is to convince them that you can show them that what they believe is questionable. This is the first and the most critical step to get their attention. Davis even suggested that it is possible to routinize this strategy such that one can systematically respond to the current beliefs of a group of people. He identified 12 dialectical relations regarding hypotheses and their antitheses (Table 9.7). For example, if everyone believes that A and B are not connected, then its antithesis argument that A and B are connected is likely to be interesting. If everyone believes that A is changing, then one would be interested in an argument that A is constant. Davis warned that if one takes this strategy too far, it may backfire. The antithesis may sound too ridiculous to retain anyone to listen. Davis' framework is in fact a classification of patterns in our knowledge. If our current belief is in one form of knowledge, an antithesis pattern may provide an alternative interpretation. If it is believed that "A is a B," then one is likely to find it interesting why "A is not a B" is even possible. Similarly, causal relations are an important type of knowledge. Which one should we believe: "HIV causes AIDS" or "HIV does not cause AIDS"?

A large degree of differences between claims on related topics may reflect a degree of uncertainty concerning the status of underlying knowledge. The higher

Table 9.5 Distributions of uncertainty cue words in scientific publications (with reference to the term knowledge)

Google scholar		ScienceDirect		Web of science		Springer		Mendeley		Pubmed		Core	
Unknown	0.990	Unknown	0.6021	Unknown	1.3294	Unknown	0.3416	Unknown	0.3875	Unknown	0.6526	Uncertainty	0.4748
Incomplete	0.755	Conflicting	0.3613	Unusual	0.5966	Conflicting	0.2516	Unclear	0.2166	Undetermined	0.6203	Unknown	0.4224
Impossible	0.7251	Surpris*	0.3309	Suspect	0.5217	Contrary	0.1998	Uncertainty	0.1806	Unclear	0.3784	Impossible	0.3895
Consensus	0.6992	Contrary	0.3308	Surpris*	0.4091	Impossible	0.1972	Consensus	0.1582	Unusual	0.3269	Surpris*	0.3685
Uncertainty	0.6952	Uncertainty	0.3225	Uncertainty	0.3527	Surpris*	0.1783	Unusual	0.1376	Consensus	0.2398	Contrary	0.3226
Unexpected	0.6394	Unclear	0.2945	Controversial	0.3373	Unclear	0.1625	Contrary	0.1187	Uncertain	0.1965	Consensus	0.2473
Surpris*	0.6016	Impossible	0.2901	Contrary	0.3341	Uncertainty	0.1552	Controversial	0.1121	Controversial	0.1747	Incomplete	0.2440
Uncertain	0.5896	Suspect	0.2426	Unclear	0.2754	Incomplete	0.1312	Incomplete	0.1020	Incomplete	0.1490	Ambigu (ity[ous})	0.2266
Unusual	0.5319	Incomplete	0.234	Conflicting	0.2712	Suspect	0.1296	Uncertain	0.0979	Contrary	0.1403	Unusual	0.2164
Contrary	0.5259	Unusual	0.2285	Unexpected	0.2401	Consensus	0.1189	Unexpected	0.0735	Conflicting	0.1200	Inconsistent	0.2039

* is a wildcard, e.g., surpris* including surprises, surprisingly, and surprised

Table 9.6 Distributions of uncertainty cue words in non-scientific publication sources

Supreme		USPTO		NYTimes		Google		NSF	
Contrary	1.3001	Impossible	1.1443	Impossible	1.0749	Unknown	0.9633	Uncertainty	0.1980
Controversial	1.0170	Contrary	1.1167	Unusual	0.8905	Supris*	0.6358	Unusual	0.1161
Dispute	0.9794	Unknown	0.7607	Dispute	0.7309	Dispute	0.6064	Debatable	0.0868
Inconsistent	0.7455	Incomplete	0.3731	Contrary	0.5375	Myster*	0.5284	Conflicting	0.0817
Impossible	0.4704	Unexpected	0.3715	Unknown	0.5128	Impossible	0.3972	Surpris*	0.0672
Ambigu*	0.2825	Supris*	0.3029	Suspect	0.3472	Unusual	0.2358	Incomplete	0.0540
Conflicting	0.2678	Unusual	0.2350	Unexpected	0.3306	Unexpected	0.1771	Uncertain	0.0516
Doubtful	0.2226	Incompatible	0.1970	Uncertain	0.3192	Suspect	0.1752	Impossible	0.0510
Unusual	0.2175	Inconsistent	0.1898	Suspicion	0.3004	Bizarre	0.1523	Unexpected	0.0400
Uncertainty	0.1844	Unreliable	0.1545	Controversial	0.2822	Controversial	0.1431	Consensus	0.0388

Table 9.7 12 dialectical relations identified by Murray Davis

Phenomenon		Dialectical relations		
Single	Organization	Structured	⟷	Unstructured
	Composition	Atomic	⟷	Composite
	Abstraction	Individual	⟷	Holistic
	Generalization	Local	⟷	General
	Stabilization	Stable	⟷	Unstable
	Function	Effective	⟷	Ineffective
	Evaluation	Good	⟷	Bad
Multiple	Co-relation	Interdependent	⟷	Independent
	Co-existence	Co-exist	⟷	Not co-exit
	Co-variation	Positive	⟷	Negative
	Opposition	Similar	⟷	Opposite
	Causation	Independent	⟷	Dependent

the uncertainty, the more discrepant results there are. The uncertainty reduces as we learn more and more about our topic. Research is driven by the uncertainty in that once a topic has revolved much of its uncertainty, the research of the topic will lose its attraction to researchers. Competing on a settled topic is pointless. A topic with much of its uncertainty resolved would become a good topic for a textbook. The knowledge is codified.

Figure 9.8 depicts the distributions of two contradictory semantic predications found in each year's MEDLINE records. The predication "HIV Causes AIDS" is overwhelming in terms of its volumes (shown in purple). The predication "HIV is not the cause of AIDS" appears almost every year, but its volume is much smaller. The co-existence of contradictory claims indicates a considerable degree of uncertainty. Active researchers are likely to be aware of such uncertainties in their areas of expertise. In fact, one can claim that the domain expertise is the knowledge uncertainty.

The predication "HIV Causes AIDS" has the second strongest burstness (38.7063) among MEDLINE records published between 1900 and 2014. In particular, the predication first appeared in 1984 and it began to burst from 1991 till 2000.

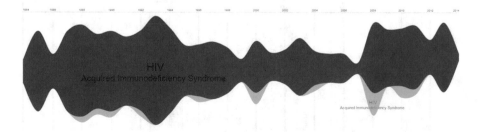

Fig. 9.8 Contradicting semantic predications extracted from MEDLINE records on causal relations between HIV and AIDS

Table 9.8 contains two opposing semantic predications regarding the causal relation between HIV and AIDS. The first four sentences contain the positive predication: HIV causes AIDS, whereas the second four sentences contain the negation of the predication: HIV is not the cause of AIDS. The two predications are contradictory. In the first example, a study suggests that "HTLV-III is the primary cause of human AIDS." The semantic predication of "HIV causes AIDS" partially preserves the meaning of the original sentence. First, the use of hedging word suggest modifies the status of the simplistic predication. Second, HTLV-III in the original sentence is abstract to the broader concept HIV. The more specific term human AIDS is mapped to the broader concept AIDS. Furthermore, the "primary cause" is simplified by the semantic type CAUSES.

If a knowledge system contains contradictory claims, then it is important for a researcher to be able to identify the status of these claims precisely. Furthermore, researchers would often need to take into account the provenance of evidence associated with each of the claims and how such evidence is validated and assessed. If Fuchs' theory is correct, resolving contradictory claims is most likely to play a central role in the work of research fronts because, as a type of competition, resolving contradictory claims would be critical for re-allocating recognition and resources. If contradictory claims appear within the boundary of a specialized area of research, resolving them is unlikely to have a greater degree of impact than that from the first scenario. The specialization effectively shields off much competition. The matter would be even less impactful if contradictory claims are limited to an area of research that is already fragmented off the main stream. To Fuchs, competition leads to scientific change.

Table 9.8 HIV causes AIDS (with the green background) and HIV is not the cause of AIDS (with the pink background)

SID	PMID	Sentence
35528335	6145881	The results strongly indicate that the antibodies to HTLV-III are diagnostic of AIDS or indicate significant risk of the disease, and suggest that **HTLV-III is the primary cause of human AIDS**
34893490	6200936	These results and those reported elsewhere in this issue suggest **that HTLV-III may be the primary cause of AIDS**
35618164	6095415	A transmissible agent especially a retrovirus (HTLV, LAV), is now widely considered in AIDS etiology.
30287966	6100647	HTLV-I is etiologically associated with adult T-cell leukemia-lymphoma (ATLL), HTLV-II has been isolated from a patient with hairy T-cell leukemia, and HTLV-III is the cause of acquired immune deficiency syndrome (AIDS). |
20897139	3399880	HIV is not the cause of AIDS.
33396961	2644642	(iii) pure HIV does not cause AIDS upon experimental infection of chimpanzees or accidental infection of healthy humans.
40872383	8906995	Furthermore, Cys-138 was found in chimpanzee immunodeficiency virus (CIV), a lentivirus that is similar to HIV but does not cause AIDS in chimpanzees.
49995531	1342726	Molecular biologist Peter Duesberg's argument that HIV is not the cause of AIDS is analyzed in light of his contention that a version of Koch's postulates has not been satisfied.

The existence of contradictory claims may potentially lead to the recognition of anomalies, which may in turn overthrow a well established paradigm. The key to determine whether contradictory claims may have the potential for a Gestalt Switch depends on why and how these claims differ. For example, if we consider HIV Causes AIDS, HIV Causes AIDS in human, and HIV does not cause AIDS in chimpanzees as different claims, then there will be no contradiction. On the contrary, if we use the same semantic predication HIV Causes AIDS or the negation of the predication to represent these claims, then our interpretations of these claims are contradictory. The 8th sentence in the table explicitly indicates that the contradiction exists at both levels of the extracted predications and the original writings.

The Reduction of Uncertainty

Table 9.9 demonstrates how the uncertainty associated with a scientific topic may be reduced over time as we learn more about the topic. In 1987, dementia is common in patients with AIDS, but its mechanism was unknown. In 1993, the cause of the AIDS dementia was still unknown, but there was some progress. Radiological and pathological studies have focused on subcortical white matter. In 2000, while the cause of neuronal damage in AIDS was still unclear, its relationships with HIV dementia remain debatable. In 2004, the search narrowed down to HIV-1 transactivating factor Tat. A sequence like this demonstrates how the uncertainty of scientific knowledge can be reduced over time.

A meta analysis is a study of studies that address a set of research questions. A meta analysis statistically normalizes various discrepancies in the findings of studies with equivalent or comparable designs. Ioannidis and Trikalinos (2005) conducted a meta meta-analysis, which means a study of meta-analytic studies. They attempted to answer two questions:

1. How is the between-study variance for studies on the same question changed over time?
2. When did the studies appear with the most extreme results?

They found that the between-study variance appears to decrease over time. They also found that the most extreme results are likely to appear at the beginning period of the research. As shown in Fig. 9.9, the results swung widely with reference to the results published immediately before them. The magnitude of the differences decreases over time.

Table 9.9 The knowledge of the cause of dementia in patients with AIDS

P-VLLD	Year	Subject	Predicate	Object	Sentence
3039662	1987	Dementia	PROCESS_OF	Patients	Dementia is common in patients with AIDS, but the mechanism by which the human immunodeficiency virus type 1 (HTV-1) causes the neurological impairment is *unknown*
3039662	1987	Acquired immunodeficiency syndrome	PROCESS_OF	Patients	Dementia is common in patients with AIDS, but the mechanism by which the human immunodeficiency virus type 1 (HTV-1) causes the neurological impairment is unknown
7689819	1993	White matter	LOCATION_0 F	Diagnostic radiologic examination	The cause of acquired immunodeficiency syndrome (AIDS) dementia, which is a frequent late manifestation of human immunodeficiency virus (HIV) infection, is unknown but radiological and pathological studies have implicated alterations in subcortical white matter
10871764	2000	HIV infections	COEXISTS_WITH	Acquired Immunodeficiency Syndrome	Neuronal apoptosis has been shown to occur in HIV infection by a number of in vivo and in vitro studies, however, the cause of neuronal damage in AIDS is still unclear and its relationships with the cognitive disorders characteristic of HIV dementia remain a matter of debate
15361847	2004	AIDS dementia complex	PROCESS_OF	AIDS patient	The HIV-1 transactivating factor, Tat, has been suspected of causing neuronal dysfunction that often leads to the development of HTV-associated dementia in AIDS patients

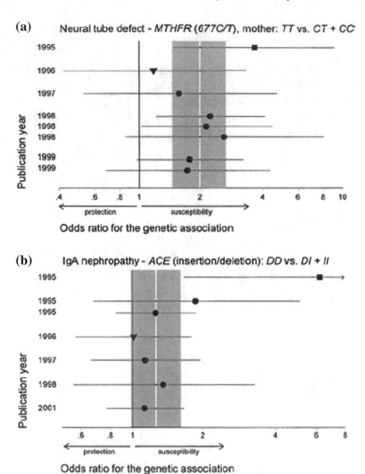

Fig. 9.9 The diversity of published claims decreases over time. *Source* Ioannidis and Trikalinos (2005)

Propositions and Their Epistemic Status

The term meta-discourse in philosophy refers to a discussion about a discussion, as opposed to a simple discussion about a given topic. It also refers to a word or phrase that serves as a guide for the reader on the importance of an example or the role of the text to follow in the discourse. Strictly speaking, meta-discourse is not the subject itself; rather, it provides additional information about the subject. Meta-discourse includes phrases such as "on the other hand," "after all," and "to our best knowledge." In scientific writing, meta-discourse may serve multiple purposes (Table 9.10). It is generally advised that technical, academic, and other non-fiction writers should use meta-discourse but bear in mind not to bury the subject itself.

Table 9.10 Purposes served by meta-discourse

Purpose	Meta-discourse
To denote the writer's confidence	May, perhaps, certainly, must
To denote the writer's intentions	In summary, in a nutshell
To give directions to the reader	Therefore, however, finally
To organize the text	First, second, therefore

Much of scientific assertions found in scholarly publications share a generic structure that consists of two parts: the core of the assertion and a modifier or a descriptor about the assertion. The modifier serves a similar role as the meta-discourse. We can think of many structures that share the same composite pattern in which one part serves the central role and the other part characterizes the central part.

- Analysis = Meta-Analysis + Analysis Proper
- Data = Meta-Data + Data Proper
- Message = Meta-Message + Message Proper
- Discourse = Meta-Discourse + Discourse Proper
- Knowledge = Meta-Knowledge + Knowledge Proper
- Statement = Epistemic Status (Meta-Proposition) + Proposition.

What we are interested is the last one on the list: a statement is seen as a proposition and its epistemic status. For instance, given the statement that *the mechanism of the disease is unknown*, the statement that mechanism is unknown conveys the epistemic status of the subject. Consider another example, *there is currently no consensus on what causes the disease*. The *what causes the disease* is the core message, whereas the lack of consensus is the epistemic status, or the meta-knowledge.

Research on representing scientific knowledge has overwhelmingly focused on the Proposition part of a statement. For instance, Semantic MEDLINE's semantic predications essentially correspond to the Proposition part of the pattern. Given a semantic predication of HIV CAUSES AIDS, none of its epistemic status nor the provenance of its evolution is preserved—the meta-knowledge is not accessible in association with the plainly expressed semantic predication. There is no trace of its original context. There is no indication how confidently the claim was made. There is no sign of any controversies involved. Thus we refer to this type of information as propositions, which form part of scientific knowledge but they are not complete in that one cannot make any meaningful inference or reasoning just based on propositions without knowing to what extent they are considered true and to what extent they are still unknown.

The Epistemic Status part of the statement is largely overlooked with notable exceptions in the study of hedging in scientific writing (Hyland 1996). The Epistemic Status part is meta-discourse in nature because it guides the reader about how to interpret the Proposition part. The use of hedging words is a sign of uncertainty at least from the position of the writer. A clause that contains suggestions of incomplete, conflicting, or contradictory information presents evidence that the certainty of a proposition is questionable.

The following MySQL query highlights the two-part structures in scientific writing. In particular, the query finds sentences that contain a specific claim and a meta-discourse that qualifies the claim. The query searches for the phrase "claim that" as the anchor and shows the text before and after the anchor phrase. This is a commonly used information search method known as the Keyword in Context (KiC) method. The table contains a text field—context—of paragraphs from scientific articles.

```
SELECT   article_id,   substring(context,   if(locate('claim   that',
        context)>30,  locate('claim that', context)-30, 1),  60) As
        KiC
FROM fulltext
WHERE project='sample' AND context LIKE '% claim that %'
LIMIT 20;
```

Table 9.11 shows examples of sentences that are joined by the anchor phrase 'claim that'—the text before the anchor is serving the role of a meta-discourse, whereas the text follows the anchor is the actual claim the authors are making. For instance, several cases are indirect quotations from published articles. In two of the examples, authors exclude a claim rather than make a claim.

Table 9.12 shows examples of the contexts in which the word 'uncertainty' are used. The level of uncertainty varies from 'entirely uncertain,' 'in part, fragmentary and uncertain,' 'at best difficult and uncertain,' to 'the extent of … is uncertain,' and 'the ultimate role of … is uncertain.'

Separating sentences into such two parts allows us to study the dynamics of uncertainty and its role in the development of scientific knowledge. The absence of the epistemic status part commonly implies that the proposition is considered true or valid. For instance, HIV causes AIDS is equivalent to a statement: *research has long established that HIV causes AIDS*. The length of the epistemic status part may serve as a simplistic indicator of the level of uncertainty—the longer the string length of this part, the higher the likelihood of the uncertainty. Of course, it is quite conceivable that one can express a high level of uncertainty concisely.

A useful device to analyze groups of words rather than individual words is a dependency graph. Since we need to effectively separate the proposition from a description of its epistemic status or other types of modifiers and wrappers, dependency graphs lend us graph-theoretic properties as well as linguistic and semantic relations. In the following examples, we will illustrate how we can identify a proposition and its epistemic status from a corresponding dependency graph. Furthermore, we will search for patterns that can be computationally processed and synthesized.

Table 9.11 Examples of claims and leading meta-discourse

#	Article ID	KiC
1	2007057	with their triple helix model **claim that** the contribution of
2	2007068	On this basis they claim that the technology reflected in th
3	1994398	of study and Hannam's (2009) claim that tourism studies is
4	2418115	give evidence to support the claim that improvements in edu
5	1930043	roper use of the term, we can claim that research broadly re
6	2131039	Moya-Anegon et al. [3] also claim that 85% of the journals in
7	2139416	1988), and Granovetter (1973) claim that individuals' person
8	2055620	We make no claim that the resulting sample is by any means a
9	2416554	Sipido et al.20 claim that the average life expectancy of pa
10	1982348	Legl23 corroborated Davis's24 claim that library instruction
11	1982356	s corroborated Davis's (2003) claim that library instruction
12	1909729	urrent discussion, we make no claim that DEA suffers from su
13	1953896	Third, we claim that the really destructive critique to the
14	2355226	iz-Baños, and Courtial (2005) claim that power laws are not
15	2346377	y, Baten and Muschalli (2012) claim that since the 1990s eco
16	2199679	(1998) claim that personality varies with structural holes a
17	2199787	Finally, we claim that the emergence of strategic roles can
18	1965013	One could claim that the quality incentive is embedded in th
19	2078654	imulations in Japan and China claim that the reduction impac
20	2078783	Many analysts claim that the use of green roofs is an effici

Table 9.12 Sentences containing the word 'uncertainty' in MEDLINE articles

PMID	Sentence
5321391	The duration of function of individual grafts is entirely uncertain at present
5940637	Severe osteomalacia of uncertain etiology was observed in a 44-year-old woman
11526856	The behavioral role of these response sequences is uncertain
11881655	All three approaches are beset with uncertainties, and it is important to state at the outset that no completely convincing evidence exists for extraterrestrial life
12056428	On the basis of these data that are, in part, fragmentary and uncertain, upper and lower limits of rad doses under different amounts of mass shielding are estimated
13118110	The extent of the uptake, however, is uncertain, again because of the liberation of chromogenic substances
13561107	The ultimate role of these agents in the treatment of major emotional disorders, such as schizophrenic reactions, still is uncertain
13684978	The value and risks of the procedure have been examined in 20 patients with obstructive jaundice of uncertain origin and in one further patient with a post-cholecystectomy syndrome
14287175	The assays indicated 1.2–2.6% RNA, similar to previously published work, but only 0.0–1.0% DNA, near enough the sensitivity limits to render the presence of DNA in the preparations uncertain
14792375	Prognosis in pancreatitis is at best difficult and uncertain

Dependency Graphs

Given a sentence, the dependent relations derived from the sentence can be represented in a dependency graph as shown in Fig. 9.10. The original sentence "A transmissible agent especially a retrovirus (HTLV, LAV), is now widely considered in AIDS etiology." is from an article published in 1984 (PMID: 6095415). The dependence graph divides the sentence into a few groups of words. For example, the semantic predication "HIV CAUSES AIDS" is extracted from the segment "(HTLV, LAV) is now widely considered in AIDS etiology."

The dependency parser from the Stanford NLP library identifies HTLV and LAV as the subject of this segment (nsubjpass). The word considered/VBN-12 means that it is a verb at the 12th position of the sentence. The text "is now widely" modifies the word considered, thus in the dependency graph, they are shown as the three nodes below the considered node. By retaining words with specific dependency types, we can computationally simplify a sentence by retaining the most salient message. For example, instead of considering the entire sentence, we can focus on the key message: HTLV and LAV are considered in AIDS etiology.

It is intuitively easy to separate a proposition from its conditional or contextual wrapper from a dependency graph because it is straightforward to identify sub-graphs that correspond to the two parts. For example, in the dependency graph shown in Fig. 9.10, the core proposition is represented by the sub-graph located at the lower right part of the graph, whereas the sub-graph on the left represents a modifier of the former sub-graph. The number [1] in Fig. 9.10's caption means that this is the first sentence in the abstract of the MEDLINE article.

The dependency graph shown in Fig. 9.11 represents a long and complex sentence from a 1984 article (PMID: 6100647). This is the 4th sentence from the abstract of the article:

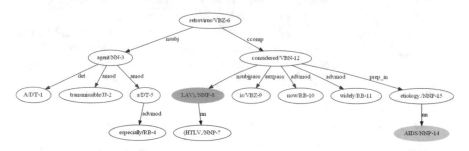

Fig. 9.10 The first appearance of the predication "HIV CAUSES AIDS" in 1984 (PMID: 6095415; SID: 35618164 [1]: "A transmissible agent especially a retrovirus (HTLV, LAV), is now widely considered in AIDS etiology.")

HTLV-I is etiologically associated with adult T-cell leukemia-lymphoma (ATLL), HTLV-II has been isolated from a patient with hairy T-cell leukemia, and HTLV-III is the cause of acquired immune deficiency syndrome (AIDS).

The sentence contains three statements. The HIV-CAUSES-AIDS predication is extracted from the last statement, which is represented by the sub-graph of the word cause/NN-25. By filtering the dependency types, we can simplify the sub-graph to a much simpler graph: HTLV-III—cause—AIDS. The complexity of the sentence is clearly reflected in the complexity of the dependency graph. The dependency graph provides a sense of context for the predication of our interest as well as other predications.

In Fig. 9.12, the predication is derived from the sub-graph at the lower right of the graph under the word cause: HTLV-III—cause—AIDS. The sub-graph as a whole is the object of the verb suggest/VBP-22, which is the verb at the 22nd position of the sentence. Words such as suggest are considered as hedging words. Writers often use hedging words to express a degree of uncertainty of their statements. A statement expressed with hedging words implies that the writer does not

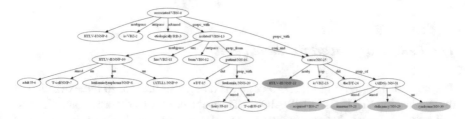

Fig. 9.11 The dependency graph of a sentence in a 1984 article (PMID: 6100647; SID: 30287966 [4]). HTLV-I is etiologically associated with adult T-cell leukemia-lymphoma (ATLL), HTLV-II has been isolated from a patient with hairy T-cell leukemia, and HTLV-III is the cause of acquired immune deficiency syndrome (AIDS)

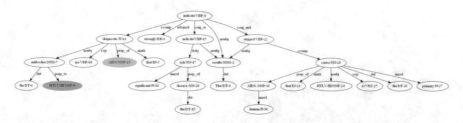

Fig. 9.12 The dependency graph of a sentence from a 1984 article (PMID: 6145881; SID: 35528335[6]). This is the 6th sentence in the abstract: *The results strongly indicate that the antibodies to HTLV-III are diagnostic of AIDS or indicate significant risk of the disease, and suggest that HTLV-III is the primary cause of human AIDS*

rule out the possibility of exceptions. Without hedging, the predication sounds like "HIV causes AIDS, period!" With hedging, it conveys that the status of the statement may be conditional on other factors, for example, "To our best knowledge, HIV causes AIDS."

The example shown in Fig. 9.13 also contains a hedging word suggest. In addition, there is another layer of hedging—may be—in the core statement: HTLV-III may be the primary cause of AIDS. It is reasonable to perceive that this sentence has a higher degree of uncertainty than the one in the previous example because of the presence of two levels of hedging. The word cause is modified by the word primary, which can be seen as another level of hedging because it does not rule out other possible causes. The three levels of hedging make the statement as precise as the writer wants to convey his/her best knowledge about this matter. The writer only needs to do that when the real status of the proposition is still uncertain. Therefore, the presence of hedging is an indicator that the scientific assertion in question is associated with a considerable degree of uncertainty.

The dependency graph shown in Fig. 9.14 is complicated. The predication in the complex sentence boils down to a short statement re-constructed from the graph: A direct role of PBM in the pathogenesis of AIDS is postulated.

The Length of Uncertain Statements

The dependency graph in Fig. 9.15 contains a segment that led to the extraction of the predication "HIV CAUSES AIDS." The subject was Barbara Hogan. The text includes a segment that she affirmed that HIV causes AIDS. The sub-graph that represents the assertion is very simple, as colored in the graph. This observation leads us to propose another way to measure the uncertainty of a scientific assertion: the longer an assertion is in terms of the total number of words, the more uncertain it is likely to be. In other words, if one has to state a claim with uncertainty, then he/

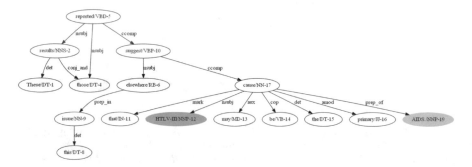

Fig. 9.13 The dependency graph of a sentence from a 1984 article (PMID: 6200936; SID: 34893490[7]). This is the 7th sentence in the abstract: *These results and those reported elsewhere in this issue suggest that HTLV-III may be the primary cause of AIDS*

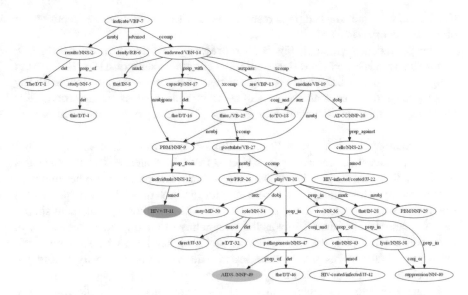

Fig. 9.14 The dependency graph of a sentence in a 1990 article (PMID: 2104787; SID: 18493183 [16]). This is the 16th sentence in the abstract: *The results of this study clearly indicate that PBM from HIV+ individuals are endowed with the capacity to mediate ADCC against HIV-infected/ coated cells and thus, we postulate that PBM may play a direct role* in vivo *in lysis or suppression of HIV-coated/infected cells and in the pathogenesis of AIDS*

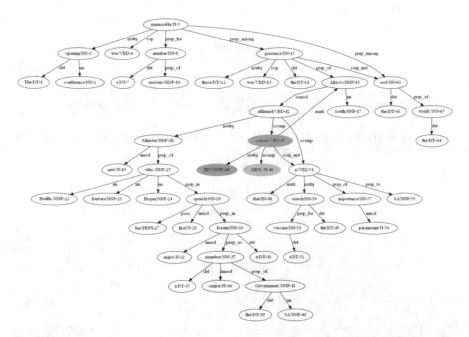

Fig. 9.15 A dependency graph of a sentence in a 2009 article (PMID: 19202348; SID: 120435934[12]). This is the 12th sentence in the abstract: *The conference opening was memorable for a number of reasons: among these was the presence of South Africa's new Minister of Health, Barbara Hogan who, in her first speech in a major forum as a senior member of the SA Government, affirmed that HIV causes AIDS, and that the search for a vaccine is of paramount importance to SA and the rest of the world*

she is likely to include hedging, restrictions, limiting conditions, exceptions, or other factors or sources of uncertainty.

The dependency graph in Fig. 9.16 represents the title of an article (PMID: 3399880). This is a very short statement: HIV is not the cause of AIDS. It is hard to imagine how it can be shortened any further. It has no hedging, no qualifying conditions, and no exceptions. As far as it is concerned as an assertion, it is absolutely certain; or, its uncertainty is zero. The author of the sentence is very confident.

The example shown in Fig. 9.17 is the 8th sentence from a 1990 article (PMID: 1980675). A negative predication is extracted from the sentence. The sentence as a whole simply states the position of Duesberg. There is no hedging or other indicator of uncertainty. How should we assess the uncertainty of an indirect quote of a statement? We have two options: 1. Assume that the paraphrasing statement should have the same level of uncertainty as the uncertainty of the original statement, or 2. Consider that the paraphrasing statement has a higher level of uncertainty. We believe the latter makes more sense because its writer is not really taking the responsibility for the core claim.

Figure 9.18 shows another example of the negation of the predication "HIV CAUSES AIDS." This sentence makes two points: CIV is similar to HIV, but CIV does not cause AIDS in chimpanzees. Thus, the predication "HIV does not cause AIDS" extracted by SemMed does not preserve the original meaning of the text. This is another type of uncertainty. It is introduced in the process of mapping to a semantic type.

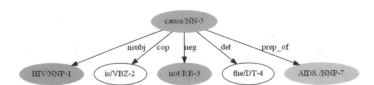

Fig. 9.16 The dependency graph of the title of a 1988 article (PMID: 3399880; SID: 20897139 [title]). The title is: *HIV is not the cause of AIDS*

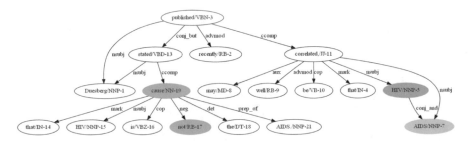

Fig. 9.17 The dependency graph of a sentence of a 1990 article (PMID: 1980675; SID: 51884237 [8]). This is the 8th sentence in the abstract: *Duesberg recently published that HIV and AIDS may well be correlated, but stated that HIV is not the cause of AIDS*

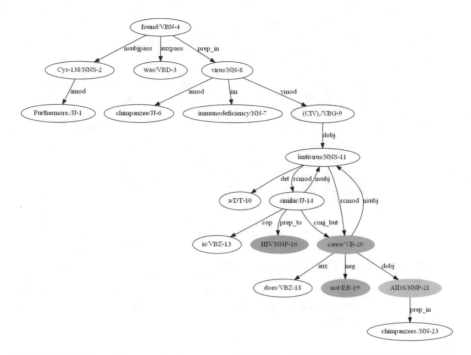

Fig. 9.18 The dependency graph of a sentence of a 1996 article (PMID: 8906995; SID: 40872383 [8]). This is the 8th sentence in the abstract: *Furthermore, Cys-138 was found in chimpanzee immunodeficiency virus (CIV), a lentivirus that is similar to HIV but does not cause AIDS in chimpanzees*

The example shown in Fig. 9.19 demonstrates another type of uncertainty. The core statement was "HIV could not cause AIDS simply through direct cytopathic mechanisms alone." Does it mean that HIV does not cause AIDS? Does it mean that HIV may cause AIDS through other mechanisms or a combination of multiple types of mechanisms? This type of uncertainty is resulted from the ambiguity that is unlikely to be resolvable at the level of individual sentences.

Figure 9.19 shows a streamgraph visualization. It depicts the volume of a stream of each semantic predication of causal relations found in SemMedDB. The width of a stream at a particular year is proportional to the number of articles in which the predication appears. Each stream is labeled by the subject and the object of the predication. The semantic type is not labeled because they are all causal relations. For example, the predication "HIV CAUSES AIDS", labeled as HIV/Acquired Immunodeficiency Syndrome in the streamgraph, emerged in 1984. It had the widest stream in 1985. In 1986, the most popular predication was "Retroviridae CAUSES Acquired Immunodeficiency Syndrome," but the predication "HIV CAUSES AIDS" became the most popular one again in 1987 and 1988. From this simple visualization, we learn that the research of HIV and AIDS was most active during 1984 and 1988 (Figure 9.20).

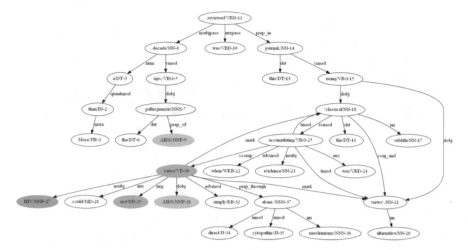

Fig. 9.19 The dependency graph of a sentence from a 2008 article (PMID: 18624032; SID: 111111060[1]). This is the opening sentence of the abstract: *More than a decade ago, the pathogenesis of AIDS was reviewed in this journal, using the subtitle 'classical and alternative views', when evidence was accumulating that HIV could not cause AIDS simply through direct cytopathic mechanisms alone*

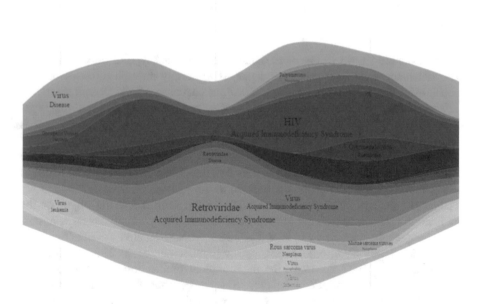

Fig. 9.20 A streamgraph visualization of semantic predications between 1984 and 1989

Summary

The understanding of the type and the degree of the uncertainty associated with a scientific proposition is about the epistemic status of scientific knowledge—it is meta-knowledge of science. Without the meta knowledge, a scientist would be like someone who only learns how to swim by reading books. Without the meta knowledge, scientists will have no way to differentiate codified knowledge from knowledge that is in the making.

The mainstream research of representing scientific knowledge has overwhelmingly focused on predications extracted from scientific literature. While representing scientific knowledge in a simplified form may serves important goals, in a long run, the omission of their epistemic status from the representation of scientific knowledge is likely to hinder the accessibility of scientific knowledge. Many problems with policy and administrative implications may not be adequately resolved. The public understanding of science may not offer the public with efficient and effective means to clarify controversies of scientific debates or reconcile contradictory results and interpretations in scientific literature.

Understanding the wide variety of types of uncertainties in science and their roles in the advance of science itself and in the broader context of everyday life is the first step towards a better understanding of how science works. A high level of uncertainty may attract more competitions because it may imply a potentially higher reward. A sudden increase of uncertainty may indicate the emergence of a new paradigm. Once the perceived level of uncertainty drops below a level, an area of research may lose its attraction. For funding agencies and peer reviewers of high-risk and high-reward programs, the perceived risk and the underlying uncertainty are the two sides of the same coin. They are integral part of innovative and competitive research. They should be treated as such.

Concluding Remarks

We began the book with Heilmeier's Catechism as a desired level of clarity and granularity in communicating scientific knowledge effectively. A competent scholar should be able to communicate complex scientific work that people without the relevant domain knowledge can understand to the extent it matters to them. For example, how many days of Ebola quarantine would be sensible? What is the key to help more people to understand controversies about climate change?

Rome was not built in one day. Many research programs' pragmatic values may not become clear for many generations. What are the arguments for or against supporting basic research as opposed to applied sciences? To put these questions in perspective, we introduced three major theories of scientific change at macroscopic levels from three distinct perspectives—philosophical, sociological, and evolutionary. The value of these theoretical visions is twofold: armed with these theories, we have a rich set of tangible properties that we can match and verify from different perspectives, and we can start to construct a theory of our own that may connect predictions made by existing theories and reconcile inconsistencies across different expectations. Macroscopic theories of science focus on holistic properties of scientific domains. The notion of a scientific domain is a generic concept of a complex adaptive system, which may exist across multiple levels of granularity. It is valuable to develop a vision at this level to see the forest of scientific knowledge as well as the individual trees.

At lower levels of abstraction, we have reviewed a series of information metrics that measures the importance of information, semantic relatedness, and scholarly impact. An important issue concerning all the quantitative indicators is how to normalize a measurement to minimize bias and makes a comparison fair. Given the ever increasing enthusiasm in ranking increasingly diverse and heterogeneous targets, it is essential to be aware of the basic principles and implications of various normalization schemes.

Text mining techniques and applications in biomedical domains in particular are introduced. Pioneering, intermediate, and recent developments are outlined to highlight the major milestones in the course of development.

Semantic MEDLINE is a very valuable resource. It helps us understand many significant properties of semantic predications extracted from unstructured text. We illustrated how to utilize visual analytic functions in CiteSpace to explore semantic networks constructed from semantic predications. We outlined the development of an ambitious plan—a Visual Analytic Observatory of Scientific Knowledge (VAO) as the first step towards representing scientific knowledge that takes the uncertainty of science into account. We demonstrated two major sources of uncertainty in scientific literature, namely hedging and contradictory information. Finally, we illustrated a series of uncertainty types through dependency graphs of sentences of various complexity.

The uncertainty associated with a research question drives the research. The unknown or the uncertainty makes a competition meaningful because a competition needs a problem to solve. As the research advances, the level of uncertainty reduces and the competition becomes less motivated. Scientists either move elsewhere to challenge themselves with new problems or they proceed with specializations by using codified and routinized knowledge that has little room for uncertainty.

References

Angeles ADL, Ferrari F, Fujiwara Y, Mathieu R, Lee S, Lee S, Tu H-C, Ross S, Chou S, Nguyen M, Wu Z, Theunissen TW, Powell BE, Imsoonthornruksa S, Chen J, Borkent M, Krupalnik V, Lujan E, Wernig M, Hanna JH, Hochedlinger K, Pei D, Jaenisch R, Deng H, Orkin SH, Park PJ, Daley GQ (2015) Failure to replicate the STAP cell phenomenon. Nature 525(7570):E6–E9. doi:10.1038/nature15513

Bibby K, Casson LW, Stachler E, Haas CN (2015) Ebola virus persistence in the environment: state of the knowledge and research needs. Environ Sci Technol Lett 2(1):2–6

Behnam B, Naeimi A, Darvishzade A (2012) A comparative genre analysis of hedging expressions in research articles: is fuzziness forever wicked? Engl Lang Lit Stu 2(2):20–38

Chaomei Chen, Ming Song, Go Eun Heo (2017) A Scalable and Adaptive Method for Finding Semantically Equivalent Cue Words of Uncertainty. arXiv:1710.08327. https://arxiv.org/abs/1710.08327

Chen C (2016) Grand challenges in measuring and characterizing scholarly impact. Frontiers Res Metrics Analytics. doi:10.3389/frma.2016.00004

Chen C, Hu Z, Milbank J, Schultz T (2013) A visual analytic study of retracted articles in scientific literature. J Am Soc Inf Sci Technol 64:234–253. doi:10.1002/asi.22755

Clark C, Aberdeen J, Coarr M, Tresner-Kirsch D, Wellner B, Yeh A, Hirschman L (2011) MITRE system for clinical assertion status classification. J Am Med Inform Assoc 18(5):563–567

Clark M, Kim Y, Kruschwitz U, Song DW, Albakour D, Dignum S, Beresi UC, Fasli M, De Roeck A (2012) Automatically structuring domain knowledge from text: an overview of current research. Inf Process Manage 48(3):552–568. doi:10.1016/j.ipm.2011.07.002

Cross N (1997) Creativity in design: analyzing and modeling the creative leap. Leonardo 30 (4):311–317

Davis MS (1971) That's interesting! towards a phenomenology of sociology and a sociology of phenomenology. Philos Social Sci 1(2):309–344

de Knijff J, Frasincar F, Hogenboom F (2013) Domain taxonomy learning from text: the subsumption method versus hierarchical clustering. Data Knowl Eng 83:54–69

Di Marco C, Kroon F, Mercer R (2006) Using hedges to classify citations in scientific articles. In: Shanahan J, Qu Y, Wiebe J (eds) Computing attitude and affect in text: theory and applications, vol 20. The Information Retrieval Series. Springer, Netherlands, p 247–263. doi:10.1007/1-4020-4102-0_19

Falahati R (2006) The use of hedging across different disciplines and rhetorical sections of research articles. In: Proceedings of the 22nd NorthWest Linguistics Conference (NWLC22), Burnaby, February 18–19, 2006

Fuchs S (1993) A sociological theory of scientific change. Soc Forces 71(4):933–953

Greenberg SA (2009) How citation distortions create unfounded authority: analysis of a citation network. BMJ 339:b2680

Harris Z (1954) Distributional structure. Word 10(23):146–162

Heffernan O (2007) Clarity on uncertainty. Nature Reports, Climate Change, p 5

Horn K (2001) The Consequences of Citing Hedged Statements in Scientific Research Articles: When scientists cite and paraphrase the conclusions of past research, they often change the hedges that describe the uncertainty of the conclusions, which in turn can change the uncertainty of past results. Bioscience 51(12):1086–1093. doi:10.1641/0006-3568(2001)051[1086:tcochs]2.0.co;2

Hyland K (1996) Talking to the academy: forms of hedging in science research articles. Written Commun 13(2):251–281

Hyland K (1998) Boosters, heding and the negotiation of academic knowledge. Text 18(3):349–382

Ioannidis JPA, Trikalinos TA (2005) Early extreme contradictory estimates may appear in published research: the Proteus phenomenon in molecular genetic research and randomized trials. J Clin Epidemiol 58:543–549

Jensen JD (2008) Scientific uncertainty in news coverage of cancer research: effects of hedging on scientists' and journalists' credability. Human Commun Res 34:347–369. doi:10.1111/j.1468-2958.2008.00324.x

Johnson BB, Slovic P (2015) Fearing or fearsome Ebola communication? Keeping the public in the dark about possible post-21-day symptoms and infectiousness could backfire. Health, Risk & Society 17(5–6):458–471

Kakuk P (2009) The legacy of the Hwang case: research misconduct in biosciences. Sci Eng Ethics 15:545–562

Kilicoglu H, Bergler S (2008) Recognizing speculative language in biomedical research articles: a linguistically motivated perspective. BMC Bioinformatics 9(Suppl 11):S10

Kuhn TS (1962) The structure of scientific revolutions. University of Chicago Press, Chicago

Lewandowsky S, Gignac GE, Vaughan S (2013) The pivotal role of perceived scientific consensus in acceptance of science. Nat Climate Change 3(4):399–404. doi:10.1038/nclimate1720

Light M, Qiu X, Srinivasan P (2004) The language of bioscience: facts, speculations, and statements in between. Paper presented at the HLT-NAACL 2004 Workshop, Biolink, 2004

Lippi M, Torroni P (2016) Argumentation mining: state of the art and emerging trends. ACM Trans Internet Technol 16(2):10:11–10:25

Malhotra A, Younesi E, Gurulingappa H, Hofmann-Apitius M (2013) 'HypothesisFinder:' a strategy for the detection of speculative statements in scientific text. PLoS Comput Biol 9(7): e1003117

McDonald S, Ramscar M (2001) Testing the distributional hypothesis: the influence of context on judgements of semantic similarity. Proceedings of the 23rd annual conference of the cognitive science society. pp 611–616

Medlock B (2008) Exploring hedge identification in biomedical literature. J Biomed Inform 41:636–654. doi:10.1016/j.jbi.2008.01.001

Medlock B, Briscoe T (2007) Weakly supervised learning for hedge classification in scientific literature. Paper presented at the proceedings of the 45th annual meeting of the association of computational linguistics, Prague, Czech Republic, June 2007

Mikolov T, Sutskever I, Chen K, Corrado GS, Dean J (2013) Distributed representations of words and phrases and their compositionality. In: Advances in neural information processing systems, pp 3111–3119

Nakao N, Yoshimura A, Morita H, Takada M, Kayano T, Ideura T (2003) Combination treatment of angiotensin-II receptor blocker and angiotensin-converting-enzyme inhibitor in non-diabetic renal disease (COOPERATE): a randomised controlled trial. Lancet 361(9352):117–124

Noorden Rv (2014) Publishers withdraw more than 120 gibberish papers. Nature. doi:10.1038/nature.2014.14763

Piffer D (2012) Can creativity be measured? An attempt to clarify the notion of creativity and general directions for future research. Thinking Skills Creativity 7(3):258–264. doi:http://dx.doi.org/10.1016/j.tsc.2012.04.009

Rizomilioti V (2006) Exploring epistemic modality in academic discourse using corpora. In: Macia EAo, Cervera AS, Ramos CR (eds) Information technology in languages for specific purposes of educational linguistics. Springer, New York, USA, p 53–71

Rzhetsky A, Iossifov I, Loh JM, White KP (2006) Microparadigms: chains of collective reasoning in publications about molecular interactions. PNAS 103(13):4940–4945. doi:10.1073/pnas.0600591103

Service RF (2002) Bell Labs fires star physicist found guilty of forging data. Science 298:30–31

Shneider AM (2009) Four stages of a scientific discipline: four types of scientists. Trends Biochem Sci 34(5):217–223

Summers-Stay D, Voss C, Cassidy T (2016) Using a distributional semantic vector space with a knowledge base for reasoning in uncertain conditions. Biologically Inspired Cogn Architectures 16:34–44

Szarvas G, Vincze V, Farkas R, Csirik J (2008) The BioScope corpus: annotation for negation, uncertainty and their scope in biomedical text. BioNLP 2008: current trends in biomedical natural language processing. Association for Computational Linguistics, Columbus, Ohio, USA, pp 38–45

Szarvas G, Vincze V, Farkas R, Mora G, Gurevych I (2012) Cross-genre and cross-domain detection of semantic uncertainty. Comput Linguist 38(2):335–367

Uzzi B, Mukherjee S, Stringer M, Jones B (2013) Atypical combinations and scientific impact. Science 342(6157):468–472

van Raan AFJ (2004) Sleeping beauties in science. Scientometrics 59(3):461–466

Vincze V, Szarvas G, Farkas R, Mora G, Csirik J (2008) The BioScope corpus: biological texts annotated for uncertainty, negation and their scopes. BMC Bioinformatics 9(Suppl 11):S9

Wager E, Williams P (2011) Why and how do journals retract articles? An analysis of Medline retractions 1988-2008. J Med Ethics 37:567–570

Wakefield AJ, Murch SH, Anthony A, Linnell J, Casson DM, Malik M, Berelowitz M, Dhillon AP, Thomson MA, Harvey P, Valentine A, Davies SE, Walker-Smith JA (1998) Ileal-lymphoid-nodular hyperplasia, non-specific colitis, and pervasive developmental disorder in children (Retracted article. See vol 375, pg 445, 2010). The Lancet 351(9103):637–641

Zhu X, Turney P, Lemire D, Vellino A (2015) Measuring academic influence: not all citations are equal. J Assoc Inf Sci Technol 66(2):408–427

Printed in the United States
By Bookmasters